Aspects of time

for Michael and Lisa
ἡ ζωὴ μιὰ δροσιὰ εἶναι, ἔνα κῦμα

edited by C. A. PATRIDES

Aspects
of time

Manchester
University Press

University
of Toronto Press

While copyright in the volume as a whole is vested in C. A. Patrides, copyright in each chapter is vested in its author or publisher, and no chapter may be reproduced whole or in part without the permission of the copyright holder.

Published by
Manchester University Press
Oxford Road
Manchester M13 9PL

MUP ISBN 0 7190 0632 5

North America
University of Toronto Press
Toronto and Buffalo

UTP ISBN 0 8020 2232 4

Designed by Max Nettleton

Printed in Great Britain by Western Printing Services Ltd, Bristol

NORMAL LOAN

44706 03/05

Contents

To the reader

though Time be the mightiest of Alarics, yet is he the mightiest mason of all.

Melville, *Mardi*, LXXV

The aim of the present venture is to provide contexts for the diverse manifestations of time in modern literature. A secondary aim is to enable students of European thought as of English literature to understand the developing views on time to the outset of the twentieth century. These developments are in the first instance outlined in the Introduction (pp. 1 ff.) and expounded more fully in Part I (pp. 19 ff.). In Part II a general survey of modern views leads to an exposition of the implications of Einstein's contributions, and thence to an account of the response of the visual arts inclusive of the cinema (pp. 67 ff.). In Part III the capital concern of this volume moves to the forefront to include, after a backward glance at Sterne, sustained discussions of fourteen modern authors (pp. 103 ff.).

The need for a collection of essays on time became apparent to me while studying the rise and decline of the Christian view of history (see Bibl. 276). Not surprisingly, I was unable to discover any comprehensive study of time within a framework at once historical, philosophical, and literary; for, vast as the subject is, it could hardly be undertaken by any single individual. As a co-operative endeavour appeared to have been indicated, I thought it would prove useful to gather the twenty-two essays here made available, and to provide not only an Introduction but a necessarily wide-ranging bibliography. The latter's importance should not be underestimated, since the concern with time in literature is intimately related to developments in other fields. The impact of the physical sciences is already a matter of record; and if the bibliography also encompasses studies of time in sociology among other disciplines, it is in order to suggest attitudes which are yet to be mirrored in contemporary literature.

The essays in this collective effort are gathered from a variety of sources as the relevant notes indicate. All but five are reprinted without changes. The exceptions are: Ricardo J. Quinones's 'Time and historical values in the literature of the Renaissance', which selects for the present collection aspects of the argument in his magisterial study *The Renaissance Discovery of Time* (Bibl. 329); William Bysshe Stein's 'Conrad's word-world of time', which is an original contribution only distantly related to his previously published work (528); R. W. Stallman's ' "The sacred rage": the time-theme in *The Ambassadors*', which is a considerably reduced version of an earlier study (525); my own essay on 'T. S. Eliot and the pattern of time', which is an amended version of another effort in the same direction (495); and John Dixon Hunt's 'The complex grammar of Auden's time', which is an original contribution specially prepared for the present collection.

My acknowledgements must necessarily begin with my appreciation for the labours of Messrs Hunt, Quinones, Stallman, and Stein. In addition, I am grateful to the following publishers and individuals who have granted their permission to reprint the material used in this volume:

Cooper Square Publishers, Inc., 59 Fourth Avenue, New York 10003, and Dr Luise Kaufmann for permission to reprint pages 125–32 from Fritz Kaufmann's *Thomas Mann: The World as Will and Representation* (1957).

Faber and Faber Ltd for permission to reprint Marcel Brion's essay 'The idea of time in the work of James Joyce', translated by Robert Sage, from *Our Exagmination Round his Factification for Incamination of Work in Progress*, by Samuel Beckett and others (1961).

Faber and Faber Ltd, Harcourt Brace Jovanovich Inc. and Random House Inc. for permission to reproduce quotations from the work of W. H. Auden and T. S. Eliot.

Films and Filming and Mr Alan Stanbrook for permission to reprint his essay 'The time and space of Alain Resnais', *Films and Filming*, x (1964), iv, pp. 35–8.

Editions Gallimard for permission to reprint Jean-Paul Sartre's essay 'A propos de *Le Bruit et la Fureur*: la temporalité chez Faulkner' from *Situations I* (1947), in the translation by Annette Michelson (see below, under Hutchinson). Copyright by Editions Gallimard 1947.

Harvard University Press for permission to reprint Sigfried Giedion's *Space, Time and Architecture*, 5th rev. ed. (1967), pp. 430–48; the same Press, as well as Professor Jerome H. Buckley, for permission to reprint chapter i of *The Triumph of Time* (1967); and the same Press, as well as Professor Ricardo J. Quinones, for permission to adapt in his essay (below, pp. 38 ff.) aspects of the argument in *The Renaissance Discovery of Time* (1972). Copyright by the President and Fellows of Harvard College.

Holt, Rinehart and Winston, Inc., for permission to reprint Dorothy van Ghent's *The English Novel* (1953, 2nd ed. 1961), pp. 84–93.

The Hudson Review for permission to reprint José Ortega y Gasset's 'Time, distance, and form in Proust', translated by Irving Singer, *Hudson Review*, xi (1958), pp. 504–13. Copyright 1959 by The Hudson Review, Inc.

Humanities Press Inc. and Professor A. A. Mendilow for permission to reprint chapter i of *Time and the Novel* (1952, repr. 1971).

Hutchinson Publishing Group Ltd for permission to reprint Jean-Paul Sartre's essay 'On *The Sound and the Fury*: time in the work of Faulkner' from his *Literary and Philosophical Essays*, translated by Annette Michelson (1955).

The Johns Hopkins University Press for permission to reprint the chapter on Valéry from Georges Poulet's *Studies in Human Time*, translated by Elliott Coleman (1956).

Mme Alice Mauron for permission to reprint Charles Mauron's essay 'On reading Einstein', translated by T. S. Eliot, *The Criterion*, x (1930–1), pp. 23–31.

Michigan Quarterly Review, for permission to adapt C. A. Patrides's essay 'The renascence of the Renaissance: T. S. Eliot and the pattern of time', *Michigan Quarterly Review*, xii (1973), pp. 172–96.

Modern Drama and Professor Richard Schechner for permission to reprint his essay 'There's lots of time in *Godot*', *Modern Drama*, ix (1966), pp. 268–76.

New Directions Publishing Corporation for permission to reprint Marcel Brion's essay 'The idea of time in the work of James Joyce', translated by Robert Sage, from James Joyce, *Finnegans Wake: A Symposium*. Copyright by Sylvia Beach 1929.

The Philosophical Review for permission to reprint Herman Hausheer's 'St Augustine's conception of time', *Philosophical Review*, xlvi (1937), pp. 503–12; and the same journal, as well as Professor Catherine Rau, for permission to reprint her essay 'Theories of time in ancient philosophy', *Philosophical Review*, lxii (1953), pp. 514–25.

Professor R. W. Stallman for permission to reprint a condensed version of pages 34–51 of *The Houses that James Built and Other Literary Studies* (1961). Copyright by R. W. Stallman.

Twentieth Century Literature and Professor Margaret Church for permission to reprint her essay 'Time and reality in Kafka's *The Trial* and *The Castle*', *Twentieth Century Literature*, ii (1956), pp. 62–9.

The University of New Mexico Press and Professor W. P. Albrecht for permission to reprint his essay 'Time as unity in Thomas Wolfe', *New Mexico Quarterly Review*, xix (1949), pp. 320–329. Copyright by the University of New Mexico Press.

The University of Texas Press and Professor Ambrose Gordon, Jr, for permission to reprint his essay 'Time, space, and Eros: *The Alexandria Quartet* rehearsed' from *Six Contemporary Novelists*, edited by William O. S. Sutherland (1962).

The University of Toronto Press and Professor John Graham for permission to reprint his essay 'Time in the novels of Virginia Woolf', *University of Toronto Quarterly*, xviii (1949), pp. 186–201.

In my Introduction, moreover, I have quoted from the following translations: Aristotle's *Physics*, trans. H. G. Apostle (Bloomington, Ind., 1969); Augustine's *Confessions*, trans. A. C. Outler (Library of Christian Classics, Philadelphia, 1955), and *The City of God*, trans. John Healey (1610), revised by R. V. G. Tasker (1942); Boethius's *Consolation of Philosophy*, trans. W. V. Cooper (1902); Justin Martyr's *Dialogue with Trypho*, trans. A. L. Williams (1930); Plato's *Timaeus*, trans. B. Jowett (4th ed., Oxford, 1953); and Plotinus's *Enneads*, trans. Stephen MacKenna as revised by B. S. Page (3rd ed., 1956). The quotation from Origen on p. 5 is from Jaroslav Pelikan's *The Shape of Death* (1962).

I should finally record my gratitude to the staffs of the British Library, the New York Public Library, and the libraries of the University of York and New York University, for their manifold courtesies and unfailing assistance; to the authorities of New York University who in appointing me Berg Professor of English Literature for the autumn of 1974 enabled me to gain access to their city's vast resources; to the Director and staff of the Institute for Medieval and Renaissance Studies at the City College of the City University of New York for inviting me to deliver a version of my Introduction as a lecture within the series 'Time: Medieval, Renaissance, Perennial'; to Mr Martin Spencer who welcomed this volume on behalf of Manchester University Press and warmly supported its publication; to Mr J. R. Banks, of the same Press, who patiently attended to numerous aspects of my manuscript; to Miss Dilys Powell, the eminent film critic, who guided me with alacrity through the ever-growing literature on the cinema; and to Professors R. F. Atkinson of the University of York, and David Burrows and Jay Leda of New York University, who advised me on philosophy, music and the cinema respectively.

In securing the rights to the essays reprinted here, I was necessarily involved in extensive correspondence. But the rewards were often substantial. In three instances my initially formal letters marked the beginning of new friendships; and on a particularly memorable occasion—the gracious letter I had from Mme Alice Mauron in connection with her late husband's essay 'On reading Einstein' (below, pp. 75 ff.)—I was granted an account of M. Mauron's last meeting with T. S. Eliot in November 1949 at the Institute Français in London.

C. A. P.

Time past
and
time present

Quid est ergo tempus? Si nemo ex me quaerat, scio; si quaerenti explicare velim, nescio.

I 'A LUMINOUS HALO'

St Augustine and Virginia Woolf are odd companions at the best of times. But they were in agreement concerning the elusive mystery of time. St Augustine phrased the embarrassing predicament forcefully enough: 'What, then, is time? If no one asks me, I know what it is. If I wish to explain it to him who asks me, I do not know' (*Conf.* XI, 14). Virginia Woolf provided the annotation:

Time, unfortunately, though it makes animals and vegetables bloom and fade with amazing punctuality, has no such simple effect upon the mind of man. The mind of man, moreover, works with equal strangeness upon the body of time. An hour, once it lodges in the queer element of the human spirit, may be stretched to fifty or a hundred times its clock length; on the other hand, an hour may be accurately represented on the timepiece of the mind by one second. This extraordinary discrepancy between time on the clock and time in the mind is less known than it should be and deserves fuller investigation.[1]

The fuller investigation had in fact been mounted already, and was to be extended with manifold variations in philosophy as in literature, in the visual as in the plastic arts, in the cinema as in architecture, and in the physical sciences as in psychology. It were indeed an understatement even to claim that the twentieth century is obsessed with time (see A. A. Mendilow, 'The time-obsession of the twentieth century', below, pp. 69 ff.). Samuel Alexander regarded time as 'the most characteristic feature' of modern thought, while Bernard Bosanquet termed it 'the central crux of philosophy'.[2] Within a religious context, it is true, man has never ceased to inquire into the relationship between time within which he exists, and eternity toward which he aspires; so much so, that religion has not inaccurately been defined as 'the expression of man's fundamental instinct to seek security from the menace of Time' (Bibl. *26*). Hebraic thought in responding to the challenge of time posited a circular time as regards the succession of generations, a horizontal time as regards the passage of years, and a vertical time that constitutes the unique Hebrew contribution: 'the direct action of the active God' (Bibl. *67*). As Christianity likewise concluded that events within the temporal dimension cumulatively signify 'a process determined by the creative act of God vertically from above',[3] history came to be regarded as an experience directed by Providence and oriented toward the future until the rivers of time are to tumble at last into the vast ocean of eternity. 'Creatures of an inferiour nature', John Donne was to say in the seventeenth century, 'are possest with the *present*; *Man* is a *future Creature.*'[4]

But the decline of the Christian world-view since the Renaissance transferred discussions of time outside the expressly religious framework. St Augustine and Virginia Woolf may alike have pondered the mystery of time, yet their ultimate aims diverge dramatically. The inquiry of the one into the nature of time was sustained by the assurance provided by his metaphysical assumptions; the inquiry of the other was motivated by an earnest desire to seek order within a context decreasingly able to provide it. For now, in the new dispensation of 'human time' (cf. Bibl. *499*), the emphasis shifted from man as a social entity intent upon the future, to man seeking in agonising solitariness to understand the inscrutable present and to recapture the erratic past. The novel process, as Beckett said of Proust, is one of decantation:

The individual is the seat of a constant process of decantation, decantation from the vessel containing the fluid of future time, sluggish, pale and monochrome, to the vessel containing the fluid of past time, agitated and multicoloured by the phenomena of its hours.[5]

In this severely anthropocentric universe primacy belongs not so much to the individual as to the human mind. For Virginia Woolf the task of the artist appeared patently obvious:

The mind receives a myriad impressions—trivial, fantastic, evanescent, or engraved with the sharpness of steel. From all sides they come, an incessant shower of innumerable atoms . . . Life is not a series of gig lamps symmetrically arranged; life is a luminous halo, a semi-transparent envelope surrounding us from the beginning of consciousness to the end. Is it not the task of the novelist to convey this varying, this unknown and uncircumscribed spirit . . . ?

'Let us', she added,

Let us record the atoms as they fall upon the mind in the order in which they fall, let us trace the pattern, however disconnected and incoherent in appearance, which each sight or incident scores upon the consciousness. [Bibl. *558*]

But can literature realistically record the falling atoms, so long as it remains confined to the linearity of the temporal dimension while 'time in

the mind'—the luminous halo of asymmetrical life—transcends time itself?
Stubbornly the commonplace distinction persists between the visual arts
which can achieve simultaneity of effect, and literature or music which are
by nature 'serial' arts.[6] T. S. Eliot, who claimed that 'genuine poetry can
communicate before it is understood', was no less conscious of the insur-
mountable difficulties attending both literature and music: for

> Words move, music moves
> Only in time.[7]

However readily we may assent to Pound's proposition that *Ulysses* possesses the form of a sonata, we resist his suggestion that Cocteau's poetry is
'cinematographic'.[8] But perhaps modern literature has so far transcended
its sequential nature that it can approximate to the potentially a-temporal
cinema, the arts, and even architecture. Perhaps a new dimension now
obtains, altogether different from the assumptions characteristic of earlier
poets and novelists.

The numerous possibilities are suggested in the last and longest section
of the present collection of essays (Part III, 'Time in modern literature',
pp. 103 ff.). But to understand the paths that led to the varieties of modern
literature it is first necessary to glance at the general background (Part I,
'Time past', pp. 19 ff.) and next to suggest the coincidence of parallel
endeavours in fields other than literature (Part II, 'Time present', pp.
67 ff.). To provide these contexts may well prove to be indispensable to
an appreciation of the frequently disconcerting oddities of the literature of
the twentieth century.

II 'ONE PERMANENT POINT'

If the Hebrews as we noted were emphatically concerned with the single
dimension of time, the Greeks were no less enthralled by the three dimensions of space. Greek interests and assumptions militated against a sustained attention to the problem of time. Assured in their vision of the
eternal recurrence of events, they rightly concluded that to philosophise on
'history' were pointless. What possible meaning could endlessly repeated
cycles possess? The stress of Plato as of Aristotle on the cyclical nature of
time was therefore unremitting. Plato's patterned myths of the ever-
recurring cycles of ages are known only too well, but it is worth noting
that even his rhetorical flourish in the *Timaeus* that time is 'a moving image
of eternity' incorporates the metaphor of the circularity of motion ('moving' *id est* 'revolving': κυκλομένου). Aristotle as was his fashion preferred
prosaic explicitness: 'time itself is thought to be a circle.'[9]

Plato's remarks on time are suggestive but tentative, largely couched in
mythological terms. Aristotle's sustained discourse on the existence of time
(*Phys.* IV, 10 ff.) is more sophisticated because more scientific but does not
necessarily deviate from the common assumptions (see Catherine Rau,

'Theories of time in ancient philosophy', below, pp. 21 ff.). Psychological time—'time in the mind'—concerned neither Plato nor Aristotle. Both sought time's frame of reference in external space, and found it in the revolutions of the spheres. Equally important, they differentiated sharply between time and eternity. To a degree, this aspect of their thought is related to the inquiries into the nature of reality already ventured by Heraclitus and Parmenides. Heraclitus at the dawn of philosophy had maintained that the fundamental reality is change and impermanence (πάντα ῥεῖ), a ceaseless flux best symbolised by fire which appears to be changeless yet changes constantly. But if change and impermanence had obliged Heraclitus to accept time as palpably real, Parmanides in affirming constancy and permanence expressly denied it. In turn, Plato's differentiation between time and eternity involves a dichotomy between the world of becoming which is temporal and subject to change as envisaged by Heraclitus, and the dimension beyond time and change as articulated by Parmenides. The dichotomy is emphasised in the rhetorical flourish already quoted that time is 'a moving image of eternity', promptly annotated as 'moving [κυκλομένου] according to number, while eternity itself rests in unity'. Plato's tentative remarks once hammered into shape by Aristotle's organising mind, it was left to Plotinus dramatically to magnify the dichotomy between time which is 'bound to sequence' (III, vii, 6), and eternity which is 'undivided totality, limitless, knowing no divagation, at rest in unity and intent upon it', 'unchanging, self-identical, always endlessly complete', 'at no point broken into period or part' (III, viii, 3 and 11). Chalcidius in his vastly influential commentary on the *Timaeus* thus summarised the accumulated interpretations:

the pattern, that is the intelligible world, is eternal. But that which is constituted after the pattern, to wit the sensible world, is temporal. And the characteristic of time is to proceed. The characteristic of eternity is to abide and always persevere identical. The parts of time are days, nights, months and years. Eternity has no parts. The modes of time, also, are past, present and future. The being of eternity is uniform in a sole and perpetual present. The intelligible world, therefore, always *is*. This world, which is the image of the other, always *was*, *is*, and *shall be*.[10]

Yet early Christian thought resisted acceptance of the mutually exclusive domains of time and eternity, in spite of their nominal relationship to St Augustine's equally dichotomous cities or states, the *civitas terrena* and the *civitas Dei*. Two aspects of the Platonic vision began to haunt generations of Christian apologists. The first was the distinction itself between time and eternity, which went contrary to the Christian claim that the Prime Mover is in love with the productions of time; and the second, the Greek insistence on the circularity of time, which negated the Christian emphasis on the linear nature of time and consequently the irreversibility and uniqueness of all events. Not surprisingly, the condemnation of the cyclical view was undertaken by the early Christians with passionate energy and impressive unanimity (see Bibl. 276). History, they argued, progresses in a straight line from the six days of creation to the single Day of Judgement; it is therefore impossible 'that things will always remain as they are, and further, that you and I shall live again as we are living now, without having become either better or worse' (St Justin Martyr, *Dial.* I, 5). Even Origen lashed out against belief in the circularity of time, since to accept it were to credit that 'Jesus will again come to visit this life and will do the same things that he has done, not just once but an infinite number of times according to the cycles' (*C. Cels.* IV, 67–8; V, 20–1; etc.). But as Origen was unable in the end to resist the appeal of the concept, the most explicit censure of the 'continual rotation of ages past and present' was reserved for St Augustine, who in opposition argued in *The City of God* that the history of the universe is 'single, irreversible, unrepeatable, rectilinear', unfolding as a 'uni-dimensional movement in time' from the creation through its centre in Jesus to the end of the world.[11]

St Augustine's view of time in *The City of God* and more particularly in his *Confessions* (XI, 11 ff.) is the first major endeavour in the history of philosophy to explain the problem (see Catherine Rau, below, pp. 26 ff., and the more detailed account by Herman Hausheer, pp. 30 ff.). The affirmation of linear time advancing purposefully in a given direction was not proposed in isolation from other factors. On the contrary, where the Greeks had related time to external space, Augustine referred it to the experience internal within our minds ('It is in you, O mind of mine, that I measure the periods of time' (XI, 27)) and thence to our perception of the totality of events. Psychological time and historical time were consequently merged. Augustine accepted the traditional differentiation between time and eternity but internalised both; for time was now regarded as the world of flux where the heart of man 'flies about in the past and future motions of created things, and is still unstable', while eternity is the state of a single 'day' which is 'not recurrent but always today'—the state, more particularly, where 'the whole is simultaneously present' (*Conf.* XI, 11 and 13). The reality of time as an experience external to man cannot be denied; yet past, present and future are also experiences within the mind of man. In Augustine's words,

it is not properly said that there are three times, past, present, and future. Perhaps it might be said rightly that there are three times: a time present of things past; a time present of things present; and a time present of things future. For these three do coexist somehow in the soul, for otherwise I could not see them. The time present of things past is memory; the time present of things present is direct experience; the time present of things future is expectation. [*Conf.* XI, 20]

The coexistence of the three 'presents' in the human mind is the reflection within the temporal order of their simultaneous presence in the eternal Divine Mind. When all is said, therefore, reality is meaningful only in relation to God acting within history and in the mind of man.

The Augustinian concept of the Eternal Present ('the whole is simultaneously present') was to pave the way for Boethius's definitive statement that eternity is 'interminabilis vitae tota simul et perfecta possessio'—'the simultaneous and complete possession of infinite life' (*De cons. phil.* v prose vi). 'God has a condition of ever-present eternity', wrote Boethius. 'His knowledge, which passes over every change of time, embracing infinite lengths of past and future, views in its own direct comprehension everything as though it were taking place in the present.' As a commentator explained in the seventeenth century,

in relation to himself there is no *Foreknowledge* in God, all things which in our inferior Capacities seem either past, or to come, are actually præsent to him, whose duration is altogether but one constant and permanent part, one Tὸ νῦν, entire in unity, and incapable of division into successive minutes.[12]

Augustine cast his mighty shadow over the entire period to the Renaissance. His contrast between limited time and limitless eternity became a commonplace, most often asserted through a number of telling metaphors variously describing time as 'the candle of eternity', else 'the point . . . of an infinite circumference', 'a sparke of fire in the midst of the vast Ocean', or 'a certain space borrowed or set apart from *eternitie*: which shall at the last return to eternitie again: like the rivers, which have their first course from the seas; and by running on, there they arrive, and have their last.'[13] Eternity itself was conceived in equally traditional terms. 'There is no Dinumeration of tyme with God', wrote Luther, for in the Eternal Mind 'all thynges are lapped up as it were in one bundle.' Other apologists similarly asserted that 'God is above all time', that his realm 'knows no distinction of tenses', that 'Diuinitie is not cloistered or confined to time, either past, or future, but commands all as present'.[14] The concept of the Eternal Present which all these statements affirm, was in English literature most brilliantly articulated by Sir Thomas Browne in *Religio Medici*:

that terrible terme *Predestination*, which hath troubled so many weake heads to conceive, and the wisest to explaine, is in respect of God no previous determination of our estates to come, but a definitive blast of his will already fulfilled, and at the instant that the first decreed it; for to his eternitie which is indivisible and altogether, the last Trumpe is already sounded, the reprobates in the flame, and the blessed in *Abrahams* bosome. Saint *Peter* spake modestly when hee said, a thousand yeares to God are but as one day; for, to speake like a Philosopher, those continued instants of time which flow into a thousand yeares, make not to him one moment; what to us is to come, to his Eternitie is present, his whole duration being but one permanent point, without succession, parts, flux, or division. [ɪ, 11]

To regard the universe from the vantage point of timeless eternity is necessarily to grant the astonishing paradox that

the world was before the Creation, and at an end before it had a beginning; and thus was I dead before I was alive; though my grave be *England*, my dying place was Paradise, and *Eve* miscarried of mee before she conceiv'd of *Cain*. [ɪ, 59]

III 'FROM THE OUTER TO THE INNER'

The Renaissance was partial to the Augustinian view of time even while it ushered in developments that transformed the understanding of time radically (see Ricardo J. Quinones, 'Time and historical values in the literature of the Renaissance', below, pp. 38 ff.). The invention of mechanical clocks from the end of the thirteenth century and the subsequent advent of horology were developments that looked beyond themselves, to a deepening awareness of the expeditious flight of time no longer annexed to an assured sense of its providentially-directed passage. Sir Walter Ralegh's brooding meditations on the state of the universe voice the ever-increasing disquiet:

as all things vnder the Sunne haue one time of strength, and another of weaknesse, a youth and beautie, and then age and deformitie: so Time it selfe (vnder the deathfull shade of whose winges all things decay and wither) hath wasted and worne out that liuely vertue of Nature in Man, and Beasts, and Plants; yea the Heauens themselues being of a most pure and cleansed matter shall waxe old as a garment . . .[15]

If as we have been told consciousness of time was a factor in the advent of Greek tragedy (Bibl. *333*), it may be that the same generalisation applies to the advent of Shakespearean drama (Bibl. *329*). In Shakespeare, certainly, time assumes an obsessive significance altogether absent from the calm espousal by the Mystery Plays of the progress of the *civitas terrena* toward its consummation in the felicities of the *civitas Dei*. Shakespeare's sonnets stand witness to a vastly intensified consciousness of the destruction effected by 'devouring time' (XIX), 'bloody tyrant time' (XVI), 'time's injurious hand' (LXIII):

> Nativity, once in the main of light,
> Crawls to maturity, wherewith being crown'd,
> Crooked eclipses 'gainst his glory fight,
> And Time that gave doth now his gift confound.
> Time doth transfix the flourish set on youth
> And delves the parallels in beauty's brow,
> Feeds on the rarities of nature's truth,
> And nothing stands but for his scythe to mow. [LX]

Much the same consciousness informs the plays, however various its manifestations and multiform its implications. *Macbeth*, for instance, dramatises the aspiration to jump beyond 'this bank and shoal of time' (I, vii, 6); but the fruitless endeavour 'to beguile the time' (I, v, 64) terminates in a catastrophically negative vision as the meaningless days are seen to advance relentlessly 'To the last syllable of recorded time' (v, v, 21). 'I wasted time,' Richard II laments on a different occasion, 'and now time doth waste me' (v, v, 49). The moral implications are lucid in the extreme; but gradually even they declined, for by the end of the seventeenth century novel views of time were to compound the difficulties increasingly posed by the phenomenon of time.

One dramatic development involved dualism. Perception of the tendencies in Descartes was not immediate, and his English correspondent Henry More indulged in lyrical praise ('ravished with admiration') before accepting that Cartesianism pointed unfailingly toward a dichotomy between matter and spirit. Another dichotomy was an extension of the traditional distinction of the three dimensions of space from the single one of time. Isaac Barrow asserted the absolute nature of time, Henry More the absolute nature of space—and Newton, of both. 'Absolute, true, and mathematical time', wrote Newton, 'of itself, and from its own nature, flows equably without relation to anything external.'[16] Locke agreed. Even while proposing his influential theory of the association of ideas in *An Essay concerning Human Understanding*, he claimed that simultaneity is universally valid for all observers wheresoever:

duration is but as it were the length of one straight line, extended *in infinitum*, not capable of multiplicity, variation or figure; but is one common measure of all existence whatsoever, wherein all things, whilst they exist, equally partake. For this present moment is common to all things that are now in being, and equally comprehends that part of their existence as much as if they were all but one simple being; and we may truly say, they all exist in the same moment of time. [II, xv, 11]

The reflection of these developments in eighteenth-century literature was a multiplicity of approaches: in the uncompromisingly ethical Pope, the firm connection of time to moral norms (Bibl. *381*); in Thomson and Fielding, the exclusive preference for the immediate present (Bibl. *360, 390*); and in Richardson and Sterne, the adjustment of the art of the novel to the requirements of psychological realism by way of a brilliant deployment of 'time in the mind'.

Richardson's transformation of the epistolary form in *Clarissa* into a pliable instrument of fiction has correctly been said to involve the reader in 'the suspense of a continuous present'.[17] Sterne was far more ambitious, and incalculably more original: *Tristram Shandy* was to have no parallel until Proust's *A la recherche du temps perdu* (see Dorothy van Ghent's observations on the two novels, below, pp. 106 ff.). The nominal theory of *Tristram Shandy* is stated by its equivocating narrator: 'my work is digressive, and it is progressive too,—and at the same time' (I, 22). The result is a comic extravaganza in which an ambitious discourse on time is so frequently defeated by time itself that it is finally abandoned (III, 18–19), while the very endeavour to 'explain' digression and progression is wrecked by the demonstration of the principle at work:

The highest stretch of improvement a single word is capable of, is a high metaphor, —for which, in my opinion, the idea is generally the worse, and not the better;— but be that as it may,—when the mind has done with it—there is an end,—the mind and the idea are at rest,—until a second idea enters;—and so on. [v, 42]

In what sense might one claim that Sterne is indeed 'the forerunner of the moderns'? In the sense, Virginia Woolf has authoritatively observed, that 'Sterne transfers our interest from the outer to the inner'—from clock time to time in the mind, from experiences consciously observed to the cumulative experiences in the internal universe of the subconscious. 'No writing', she added, 'seems to flow more exactly into the very folds and creases of the individual mind, to express its changing moods, to answer its lightest whim and impulse, and yet the result is perfectly precise and composed. The utmost fluidity exists with the utmost permanence.'[18]

The multiplicity of approaches to time during the eighteenth century eloquently testifies that, metaphysical assumptions like Augustine's once denied, the secular theories thereafter proposed were as numerous and diverse as their exponents. The Romantics may well appear to have developed a consistent view of time if only because their obsessive attachment to nature inevitably obliged them to regard the movement of time as analogous to the cycles of the seasons. Certainly Shelley espoused the cyclical view of time most explicitly in the 'Ode to the west wind', extending it with characteristic enthusiasm in the last choric song of his lyrical drama *Hellas*:

> The world's great age begins anew,
>   The golden years return,
> The earth doth like a snake renew
>   Her winter weeds outworn:
> Heaven smiles, and faiths and empires gleam,
> Like wrecks of a dissolving dream.[19]

Yet Shelley was not confined solely to this unexceptional view of time, aware that any attempt to comprehend 'the shadowy flux of Time'—as Coleridge remarked in the conclusion of the *Biographia Literaria*—must involve a recognition of 'Eternity revealing itself in the phaenomena of Time'. In *Prometheus Unbound*, accordingly, the action is largely internalised, unfolding in terms of the protagonist's eager anticipation of the 'retributive hour' eventually manifested in the form both of Demogorgon/Eternity and of the Spirits of the Hours which in bearing 'Time to his tomb in Eternity' herald the advent of 'a diviner day' (I, 406; III, i, 52; IV, 14, 26). In Blake's fertile imagination, on the other hand, time merges with eternity triumphantly to proclaim the holiness of all existence (cf. Bibl. 376). In Keats, a tentative effort to posit a cyclical view of time in the first *Hyperion* was displaced by the explosive paradoxes centred on time and eternity in the 'Ode on a Grecian urn' and 'To autumn'. In Wordsworth, the central position allotted to the concept of time in *The Prelude* (see Bibl. 373) involves in particular the displacement of the Christian myth of paradise, its loss, and its eventual restoration, by the myth of the paradise of childhood, its loss, and its eventual restoration. The action is again largely internalised, for even if all external events are alike propelled toward a *telos* (XIV, 110–11), time is understood to transcend itself when the apocalyptic serenity of a summer's night, breezeless and majestically silent, impels the mind to comprehend the mystery of things (XIV, 11 ff.).

The Victorians also sought the mystery but found no end in wandering mazes lost. One development in particular proved crucial, that by the middle of the nineteenth century the evidence was at last provided with which to controvert the Christian world-view whose veracity had already been spasmodically doubted. As a geologist triumphantly proclaimed in 1858,

The leading idea which is present in all our researches, and which accompanies every fresh observation, the sound of which to the ear of the student of Nature seems continually echoed from every part of her works, is—
      Time!—Time!—Time![20]

The sound was not universally appreciated. Rushkin, for one, was irritated:

If only the geologists would let me alone, I could do very well, but those dreadful Hammers. I hear the clink of them at the end of every cadence of the Bible verses.[21]

So far as the theory of evolution appeared to reinforce the idea of progress, the Victorians were to remain optimistic. But their optimism was greatly qualified by the concurrent anxiety engendered by the possibility of decadence. Beliefs so mutually exclusive were bound to produce diametrically opposed reactions (see Jerome H. Buckley, 'The four faces of Victorian time', below, pp. 57 ff.). Tennyson struggled valiantly in *In Memoriam* to quell his mounting despair; Arnold, less resolutely, was utterly engulfed by the shroud of his despondency. As many also turned inwards upon themselves to explore (in Pater's phrase) 'the narrow chamber of the individual mind',[22] it could be said that the nineteenth century terminated in claustrophobia.

IV   'NO ABSOLUTE STANDARD'

The twentieth century inherited both the despondency of the Victorians, and their massive optimism. But presently, to the vision of a universe already vastly expanded in time by Darwin and his immediate predecessors, Einstein added a further immense expansion in the joint four dimensions of space-time (see Charles Mauron, 'On reading Einstein', below, pp. 75 ff.). The theory of the relativity of simultaneity shattered the universally valid absolute time of the Newtonian universe ('this present moment is common to all things that are now in being' [above, p. 8]). Not that Einstein's theory asserts that events in the physical world are relative. Its claims centre rather on the observer—a man, a camera, a clock. Of two observers in given positions, one might conclude that a given event is true, the other that it is false. According to the theory of relativity, however, both judgements are inaccurate even while they may be jointly correct: for 'true' and 'false' are claims relative not to the events but to the observers, in that to alter one's position is necessarily to alter the conclusion reached. Moreover, as Minkowski remarked, 'Nobody has ever noticed a place except at a time, nor a time except at a place',[23] so that temporal considerations are as vital as spatial ones and equally relative to the observer. In consequence, an event can be fixed objectively solely through measurements taking account of both space and time—the 'four-dimensional space-time continuum' of Einstein's formulation (Bibl. *102–3*). This is not to say, however, that time and space have been negated. Each continues to be a reality, but a reality strictly relative to the given observer whose views are 'true' where he himself is concerned, 'false' as regards another observer.

The advent of Einstein's theory ran parallel to other developments. They included Bradley's extension of the Parmenidean concept of constancy and permanence, the reality of time now dissolved in the welter of appearances; Bergson's pursuit of the Heraclitean belief in change and impermanence, where the only acceptable reality is the psychological 'time' of a durational flow, the ceaseless flux of *la durée réelle*, 'psychical in its nature and psychological in its order' (Bibl. *93*); and McTaggart's ultimate claim that time is after all illusory.[24] Did these developments influence art and literature? Naum Gabo answered categorically enough: 'the philosophic events and events in science at the beginning of this century have definitely made a crucial impact on the mentality of my generation.'[25] The impact cannot be measured with accuracy, however, since artists never respond to one influence only. Einstein, Bradley, Bergson— the cumulative power of the moment—might therefore be said to have exerted an influence rather diffuse than palpable. True, large claims have been advanced, particularly as regards Bergson. We have been assured that 'most of Proust's work is an exposition of Bergson's philosophy' (Bibl. *527*), and much the same generalisation would appear to apply to the work of Joyce, T. S. Eliot and Virginia Woolf among others (*464*). On the other hand, we must accept that the differences between Bergson and Proust outweigh their similarities by far (*407, 552*), that Joyce's fictional world is not so much a reflection as 'a parodic representation of Bergsonian unreality' (*460*), that Eliot's enthusiastic conversion to Bergsonism in 1911 was short-lived (below, p. *162*)—and that Virginia Woolf had not even read Bergson. As with Bergson's influence, so with Einstein's (cf. *400*). Take the abolition of perspective by the cubists. The intention was that objects should henceforth be viewed relatively—'that is, from several points of view, none of which has exclusive authority' (see Sigfried Giedion, 'The new space conception: space-time', below, pp. 81 ff.; cf. Bibl. *463*). As Moholy-Nagy observed,

Contemporary painters, confronted with the static, restricted vision of a fixed perspective, countermarched to color and produced on the flat surface a new *kinetic* concept of spatial articulation, vision in motion.[26]

'Vision in motion', he added, 'is a synonym for simultaneity and space-time'. The language may be Einstein's but the difference between theoretical statement and artistic achievement defines the mystery of creation. Significantly, the Futurists, who deployed a terminology equally reminiscent of Einstein, exhibit also the influence of Bergson. 'Time and Space died yesterday', declared the improbable Marinetti in 1909. Glorying in speed, motion, change, a ceaseless flux, he assured an astonished world that 'a running horse has not four legs, but twenty, and their movements are triangular'. The ultimate ambition was stated in no uncertain terms. 'The simultaneousness of states of mind in the work of art: that is the intoxicating aim of our art.'[27]

The intoxicating aim is much in evidence in their few substantial works of art, notably Boccioni's bronze sculpture of a walking figure significantly entitled *Unique Forms of Continuity in Space* (see Bibl. *218*). More recently, Futurism has had in kinetic art a descendant equally committed to the exhibition of motion—'the changing of the network of relationships which defines the structure in space and time'.[28] Fundamental to this conception is the insistence of one artist (Günther Uecker) that one should 'let the river flow and himself to see, enquire into space and give it significance in time'. 'Seek no rest,' he advises, 'keep on the move'—'fight,' as another artist (Julio Le Parc) urges, 'fight any tendency toward stability, permanency, lastingness.' Thus will paintings and sculptures 'breathe and perspire, grow and decay' (David Medalla), the line be 'transformed by optical illusion into pure vibration, the material into energy' (Jesus-Rafael Soto), resulting in 'new dynamic multi-dimensional visual means of expression' (Bruno Munari) which alone can create 'a fourth visual dimension' (Fletcher Benton). Clearly disinclined to repudiate reality, kinetic art reflects the pervasive consciousness of the flux of time: 'Nowhere does time stand still . . . a restlessness has come upon us' (Benton). The note of disquiet sounded here reminds us forcefully of the uncertainties inadvertently induced by Bergson's durational flow as by Einstein's space-time continuum; for now, as Bertrand Russell said apropos the latter, 'there is no absolute standard of rest in the universe.'[29]

The aspirations of kinetic art coincide with those of its etymological relative, the cinema, itself eminently qualified fully to respond to the contemporary fascination with time. Its first steps were hesitant, taken largely in connection with comedy, as in Mack Sennett's unexpected violation of the expected time-sequence. With the advent of Sergei Eisenstein, however, dislocation of time became of central importance to cinematic art. This dislocation is achieved, of course, by that juxtaposition of heterogeneous images known as montage—'the filmic fourth dimension', as Eisenstein significantly called it (Bibl. *232*). In literature, Eisenstein added, the single dimension of time was circumvented only by Joyce, though at a price, namely 'the entire dissolution of the very foundation of literary diction, the entire decomposition of literary method itself'. The potential of the cinema is on the contrary immense, for a film is 'an expansion of the strict diction achieved in poetry and prose into a new realm where the desired image is directly materialised in audio-visual perceptions'.[30] Responses abounded, not least by Resnais in both *Hiroshima, mon amour* and *L'Année dernière a Marienbad* (see Alan Stanbrook, 'The time and space of Alain Resnais', below, pp. 92 ff., and Bibl. 238). Lately, Stanley Kubrick's *2001: A Space Odyssey* boldly explored time not through montage but through the implied thematic patterns. The story as a story is nominally confined to linear time. The film encompasses historic time from the dawn of man through the present and the immediate future to an apparently linear dimension ('Jupiter and beyond the infinite') that promptly folds upon itself to intimate uncharted regions not so much outside the solar system as within the mind of man—'the narrow chamber of the individual mind'. Simultaneously, the film's concluding movement is couched in terms suggestive of the abolition of time as the astronaut journeys at a geometrically advancing speed suggested through a phantasmagoric explosion of diversely coloured and shaped images. By way of justification Kubrick appealed to Einstein's theory of 'time dilation', *id est* 'as an object accelerates toward the speed of light, time slows down'.[31] If so, the much-discussed monolith periodically manifested in the film would appear to be of crucial importance. Omnipresent and changeless, it is the foremost image of timeless permanence in a world of flux.

V   'OUT OF THE STRAIGHT LINE'

Eisenstein as we have noted called montage 'the filmic fourth dimension'. He also described the film as 'a free stream of changing, transforming, commingling forms, pictures, and compositions'.

'A free stream': we are reminded of 'the stream of consciousness', the celebrated phrase first ventured by William James. Not adequate to describe reality—what single phrase is?—its metaphoric burden was gradually understood to suggest the dislocation of time in the modern novel. Consciousness, James had maintained,

does not appear to itself chopped up in bits. Such words as 'chain' or 'train' do not describe it fitly . . . It is nothing jointed; it flows. A 'river' or a 'stream' are the metaphors by which it is most naturally described . . . *let us call it the stream of thought, of consciousness, or of subjective life.*[32]

Modern literature deploys the stream of consciousness in full awareness of the formidable literary antecedents in Richardson and especially Sterne but also in nineteenth-century literature's exploration of 'the most mysterious movements of psychic life'.[33] Extra-literary antecedents include not only the intellectual developments already sketched but also the influence demonstrably exerted by Freud on the 'interior monologue' of an expressly psychological novel like *Ulysses*.[34] But whatever the antecedents, the vital consideration has been to comprehend the elusive mystery of time, whether thematically as by Proust and Thomas Wolfe, or in terms of theme *and* form as in the ambitious endeavours of Joyce, Virginia Woolf and Faulkner. The motivation is clear. It is the aspiration to account for the 'extraordinary discrepancy between time on the clock and time in the mind' which Virginia Woolf voices (above, p. 1), and consequently the dissatisfaction with the unrealistic time-as-sequence which William Golding's narrator censures at the outset of *Free Fall*:

time is not to be laid out endlessly like a row of bricks. That straight line from the first hiccup to the last gasp is a dead thing. Time is two modes. The one is an effortless perception native to us as water to the mackerel. The other is a memory, a sense of shuffle fold and coil, of that day nearer than that because more important, of that event mirroring this, or those three set apart, exceptional and out of the straight line altogether.

But in literature, as already suggested, time stubbornly refuses to be 'spatialised'.[35] Consider Gertrude Stein. She alone attempted a straightforward description of the process of time's 'spatialisation' but in the end her inimitable style coincided with her argument only fitfully:

... I at once went on and on very soon there were pages and pages and pages more and more elaborated creating a more and more continuous present including more and more using of everything and continuing more and more beginning and beginning and beginning . . .[36]

The present collection of essays is devoted to the delineation of fourteen approaches to the problem. The strategies of ten novelists are allotted pride of place in just recognition of the enormous contributions made by fiction. The chronological starting point is 1895, when the publication of Conrad's novels commenced with *Almayer's Folly* (below, pp. 114 ff.). We next pause in 1903, the year of the publication of Henry James's *Ambassadors* (pp. 126 ff.). There follow discussions of Proust (pp. 136 ff.), Joyce (pp. 153 ff.), Mann (pp. 172 ff.), Kafka (pp. 179 ff.), Virginia Woolf (pp. 187 ff.), Faulkner (pp. 203 ff.), Thomas Wolfe (pp. 210 ff.), and Lawrence Durrell (pp. 238 ff.). Three poets are also discussed—Valéry (pp. 144 ff.), T. S. Eliot (pp. 159 ff.), and W. H. Auden (pp. 225 ff.)—of whom the second stands in splendid isolation from his contemporaries in that he merged his individual talent with the tradition inherited from St Augustine. Dramatic literature is represented solely by Beckett (pp. 217 ff.), in acknowledgement of the apparently general tendency to regard the two protagonists of *Waiting for Godot* as poised to await redemption 'from the evanescence and instability of time, and to find peace and permanence outside it'.[37]

But other dramatists, poets and novelists might also have been included, for the concern with time extends far beyond the fourteen literary figures here discussed to include D. H. Lawrence. True, Lawrence failed to be seduced by time as either theme or form. Yet his negative attitude is itself positive in that he represents the most extreme instance of a passionate commitment to 'the immediate present'. 'In the immediate present there is no perfection, no consummation, nothing finished', he wrote in 1920. 'The living plasm vibrates unspeakably, it inhales the future, it exhales the past, it is the quick of both, and yet it is neither. There is no plasmic finality, nothing crystal, permanent'. He added:

Life, the ever-present, knows no finality, no finished crystallization. The perfect rose is only a running flame, emerging and flowing off, and never in any sense at rest, static, finished. . . . Tell me of the incarnate disclosure of the flux, mutation in blossom, laughter and decay perfectly open in their transit, nude in their movement before us.[38]

But a Heraclitean awareness of life's flux does not preclude a Parmenidean desire to still it. The two have often converged in fiction—especially in science fiction—even where the effort to understand time ends in the phantasmagoria of 'an infinite series of times, in a dizzily growing, ever spreading network of diverging, converging and parallel times'.[39] But even here attempts have varied. Charles Williams, in basing his *Many Dimensions* on the unlikely existence of a Stone whose virtues are 'neither here nor there but allwhere', aspired to comment on men's futile efforts to reduce a supernatural dimension to the trite level of their monomaniac greed. Opposing 'the normal corporeal sequential' to a dimension beyond time—'Time is in the Stone, not the Stone in Time' (chapter ix)—Williams invoked the Boethian concept of the coexistence 'equally' within the Divine Mind of past, present and future (chapter xvi). The moral pattern is finally articulated as a call, a mysterious sound—'which, native to heaven, can on earth be vocal and audible only between spirits already disposed to heaven' (chapter xvii). On another level altogether, a parallel intimation of the bisection of time by the timeless is also evident in the novels of Iris Murdoch. The exceptional individuals who in one form or another are in contact with demons, alone have the questionable privilege to experience the sudden annihilation of the temporal dimension. The vision of Nigel in *Bruno's Dream* is representative: 'Out of the dreamless womb time creeps in the moment which is no beginning at the end which is no end' (chapter iii). Yet the mystery is not thereby resolved; it is on the contrary rendered more profound, and disconcerting.

Poets other than Valéry and T. S. Eliot and Auden, and dramatists other than Beckett, also exhibit a sustained concern with time. In dramatic literature, for instance, Cocteau insistently emphasised the discrepancy between our sense of sequential time, and timeless time as it might be said to exist in the mythical realms beyond appearances. As Anubis proclaims in *The Infernal Machine*,

A man's time is a fold of eternity. For us it does not exist. From his birth to his death the life of Œdipus is spread flat before my eyes, with its series of episodes.[40]

The approaches of poets are equally varied. Walter de la Mare's concern with time centres on the wide-ranging melancholy meditations ('Time in triple darkness hid') that constitute *Winged Chariot*, his longest poem but also—as Auden would have it—his greatest.[41] In more elevated regions, as we shall see later, Yeats adapted the Graeco-Roman view of time's ever-recurring cycles (below, p. 163). Rilke's universe within—the 'Weltinnen-raum'—involves not so much time as space: time-as-sequence is suspended, to be displaced by the timeless subsistence of all art in a continuous spatial present so that, as has been said, 'all things which are real (that is, which can be felt) subsist side by side in a kind of magic contemporaneity'.[42] Pound, on the contrary, resolutely eschewed the abolition of time: the *Pisan Cantos* mount a quest that cumulatively explores 'the unfolding of the human spirit in the medium of time' (see Bibl. *496*). The quest is centred on the opposition between 'linear mechanical time' which Pound in associating with Christian dualism regards as destructive and therefore negative, and 'cyclical organic time' which in its relationship with the orderly seasons and primitive man's response to them is viewed as constructive and therefore positive. The dualism of time and eternity once categorically dismissed, the *Cantos* advance forward in time as backward in memory to attain the paradisal vision intimated through the multiform presence of Aphrodite (*ibid.*).

In the end St Augustine's question lingers still: 'What, then, is time?' But an epilogue were folly to attempt, and an extravagance even to contemplate. Where science outpaces science fiction in suggesting dynamic realms beyond the fourth dimension—never-ending cycles of rebirth shaping the restless universe anew, with space inverting itself to behave like linear time—it were best to walk circumspectly.

At least, 'for the time being'.

NOTES

[1] *Orlando* (1928), p. 91.

[2] Both quoted by J. A. Gunn (Bibl. 41), p. 241.

[3] C. H. Dodd, *History and the Gospel* (1938), p. 181. See Bibl. *276*.

[4] *Sermons*, ed. E. M. Simpson and G. R. Potter (Berkeley, Calif., 1956), VIII, p. 75.

[5] *Proust* (1931), pp. 4–5. The translation of *A la recherche du temps perdu* as *Remembrance of Things Past* is unfortunate in that it diverts attention from the thematic implications actually proclaimed by its title, i.e., *In Search of Lost Time*.

[6] The term is persuasively deployed by Paula Johnson, *Form and Transformation in Music and Poetry of the English Renaissance* (New Haven, Conn., 1972), chapter I.

[7] From 'Dante', in *Selected Essays* (1950), p. 200, and 'Burnt Norton', ll. 137–8, respectively.

[8] 'James Joyce et Pécuchet' (1922), in *Polite Essays* (1937), p. 89, and *The Dial*, LXX (1921), p. 110; respectively.

[9] *Timaeus* 37–8 and *Physica* IV, 14 (ὁ χρόνος αὐτὸς εἶναι δοκεῖ κύκλος τις).

[10] Trans. P. H. Wicksteed (Bibl. *71*), p. 91. Cf. the exposition of Proclus, in *The Commentaries of Proclus on the Timaeus*, trans. Thomas Taylor (1820), II, pp. 181 ff.; also Bibl. *64*.

[11] A. H. Armstrong, *An Introduction to Ancient Philosophy*, 3rd ed. (1957), p. 215, and Ernst Hoffmann, 'Platonism in Augustine's philosophy of history', in *Philosophy and History*, ed.

Raymond Klibansky and H. J. Paton (Oxford, 1936), p. 174, respectively. See also the studies cited in the Bibliography, *28–30, 35, 40, 52,* etc.; Gale (*113*) reprints the principal statements of Aristotle, Plotinus and Augustine.

[12] Walter Charleton, *The Darkness of Atheism* (1652), p. 118.

[13] *Seriatim:* Daniel Featley, *Clavis mystica* (1636), p. 401; Thomas Granger, *Syntagma logicum* (1620), p. 75; Henry Church, *Miscellanea philo-theologica* (1637), I, 25; and John Swan, *Speculum mundi* (Cambridge, 1635), p. 45.

[14] *Seriatim:* Luther, *A Commentarie . . . vppon the twoo Epistles Generall of . . . Peter, and Jude,* trans. Thomas Newton (1581), fol. 158; Henry Ainsworth, *The Orthodox Foundation of Religion* (1641), p. 18; Charleton, as above (note 12); and Humphrey Sydenham, *The Arraignment of the Arrian* (1626), p. 3.

[15] *The History of the World* (1614), ed. C. A. Patrides (1971), p. 144.

[16] *Principia mathematica,* Definitions: Scholium 1; trans. Andrew Motte, rev. Florian Cajori (Berkeley, Calif., 1962), I, 6. On these developments consult J. T. Baker and Burtt (Bibl. *77, 89*); and on the Descartes–More relationship: C. A. Patrides, ed., *The Cambridge Platonists* (1969), pp. 29 ff.

[17] Anthony Kearney, '*Clarissa* and the epistolary form', *EC*, XVI (1966), p. 44.

[18] *The Common Reader,* 2nd series (1932), pp. 79–80, 81. On the Sterne–Woolf literary continuity, see R. C. Brown (Bibl. *406*) and the sweeping remarks by A. E. Dyson, *The Crazy Fabric* (1965), especially pp. 34 f., 47; but in opposition Elizabeth Drew holds that Tristram's confusion of clock time and mental time is merely a comic device: 'He is neither reporter nor analyst of the inner psychic life of emotional associations in himself or in others' (*The Novel* [1963], pp. 83 ff.). Sterne's conception of time is said by Baird (Bibl. *354*) to have had its source in Locke's *Essay,* while Sallé (Bibl. *386*) attributes it to the 'liberal interpretation' of Locke's thesis by Addison (Spectator, 18 June 1711). But cf. Bibl. *359, 377.*

[19] *Hellas,* ll. 1060–5. Behind these lines looms Virgil's celebration of the recurring centuries (*Ecl.* IV, 5), while the symbolism of the 'Ode'—as Shelley himself observed in a note to *Queen Mab* (V, 1–6)—derives from Homer's comparison of the generations of men to leaves (*Il.* VI, 146–9) as well as from the meditation of Ecclesiastes on the interminable cycles which other Hebrew writers had categorically denied.

[20] G. P. Scrope, *The Geology and Extinct Volcanoes of Central France* (1858), p. 208; quoted by Haber (Bibl. *43*), p. 291.

[21] Quoted by Buckley (Bibl. *357*), p. 28.

[22] See John Dixon Hunt, *The Pre-Raphaelite Imagination* (1968), chapter III.

[23] In his celebrated 1908 lecture 'Space and time' (Bibl. *152*), p. 76. For an introduction to the theory of relativity consult Whitrow (Bibl. *16*), chapters V–VI; for the most lucid exposition yet written: Bertrand Russell (*167*); and for three technical but comprehensive accounts: Einstein himself (*105*), Borel (*82*), chapters VI–VII, and Whitrow (*190*), chapter IV.

[24] See Bibl. *78, 84, 148.*

[25] *Gabo,* with essays by Herbert Read and Leslie Martin (1957), p. 159. In turn, of course, developments in art influenced developments in literature. See, for instance, L. C. Breunig, 'Picasso's poets', *YFS*, XXI (1958), pp. 3–9, and Korg (Bibl. *463*).

[26] *Vision in Motion* (Chicago, Ill., 1947), p. 153.

[27] From the manifestos in Joshua C. Taylor's *Futurism* (1961), pp. 124 ff.

[28] Gerhard von Graevenitz, in *Kinetics,* being the catalogue of the exhibition at the Hayward Gallery, London (1970). The remaining statements are from the same catalogue, else from *In Motion: An Arts Council Exhibition of Kinetic Art* (1966).

[29] *The ABC of Relativity,* 3rd rev. ed. (1969), p. 104.

[30] *Film Form,* trans. Jay Leda (1951), pp. 182, 185. See also Panofsky's remarks on the cinema's 'spatialisation of time' (Bibl. *240*).

[31] In *The Making of Kubrick's '2001',* ed. Jerome Agel (1970), p. 341. For two different approaches one might also glance at Alesandro Jodorowsky's *El Topo* and Andrei Tartovsky's *Solaris.* The approach of the first is, according to its director, not linear but structural, with 'no sense of time' (interview in *The Guardian,* 17 November 1973); the approach of the second suggests, in accordance with the original novel by Stanislaw Lem (trans. J. Kilmartin and S. Cox, 1971), the concept of 'an evolving god, who develops in the course of time, grows, and keeps increasing in power while remaining aware of his powerlessness'.

[32] *The Principles of Psychology,* rev. ed. (1910), I, p. 239 (the italics are James's); first published in 1890. For a survey of the various interpretations of 'the stream of consciousness', consult James Naremore, *The World Without a Self* (New Haven, Conn., 1973), chapter IV.

To the extent that the metaphor of 'flow' echoes Newton's terminology (above, p. 8), it was correctly deployed by James; but it is a dangerous term for the post-Einstein era in that it tends to hypostasise time—as if time were a 'thing'. Cf. Bibl. *173*.

[33] Chernyshevsky's observation on Tolstoy (Bibl. *282*) is variously applicable to novelists from Jane Austen to Flaubert to Dostoevsky—and of course to Tennyson's *Maud* and Browning's dramatic monologues.

[34] See Hoffman (Bibl. *264*), chapter v, 'Infroyce'. Most discussions of the 'interior monologue' glance, whether in partial agreement or total exasperation, at Édouard Dujardin's *Le Monologue intérieur: son apparition, ses origines, sa place dans l'œuvre de James Joyce* (Paris, 1931).

[35] Above, p. 3. Joseph Frank's important essay 'Spatial form in modern Literature' (Bibl. *427*) overstates the thesis that the work of Proust, Joyce, *et al.*, is meant to be apprehended 'spatially, in a moment of time, rather than as a sequence'. True, their ambition to achieve simultaneity is crucial; but to the extent that literature like music remains confined to time-as-sequence, the term 'space' flies in the face of reality. 'Space-time' were more accurate, for it acknowledges the reality of time-as-sequence even as it suggests the broader dimension intimated by Frank. Cf. Bibl. *284*.

[36] *Composition as Explanation* (1926), p. 19.

[37] Martin Esslin, *The Theatre of the Absurd* (1961), p. 19. See also Bibl. *395, 504*.

[38] *Phoenix*, ed. E. D. McDonald (1936), pp. 218–19. Predictably, Lawrence disapproved of the novel's new directions; see William Deakin, 'D. H. Lawrence's attacks on Proust and Joyce', *EC*, VII (1957), pp. 383–403.

[39] Jorge Luis Borges, 'The garden of forking paths', reprinted in various collections of his short stories as well as in *The Traps of Time*, ed. Michael Moorcock (1968; Penguin ed., 1970), p. 156. Even the oft-used time-machine has on occasion been adapted to advantage, as in David Gerrold's *The Man Who Folded Himself* (1973), where an initially amusing 'temporal transport device' yields to a nightmare vision of interminable cycles created by man himself. Lesser writers, on the other hand, appear content merely to recall Hamlet's relevant phrase; but the result— Philip K. Dick's *Time out of Joint* (1959)—is science fiction at its most forgettable.

[40] *Orpheus*, etc., trans. Carl Wildman (1962), p. 121.

[41] *The Complete Poems of Walter de la Mare* (1969), pp. 551–96. Auden's judgement appeared in *The Observer*, 29 October 1972.

[42] Hans E. Holthusen, *Rainer Maria Rilke: A Study of his Later Poetry*, trans. J. P. Stern (Cambridge, 1952), p. 24. See also Bibl. *432, 494*, as well as Rilke's important letter to Witold von Hulewicz (1925) in *Selected Letters*, trans. R. F. C. Hull (1947), pp. 393 ff.

I

'Time past':
the background

Catherine Rau†

Theories of time
in ancient philosophy

Students of ancient philosophy are aware of a continuous tradition in the treatment of the subject of time; even in cursory examination of the texts, it is apparent that the tradition was firmly established and steadily persistent. The dominant thinkers in this field were, as one would expect, Plato, Aristotle, Plotinus, and Saint Augustine. Comparison of Augustine's discussion of time with Plotinus' discovers many similarities, including even the use of the same rhetorical devices. Plotinus intended to develop Plato's conception of time and eternity (or, at least, what he took that conception to be) and to refute Aristotle's account of time. Let us go back, therefore, to the beginning of recorded theory of time in our western tradition.

Plato's metaphysical commitments prevent him from taking time seriously. According to his famous Doctrine of Ideas (to which he recurs all through his writings, and which is given its fullest expression in the *Republic* through the description of the Line of Knowledge at the end of book VI and through the Allegory of the Cave at the beginning of book VII), the sensible world, temporal and changing, is not fully real and has no more than semblances of moral and aesthetic values, whereas the intelligible world, eternal and changeless, has true being and absolute value. Plato never discusses time in complete earnest; he even indulges, in the *Statesman* (268e–74e), in a childishly fantastic myth of periodic reversals of the course of time in the universe. His account in the *Timaeus* (27e–39d) of the creation of the universe and the beginning of time is also mythology; indeed, he gives explicit warning that the views on the material world expressed in this dialogue are to be taken as only probable.

In Plato's admittedly unphilosophic account, then, the sensible cosmos was created by the Demiurge according to the perfect pattern of the eternal world of intelligible essences; though the divine creator wished to make the cosmos as much like its ideal model as possible, he could not make it eternal, for no created thing can be eternal. Therefore, establishing order in the heavens, he made of eternity 'abiding in unity' an image 'going

† Reprinted from *Philosophical Review*, LXII (1953), pp. 514–25.

according to number'; and this image is time. Days, nights, months, and years are parts of time; past and future are created forms of it. The sun, moon, and stars were created in order to show plainly the measure of time.

This heady mixture of myth and metaphysics is not at all clear, and the definition of time as the moving image of eternity is merely agreeable rhetoric; yet a couple of points emerge as important for the thinking of both Plotinus and Augustine on this subject. Plato discusses time in connection with the creation of the universe: it is doubtful that this is a religious notion for him, but it acquires a distinctly mystic aspect for Neo-platonism, as everyone knows; and it is, of course, one of the central concepts of Christian theology. Furthermore, Plato opposes eternity to time, making the latter inferior to the former in ontic status and value. We shall see in what different ways this complicated concept is developed by the Hellenistic thinkers on one hand and by the Church Fathers on the other.

Is there in the *Timaeus* any sober philosophic view of time? A case could perhaps be made out for it. Plato seems to conceive time as a flow within which events occur, probably just as further on in this same dialogue (48e–53b) he speaks of space as the matrix or receptacle of material bodies. Platonic time is a single, continuous, evenly-flowing stream, which carries along mundane events and whose surface is marked at regular intervals by astral events; since it pervades the whole universe and is homogeneous throughout, it is absolute—but not in any precise scientific sense, obviously.

Aristotle, before giving his own account, cites and criticises (*Physics* IV, 10) views on time and its nature which have been 'handed down'. He mentions the notions that time is the motion of the heavenly sphere and that it is the sphere itself. The extant doxographers attribute the first to Plato and the second to Pythagoras; probably Aristotle had done so before them. As we have no reliable records of Pythagoras' teachings, he must be disregarded in a serious consideration of theories of time. And if the mention of the definition of time as the revolution of the heavens were intended as an indication of what Plato says in the *Timaeus*, it would be a gross misrepresentation. Also in the same passage of the *Physics*, immediately after the opinions which had been 'handed down', Aristotle examines and rejects the view then current that time is motion in general. The practice of reviewing and criticising the forerunners in this field (as in others) became traditional in the schools of philosophy; indeed, in the remaining compilations of the doxographers, the very order of the names and the wording of the summaries of opinions are stereotyped. The tradition must have grown under Aristotle's authority, for his material keeps reappearing. Plotinus follows this established practice, even to taking advantage of Aristotle's criticism of his precursors, though he opposes vigorously Aristotle's own account of time; and Augustine makes very effective use of some of the traditional material.

Still in the same chapter of the *Physics*, Aristotle raises the problem of the existence of time. The past has been and no longer exists; the future is going to be and does not yet exist. The present, which is no more than the dividing line between past and future, is continually shifting, that is, ceasing to exist. How then can time be? This piece of analysis makes it obvious that Aristotle is concerned with empirical time; his point of view is strictly scientific. Augustine opens his discussion of time in a closely similar manner and is also preoccupied with empirical time, though for very different motives, as we shall see.

After these preliminaries, Aristotle proceeds to give his own account of time (chapters 11–14). Though time is not motion, it is not independent of motion; for when we are not conscious of change of any kind, we are not aware of the lapse of time; and, conversely, when we feel no lapse of time, we notice no change. By Aristotle's definition, literally translated, time is 'number of motion according to before and after'. This rendered in modern English and modern concepts would go: time is measurement of motion in temporal order. Aristotle goes around another circle in affirming that we measure motion by time and time by motion. Yet in spite of these confusions (which Plotinus and Augustine deal with, each in his own way), Aristotle, by his insight that all events are temporal and that there is no time without events, was on the track of a more adequate theory of time as the order of events.

He states further that time is made continuous by the present and is divided by it into past and future; the present corresponds to a point which is the end of one line and the beginning of another; one point connects the two lines. Later in the *Physics* (vi, 3), he repeats this and then shows, by a neat piece of dialectic, that the present must be a mathematical point. To return to the fourth book, though the instantaneous present is indivisible, time, being a continuum, is infinitely divisible. It is, furthermore, the same everywhere at once, for simultaneous times are identical though various events are occurring in them. Time is never slow or fast, but proceeds at an even rate. He says here and repeatedly throughout the three physical treatises and again in the *Metaphysics*, that time is infinite: it had no beginning and will have no end; it cannot fail as long as motion exists, and the motion of the universe, being circular, is unvarying and unending.

Aristotle, in attributing to time complete continuity, universal homogeneity, and invariable rate of flow, is conceiving it as absolute. Furthermore, the concept of the instantaneous present is reached by analysis, whereas the present as immediately felt has duration. He wavers between a physical point of view and a psychological one: he has run into the difference between so-called 'real' time and phenomenal time, without recognising it. He opened his account with a consideration of time as we experience it in our own consciousness, and he concludes by raising the questions as to how time can be related to the soul and whether time would exist if the soul did not. We must look for no subtle insight here, however, for he explains with naïveté (of a sort accessible only to a great mind) that since nothing but soul and mind can count, if soul did not exist, time would not. Merely that of which time is an attribute, i.e., motion, would exist—if motion could exist without soul. In making this last conditional statement, he is entangled in the assumption, which runs right through Greek philosophy, that all motion in the universe must be originated, directly or indirectly, by souls.

A word should be said about the Epicureans and Stoics merely for the sake of thoroughness, since they made no important contributions to the theory of time, as far as we can judge, and seem to have had no appreciable influence upon other writers in this field. For Epicurus (Diogenes Laertius x, 72–3), time is a character of events; we experience them directly as having duration. Lucretius makes only brief passing references to time: it does not exist in itself, but is abstracted from events by the mind (I, 460–4), and the smallest unit of time perceptible by sense can be divided by reason (IV, 795–7). Epicurus had made this distinction between the shortest time perceptible by sense and the minute divisions distinguishable by the mind (x, 47, 62). There are suggestions here of a critical and competent treatment of the subject, but none of the ancients, to our knowledge, followed them out.

Zeno and his early successors in the Stoa apparently concur with Aristotle in defining time as the measure of the motion of the cosmos and in declaring past and future to be limitless (Diogenes Laertius VII, 141). All that they add is the notion of the punctuation of the infinite stretch of time by periodic world conflagration or dissolution, and this notion coalesces with their pantheistic doctrine of the alternating production of the universe by God from himself and reabsorption of it into himself (VII, 136–8, 142). The later Stoics show no interest in the problems of time. Epictetus speaks once of the conflagration of the universe (*Discourses* III, 13). Marcus Aurelius refers frequently to time, which had no beginning and will have no end; but his references are merely brief passing ones, with a moralising tone, of course.

In Plotinus' metaphysical system, it will be recalled, the One comes first; it is indefinable, for it transcends all Being; sometimes he identifies it with the Good, while at other times he maintains that it precedes both the Good and the Beautiful; in some passages, he names it God. From the One proceeds Intelligence, which is twofold as knower and known: as the knower, it is the self-vision of the One; as the known, it is the world of Ideas. From Intelligence in turn proceeds Soul, which contains all individual souls and which produces Nature and the sensible world. Soul has two parts, an upper one turned toward Intelligence, and a lower one attracted by sense. Plotinus sets up a hierarchy of reality and dignity from the supreme One down through the grades of Being and Nature to passive matter, which is non-Being. It is only in the context of this elaborate metaphysical-religious construction that his discussion of time and eternity (III, 7) can be followed.

Eternity and time are two different things, he begins; we think that we understand them clearly, and we talk about them commonly; yet when we reflect upon them, we find ourselves at a loss. So we turn to the opinions of the ancients, but we have to inquire which of the philosophers have discovered the truth. We should first seek after eternity, according to those who hold it to be different from time. Our search could, however, proceed in the opposite direction, for (repeating Plato's phrase) time is the image of eternity; we could ascend from study of the sensible and temporal to understanding of the intelligible and eternal.

Plotinus proceeds through five chapters of rhetorical developments on eternity, quoting several phrases from the *Timaeus* and mentioning Plato by name twice. Eternity is infinite life always identical, always present to itself in its totality, always an indivisible perfection, admitting no change. Eternity is of the essence of the intelligible nature; the latter, being complete and perfect, lacking nothing, has no future; as it is ungenerated and unchanging, it has no past; this continual present is its eternity. He is careful to explain here (chapter 6) that the sensible universe had no beginning in time: Plato's Demiurge is anterior to it as cause, not temporally. And elsewhere in the third *Ennead* (5, chapter 9), Plotinus insists that the account of the creation of the cosmos in the *Timaeus* is a myth which spreads out in time what it recounts, and separates beings which exist together; Plato does not intend what he says to be taken literally.

In the seventh treatise of the third *Ennead*, Plotinus, descending from contemplation of eternity to inquiry into time, finds it necessary to review the opinions of the 'ancient and blessed men'; he runs through the traditional material on time, without mentioning names, however, and ignoring for the moment the fact that one of the items on the list is attributed to Plato. He follows Aristotle closely, taking advantage of his criticisms and developing some of his arguments. Here, as all through the *Enneads*, he owes more to Aristotle than he is willing to admit. Then he devotes an entire chapter (chapter 9) to what he intends as an annihilating criticism of Aristotle's own theory of time. He very nearly succeeds. With acuteness, he points out in detail ambiguities, confusions, and lacunae in Aristotle's account, though falling into some confusion himself, especially in his mathematical notions. He vigorously rejects a subjective view of time: Why should time have to be measured in order to exist? That is like saying that the size of something remains indeterminate if there is no one to measure it. And why would time not exist before there is a soul to measure it—unless one says that a soul generates it?

When, after these preliminaries, Plotinus comes to telling us 'what we should think time to be', he gives us more metaphysical rhetoric. Time was not generated after intelligible Being, but is posterior to it logically and by nature. Soul, creating the sensible world in the image of the intelligible one, made time in place of eternity; then it subjected to time the world it created and placed it entirely within time. Whereas eternity has rest, unity, identity, permanence, and indivisibility, time has motion, multiplicity, diversity, change, and divisibility; but by its continuity and progress toward infinity, time is the image of eternity. In creating time, Soul made itself temporal. Plotinus reiterates here and elsewhere that time is in Soul as eternity is in Intelligence. Time, which is a progressive lengthening of the life of Soul, consists in motion from one state to another, always toward a future.

In a passage of some length (chapter 12), Plotinus clearly defends Plato (referred to as 'himself') against the traditional attribution to him of the doctrine that time is motion of the heavenly bodies; he quotes (inaccurately) several phrases from the *Timaeus* and explains them, or so he thinks. He adds in the next and final chapter that, though the revolution of the sun takes place in time, time does not exist in anything else; it is what it is in itself. Within its regular and steady flow occur the motion and rest of other things. Then he renews his attack on Aristotle, charging him sharply with self-contradiction and circularity in his statement that motion is measured by time, and time by motion. Plotinus invokes Plato by name, quotes again from the *Timaeus* (this time, correctly), and concludes with a remark on the soul: time has its being in the universal soul; since the latter is everywhere, time is everywhere. It is in our souls, too, for they all form but one in the universal soul.

For Augustine, as for Plotinus, eternity is a different state of being from time; but, whereas Plotinus disdains time, since for him it 'fell' from eternity, Augustine is obliged by his theology to regard time as important. God is eternal and the material world, temporal; therefore time is inferior to eternity, as the creature is below the creator; yet it is in regard to temporal life on earth that human destiny is determined for all eternity. Each soul attains salvation or is damned in its own temporal span, while in the total history of the race a divine plan unfolds. Augustine gives a complete Christian history (past, present, and future) from creation to the Last Judgement, in the *City of God*.

Whether one accepts Augustine's theology or not, one can appreciate its moral superiority over Greek metaphysics in regard to its view of human personality. For Christianity, the worth and dignity of the human being are such that it deserves an individual destiny. In all Greek philosophy, the soul is assigned no more than a cosmic function: souls are the originators of all motion in the universe (for matter is inert), but the souls of men are allowed no personal destiny beyond life on earth. Moreover, Pythagoras, Plato, and Plotinus teach a dreary succession of rebirths, not always even human; and more demoralising still is the Stoic belief that, after every world conflagration, each human life along with the whole course of the universe is repeated, the repetition continuing endlessly. When, on the Christian view, men have but one life to live, and when they are collectively acting out a tremendous drama the outcome of which is to endure for eternity, time becomes valuable.

Augustine has his own reasons for being concerned with the problems of time and eternity. His life was a singularly violent conflict between the lusts of the flesh and the longing of the soul; the opposition of matter and spirit is, for him, the tension between time and eternity. In his *Confessions*, a narrative of his search for truth and of his entering upon the way of salvation, he adverts to the problems which had agitated him along his route. Chief among these are the nature of God and the justification of creation.

Again we find a theory of time in connection with an account of the creation of the world, here according to Moses. The last three books of the *Confessions* are a commentary on the opening chapter of Genesis. In this commentary, the discussion of time (xi, 14–28) appears as a digression. Augustine explains (chapters 7–13) that God created the heaven and earth out of nothing by His word, and with them created time. It stands in contrast to eternity: time, in which there is a succession of events, has past, present, and future; in eternity there is no change or passing away, the whole being present. An ever-present eternity precedes the past and follows the future. After this exegesis, the theologian temporarily gives way to the philosopher; Augustine, fascinated by the problem of empirical time, gives his keen wits free rein.

What is time? he begins. We talk about it familiarly, yet when challenged to analyse our concept of it, we cannot. Plotinus had begun his treatise in just this way in regard to eternity and time. Augustine appeals at once to experience: we know that past events have occurred, and future ones will occur, and present ones are occurring. Then he raises Aristotle's question as to the existence of time. How can past and future exist, when the past no longer is, and the future is not yet? The present, moreover, is continually passing away. His argument swings steadily between these two points of the indubitability of immediate experience and the logical necessity of analysis.

We speak of 'a long time' or 'a short time', he continues. What is long? Not the past or the future, for neither exists. 'Let us see then, human soul, whether present time can be long, for it is given to you to perceive stretches of time and measure them' (chapter 15). How long is the present? He narrows it progressively from one hundred years to a year, a month, a day, an hour, an instant. The present is reduced by analysis to a mathematical point; it has no duration; if it did, it would be divided into past and future. Here he follows Aristotle again. But we do perceive intervals of time and measure them, he hastens to assert; we measure them as we are experiencing them. Augustine is well aware of the difference between the phenomenal present and the mathematical present; indeed, he gives a sample of the analysis wherein the former is reduced to the latter.

Now past and future events must exist, he maintains. If not, how could we talk about them at all? Then where do they exist? Wherever they are, they must be present. When we talk about the past, e.g., our childhood, we have in our memory, not the events themselves but images of them; and when we predict future happenings, e.g., the sunrise, we do not perceive the events themselves, but their signs or causes which lead us to expect them.

it is not properly said that there are three times, past, present, and future. Perhaps it might be said rightly that there are three times: a time present of things past; a time present of things present; and a time present of things future. For these three do coexist somehow in the soul, for otherwise I could not see them. The time present of things past is memory; the time present of things present is direct experience; the time present of things future is expectation. [as above, p. 5]

Having reached this conclusion as to the nature of time, Augustine turns to the problem of its measurement. It is a fact of experience that we measure stretches of time. Not past or future ones, for we cannot measure what does not exist. Present time, then? But how, since it has no duration? But we continually measure time, he insists; the process is familiar, yet obscure. Then he introduces the traditional material with engaging casualness: 'I heard from a certain learned man that the motions of the sun, moon, and stars are times themselves; but I did not agree' (chapter 23). He criticises this opinion in careful detail and with originality, going far beyond both Aristotle and Plotinus. He demonstrates clearly and concisely that by time we measure the motions of bodies, so the motions of the heavenly bodies are not time. As Plotinus had quoted Plato, he quotes Moses: the celestial lamps are *for signs and for seasons, and for days and years*. He explains further that time is not the motion of other bodies either. They move at various and varying speeds, or they remain at rest; their motions and their rest are all measured by time. His explanation is given from a consistently subjective point of view; there is no recourse to an objective metric, and no surreptitious introduction of absolute time. He points out that we measure the motion of a body by time, and we measure the time itself; indeed, we cannot measure the motion unless we measure the time it occupies. Here he cuts through Aristotle's confusions with easy penetration.

Augustine returns to his problem: How do we measure time itself? Do we measure a longer time by a shorter one, as we do space? How can we measure, say, the recitation of a poem? Analysis demonstrates that we can measure neither future, nor past, nor present, nor passing time; yet experience convinces us that we do measure stretches of time. How, in reciting a verse, do we judge some syllables long, and some short? Since they succeed each other, they cannot be juxtaposed. Still we do compare them. We do not measure the syllables themselves, but something imprinted in our memory. 'In you, my mind, I measure my stretches of time. . . . I measure that present impression which passing events make in you, and which remains when they have passed, not the events which, in passing away, have made it; this is what I measure when I measure stretches of time. Therefore either these are time, or I do not measure time' (chapter 27).

Thus, in conclusion, Augustine returns to the nature of time, taking as an example the recitation of a psalm which has been memorised. While the recitation is in progress, some of the psalm has sounded; the rest will sound. Present perception conveys the future into the past, the past growing by diminution of the future, till the future is consumed, and all is past.

But how is the future, which does not yet exist, diminished or consumed, or how does the past, which no longer exists, grow, unless the three are in the operation of the mind? For the mind expects, attends, and remembers, so that what it expects passes through what it attends to, into what it remembers. . . . And what occurs in the whole psalm does also in its individual parts and single syllables, and also in the longer action of which this psalm may be a part, in the whole life of a man, of which the parts are all the man's actions, and in the whole era of the sons of men, of which the parts are all human lives. [chapter 28]

Though one may not agree with Augustine's theory of time, its unusual merits command admiration. Notwithstanding its ancient paraphernalia, it has freshness of viewpoint and originality of treatment. Of far more importance is the great advance it makes over all previous efforts in this field. Plato gives us a handsome myth; Aristotle, inadequate physics confused by a little psychology; Plotinus, rapturous metaphysics. Augustine gives a clear, adequate, fully argued, critical theory—one not limited to solipsism, of course, for 'private' times can be correlated to construct 'public' time. Furthermore, Augustine is interesting because his is the first serious attempt at an account of time in seven centuries, and the last for fourteen more. He anticipates, everyone knows, Kant's subjective view of time as the a priori form of sensible intuition, but with this difference that he is far more lucid, coherent, and consistent than Kant. And, finally, though St Augustine now belongs to eternity, he has anticipated the relativism of the theories of time of twentieth-century physics.

Herman Hausheer†

St Augustine's conception of time

With an elevation of thought and a poetry not unlike that of Plato, and with a nicely discriminating analysis that places him among the greatest of psychologists, Augustine investigates the nature of time. His subtle and profound mind found a peculiar attraction in the contemplation of the mystery of time, which is essentially bound up with the mystery of created being (*De Civ. Dei* XII, 15). Few men have been as intensely sensitive to the pathos of mutability, of the rapidity, transitoriness, and irreversibility, of time.

Following his inclination to subjectivism, Augustine asks himself how time represents itself to the mind. He first seeks to render the idea of time clear by a brief, provisional definition, based upon the usual idea that time has three parts. While one meets nothing but riddles in an investigation of the nature of time, nevertheless so much is certain, that if nothing were passing, there would be no past; if nothing were to come, there would be no future; and nothing would exist, if there were no present. The past is that which is no more; the future that which is not yet. And if the present were perpetually present, there would be no longer any time, but only eternity. For the present to belong to time it must pass. Hence time only exists because it tends to not-being.

A logical analysis of the various conventional time-intervals discovers that the present is an instant of time which can no further be divided into smaller particles. The time-atom flies with such speed from the future to the past that it cannot be lengthened. This time-particle or present has no space. Thus, the present being the only real time, it is diminishing to an inextensive point. Such a conception would be in the tradition of the mathematical conception of time. 'If any fraction of time be conceived that cannot now be divided into the most minute momentary point, this alone is what we may call time present. . . . But the present has no extension

† Reprinted from *Philosophical Review*, XLVI (1937), pp. 503–12. Quotations from Augustine's *Confessions* are from the translation by Albert O. Outler (1955); and from *De Civitate Dei* (*Of the City of God*): by John Healey (1610) as revised by R. V. G. Tasker (1945).

[*spatium*] whatever' (*Conf.* XI, 15). Obviously this conception of time is the same as that of Descartes. Doubtless Augustine is far from attaining a formula as clear as the enunciation of a geometrical theorem. The principal thing is that he recognised the possibility of a mathematical analysis of time. Even though he does not know what time exactly is, he at least states what it is not, often the sole solution of many problems.

On the other hand, Augustine presupposes that the present is only inextensive if subjected to a logical analysis, that in reality it is still felt as duration. In general he admits that the present has no extension in abstraction. It cannot remain for long as an indivisible instant; for, however small the extension in duration, the present instantly turns itself into a past which is no longer and a future which is not yet.

The three dimensions that we customarily distinguish thus reduce themselves to one, the present, in which the past survives in memory and the future pre-exists in some way in the form of an anticipation. But the indivisible present does not cease to vanish, neither is it in reality entirely devoid of any extension of duration. The individual durations dovetail, so to say, because they have diverse contents. The number of isolated intervals can be readily noted, and thus one is in possession of a remembered or an expected total durational present.

Time thus reduces itself to the impermanent, being made of a succession of indivisible instants. It has therefore no relevance to the stable immobility of divine eternity: 'time, being transitory and mutable, cannot be co-eternal with unchanging eternity' (*De Civ. Dei* XII, 15). Between God and the creature is the same difference as between a consciousness in which all the notes of a melody are simultaneously present, and a consciousness which perceives them only in succession. In its normal operations the human mind through memory in some measure transcends time, as, for example, when we apprehend as a whole a metre or a melody, though the individual notes and sounds are successive not simultaneous (*Conf.* XI, 33).

The difficulty is not only to account for eternity, which escapes us; for time itself, which sweeps us off our feet, is a mysterious reality. The essence of time is the indivisible instant of the present, which knows itself to be neither long nor short. How then can we speak of a longer or shorter time, or even of a time double the other? However, we measure time. That is a stubborn fact. But how can we measure the length of a past which is no more, of a future which is not yet, or of an instantaneous present? What we measure is the absence of the present. It is therefore not correct to say that the past or the future is long. We rather say of the past that it *was* long, and of the future that it *will* be long. But can one truly say of the present that it *is* long? Can it be measured? A century cannot be present, neither a year, nor a month, nor a day, nor an hour. Time is never simultaneously present in all its parts, but only in an indivisible instant. Aristotle already said: 'Nothing exists of time except the present which is indivisible'

(*Physics* II, 2). Therefore neither the present nor the past nor the future can be called long or short.

Nevertheless, the fact remains that we measure time, that we make comparisons between the intervals of time. Is it nothing we measure? Have past and future no real existence? How is it that out of an unrealised future, out of not-being, the present emerges, and that the present in turn instantaneously submerges into the past, where it is annihilated? If the past has no real existence, then all history would be false, and if the future has no real existence, prediction would be impossible. They both have an objective existence in the sense that they are being discerned in the mind. While I neither perceive the past nor the future, I know where they are. For when we recall the past, we do not recall the actual events, which are no more, but the thoughts and images these have left in our mind. Our infancy vanished into the past, but we see present its image when we revive it in our memory. But if the future cannot be foreseen by means of images, how is it predicted? Just as we infer the future sunrise from the aurora by means of signs, so we learn to know the future. Prediction or prevision is a refined inference from cause-and-effect relations (*Conf.* XI, 19).

Augustine is quite aware that one of the most intricate difficulties of the problem of time is the question how time is measured. This phase of the problem raises more riddles than solutions. No one shows a keener appreciation of the contradictions involved in the proof of the objectivity of time. If time is nothing, if the past and future have no real existence, how can one measure them? For in order to measure anything there must be something. No one measures the non-existent. While no part of time is, we yet measure it. The solution of the paradox is that time is present in and measured by the mind. There are thus properly not three times—a past and a future which are not, with an immediate present which is a mere point of transition between two non-entities; but there are three presents, a present of things present, a present of things past, and a present of things future. The present of the present is experience, the present of the past is memory, and the present of the future is expectation. This triple mode of the present exists in our mind, or not at all. The only answer Augustine can give to the one who asks him how he measures the non-existing and non-spatial times is: 'I know . . .' In other words, the question still is, is time measured if it is not space? It is a profound enigma.

With renewed zest Augustine attempts once more to give verbal precision to the nature of time before he tells how we measure it. To resolve the problem we may identify time with motion. To grant such a solution seems an excessive simplification of Aristotle. For if time is not motion, it must be its own measurement. Thus we can measure time with time, motion with motion. However, the motion of a body is essentially its displacement between two points situated in space. This spatial displacement continues to be identical whatever the time consumed by the body. Moreover, if the body remains immobile at the same point, there is no motion whatever; yet I can still estimate with more or less rigorous exactitude the time of its immobility. Thus the motion which time measures is one thing, and the time which time itself measures is still another thing. Time is thus not the motion of bodies (*Conf.* XI, 24).

Feeling that the mind in some sense transcends the process of time it contemplates, Augustine could not rest satisfied with the naïve objectivism of Greek science, which identified time with the movement of the heavenly bodies. For if the movement of bodies is the only measure of time, how can we speak of past and future? A movement which has passed has ceased to exist. There remains only the present of the passing moment, a moving point in nothingness. Therefore, Augustine concludes, the measure of time is not to be found in things, but in the human mind.

But how do we measure time itself? Do I measure it by comparing a larger movement with a more limited movement? If it is with time that I measure motion, with what do I measure time? With time? In a certain sense yes; for I can measure the duration of a long syllable with that of a short one, or that of a poem with the number of verses it contains, which verses measure themselves in their turn by the number of feet, the duration of their feet by that of their syllables, and those of their long syllables finally by those of the short ones. But what can I say about it? If it is a question of their length on paper, that is space, not time, that I measure. If it is a question of verses pronounced by the voice, the dissociation of time and motion reappears under another form; for a short verse can be pronounced so that it lasts a longer time than a long verse, and *vice versa*. It is the same with a poem, a foot, a syllable.

Measurements of this kind are spatial, not temporal. Thus Augustine does not ignore the fact that time is not only a function of the amplitude of motion, but also of speed. It is above all in ourselves that we must seek the measurement of time.

In order to discover the connection between the permanent and the transitory, which for Augustine is after all the whole problem, he has recourse to a metaphor according to which he conceives of time as something analogous to space, as a kind of distension of the mind, which alone renders possible the coexistence of the future and the past in the present. Such a solution is characteristic of Augustine. In every question he finds the trial within. Here it is in memory and thought that he catches sight of his quarry. Not unlike Bergson he defines the mind in terms of attention. And as the human mind is but a dispersed image of the One, it is natural that it should have to stretch itself out in recollection of the past and strain to the future.

The distension of the mind enables one to perceive duration and makes possible the measurement of time. It is impossible to measure what does not endure and what has ceased to exist. Augustine means by mental distension the faculty of the mind to know successively the past by memory, the future by prevision, and the present by actual perception, to dilate itself, so to say, by prevision and memory from the remotest future to the most distant past.

Augustine is still not satisfied with the proffered solution of the problem at hand. If the non-existing future and past together with the instantaneous present are not amenable to measurement, neither can the uninterrupted passage of an event be measured, for measurement implies the conjunction of a beginning and an end. That is, the mind has to know at least two terms which are simultaneously in the present in order to be able to measure time. The solution of the problem is in showing what the connection between the beginning and the end, between the two terms, is. Not the transition of things measures time, but the impression they have left in the mind. Time is nothing but an impression, a mode of thought, a reflex of things passed and passing, and in particular a function of memory. The non-existent past is measured in memory. The impression which preserves the transitory survives the things themselves, and comparing them a certain measurement of their intervals or successions is made possible. What is true for the memory of the past is also true for the anticipation of the future.

Time no longer divides itself into a present, past and future existing outside of us. Its three dimensions coincide, although the present is the only one which is real and invisible. They coincide by the grace of the mind. The enduring attention of the mind provides the coincidence of the three dimensions of time. Memory, 'the light of the intervals of duration', is the subsisting distension of the present into the future and into the past. It is interesting to note the analogies which Bergson's and Augustine's psychology of duration have in common.

Finally, Augustine compares the time-process with the recitation of a poem which a man knows by heart. Before it is begun the recitation exists only in anticipation; when it is finished, it is all in memory; but while it is in progress, it exists, like time, in three dimensions. And what is true of the duration of a poem, holds equally good of the duration of each line and syllable of it. It is equally true for the whole life of man, whose actions are its parts; and, finally, it holds good for the whole human race, which is the sum of individual lives (*Conf.* xi, 28).

If this is so, what meaning is there to the question what did God do before the creation? For human consciousness in bringing the future and the past together in attention, the words before and after have no longer any significance. What tremendous effort does it take to attain a tolerable comprehension of the relation of created time and creative eternity! Man can only succeed on condition that he withdraws his thought from the flux of time, and integrates in a permanent present the totality which is no more and which is not yet. Thus alone may he and now pass from time to eternity (*Conf.* xi, 11, 29, 31).

Thus the metaphysical alone in the end provides the solution of the psychological problem of time. True, man knows by analysis as well as by intuition. In analysis time is succession. In intuition time is no more. It is eternity. Time is the distension of the eternal; eternity is an immutable present, which is neither preceded nor followed by another moment. Man's weakness in perceiving things simultaneously in the unity of an indivisible act, prevents things from existing simultaneously in the unity of a fixed permanency. Whatever succeeds each other is incapable of coexisting (*De Civ. Dei* XI, 6; XII, 15). Whereas men know things temporally, that is, in succession, God knows eternally, that is, simultaneously. Whereas human consciousness always knows exactly at which point of its unwinding activity it is, divine consciousness is unchangeably self-subsisting on its level. Having started with eternity in his study of time, Augustine also ends with eternity.

For man life is wasted because it flows, because it dissipates and consumes itself in time. The sense of this is itself due to the presence and operation of something which does not pass. For Augustine that something was no lifeless abstraction, but a concrete fullness of life, ever the same because it contains in itself all the values produced at each passing moment of time. This apprehension of eternity was one of the major factors that moulded the philosophy of Augustine.

It is thus in keeping with the heritage of Neoplatonism that Augustine seeks to preserve the dignity of God by placing him outside of time and space. Hence eternity and time are absolutely incompatible. Their differences are absolute. Time implies change, movement, transition, succession, imperfection, and improvement. Eternity is all that time is not. It is the immutable, quiescent present, the simultaneous unison of that which unfolds in time. Time and eternity are incommensurable. They are not of the same dimension. There is no comparison between an ever fixed eternity—*semper stans aeternitatis*—and a time that is never fixed (*Conf.* XI, 11). Being a totally realised perfection, God is wholly independent of time. He is an immobile eternity. His life is not an ascent to still higher perfections. Neither is it a descent to a lower world. It is a process without external aims. Its process is self-concentrated, circulating on its own horizontal level.

There are two unique peculiarities in the nature of time, which in their contrast constitute an antinomy. They are, first, the self-finality of the present; and, second, the irrevocable irreversiblity of its sequence. Only the experienced instant is given. In every instant a whole world perishes, and in every instant a whole world emerges out of nothing. Infinite past and infinite future do not exist. Moreover, because the present condenses itself to an inextensive point, it seems to dissolve all existence into emptiness. The paradox of the evaporation of the moment and the annihilation of the present is a profound abstraction. How is one to resolve the paradox of the annihilation of the intervals of time, the non-reality of the past and the future? Nothing is ever destroyed. Neither is nothing ever magically produced. The infinite moments of time, while perishable for man, coexist in God's eternal present. They abide in the *nunc stans* of the scholastics. The souls of men pass through these perishing intervals of time until they come to rest in God.

This is indeed a solution. However, it opens up new problems. Augustine leaves the question of the variability of the experience of the present— from individual to individual and within the lifespan of the individual—untouched. Is there a present which encompasses all men? What relation if any exists between God's infinitely enduring present and the varying consciousness of the present of men? Metaphors alone seem to serve here as tools of interpretation.

Based upon the presuppositions of his system of ideas, Augustine might have answered the problem of the relationship of God's eternal present and man's varying experience of the present as follows: There exists a similarity as well as a difference between God's eternal present and man's consciousness of the present. They both are real. While there are infinitely many things timelessly together in God's eternal present, there are only minute segments of eternity in man's limited consciousness of the present. The distribution of the realities of the present among men is due to their finiteness; the passage of the souls of men through the divine coexistence is an arrangement intended to procure for the finite souls the greatest possible enrichment. Augustine may give an intimation of all this in his expression 'we pass through God's today'. God thus encompasses all souls.

As there is no time in God, he does not create successively in time. Augustine realises the difficulty of how God could decree eternally that there should be a finite creation of a few thousand years. Since creation had a beginning with time, it also will have a dramatic end with time. If therefore time has no significance for God, how can God eternally determine a finite period of creation? Augustine struggles with the problem, but is unable to solve it by the tools of the Greek speculative tradition. He could have made it plausible by including time in God, that is, by using the tools of the Hebrew tradition. In the latter heritage eternity meant that which endures through all time. Augustine often uses the language of the Hebrew tradition. 'Thou art the same and thy years fail not' (Psalm 102, 27). But by merely alluding to it, he failed to work it out, as it was contrary to his basic assumptions.

The inexorable irreversibility of temporal sequence is an indisputable fact. Reality is perpetually clipped off from the duration of the present. The non-existent gnaws itself from the past into the future. The present endlessly assimilates reality to the non-existing future. How is one to escape from this absolute fact? One may assume an eternally coexisting manifold which implies all the possible momentary worlds. Augustine left this problem also unanswered. In accordance with his ideas he could have maintained that God has fixed the unilateral dimension of time and that the passage through the divine now was identical for all men.

The recognition of the uniqueness and irreversibility of the temporal process is one of the most remarkable achievements of Augustine. Hence time is not a perpetual revolving image of eternity, but is irreversibly moving in a definite direction. It has an organic finality. Creation has had an absolute beginning and travels to an absolute goal. There can be no return. That which is begun in time is consummated in eternity. Augustine was therefore actually the first man to discover the meaning of time, in spite of the fact that Plato, Aristotle, and Plotinus, had written about it. While their endeavour was primarily to explain it away, a typical Greek characteristic, Augustine explained the time-process itself. He was the first thinker to take time seriously.

Ricardo J. Quinones†

Time and historical values in the literature of the Renaissance

In Thomas Mann's great *Zeitroman, The Magic Mountain,* the spokesman for the dynamic mode of time in the west is the Italian Settembrini. His Mediterranean origin is no accident, and indeed as we come to know this garrulous exponent of the Word his own past exfoliates until he stands not only for a dying nineteenth-century liberalism but for its origins in the Italian cities of the Renaissance and for the very traditions of Italian humanism as well. Mann quite rightly makes Settembrini, ever the self-conscious pedagogue, trace his own intellectual provenance. For this engaged humanist, Dante was the citizen-poet of a great city-state whose mentor Brunetto Latini taught rhetoric, the art of language, and the rules of politics. Petrarch's critical intelligence established the bases for enlightenment and progress: he emerges as the 'father of the modern spirit', sire of the line of Erasmus and Voltaire, whose wry smiles Settembrini affects. And when he discovers that Hans Castorp possesses an engineering degree, and is thus devoted to the 'world of labour and practical genius', this only enchances their rapport which typologically expresses the earlier Renaissance consonance between humanist and merchant, a relationship especially intense, Hans Baron tells us, where time was the concern.[1] And indeed in *The Magic Mountain* the issue is time. When Settembrini quickly asks Hans his age, the latter fumbles, indicating a kind of unconscious receptivity on his part to the strange loss of the western time-sense that occurs at this enchanted height. The sanatorium is a medieval institution, where man only sleeps and feeds, and against whose allurements Settembrini can only ineffectively fulminate. So enervated and unappealing has the dynamic of the west become by the end of the nineteenth century that Hans reveals some aptness for the new consciousness. Similar to Yeats in 'The statues', Mann sees this new infusion as Asiatic, and Hans Castorp is the

† Most of the references in the notes are to works from which I quote more or less at length. For a fuller picture of the material here presented as well as for a more detailed bibliography see my *Renaissance Discovery of Time* (Cambridge, Mass., 1972) where the themes here discussed and others are indexed.

type of the German soul situated spiritually as well as geographically *in der Mitte*, between the attracting poles of east and west—with the world of the west since the Renaissance losing its positivity (an insight, incidentally, shared by D. H. Lawrence in his 'Letter from Germany', written only a few months before *Der Zauberberg* first appeared).[2] The gist of Settembrini's warning is that temporal lavishness is in the Asiatic style:

Have you never remarked that when a Russian says four hours, he means what we do when we say one? It is easy to see that the recklessness of these people where time is concerned may have to do with the space conceptions proper to people of such endless territory. Great space, much time—they say, in fact, that they are the nation that has time and can wait. We Europeans, we cannot. We have as little time as our great and finely articulated continent has space, we must be as economical of the one as of the other, we must husband them, Engineer! Take our great cities, the centres and foci of civilization, the crucibles of thought! Just as the soil there increases in value, and space becomes more and more precious, so, in the same measure, does time. *Carpe diem!* That was the song of a dweller in a great city. Time is a gift of God, given to man that he might use it—use it, Engineer, to serve the advancement of humanity.[3]

The intellectual richness of *The Magic Mountain* lies precisely in its historical sense. Mann quite accurately sees the role of time in the origin and the end of the humanistic world that Settembrini represents, and quite rightly presents the cluster of concepts that emerged in the Renaissance and was acutely felt to be threatened with disintegration in the early twentieth century. In both periods—one where the tempo of the west was quickening, and the other where it began to slacken—time as a concept came to the front in literature. When Joachim rebukes Hans's tendency to subjectivise about time, disturbing its objective and structured order, he tells him, 'Und mit der Zeit hattest du es gleich am ersten Tage zu tun' ('And from the very beginning you began bothering about time'), thereby providing us with a phrase that could serve as the motto for a discourse on Renaissance as well as on modernist literature.

Whatever qualification we might wish to make of Mann's (or is it Settembrini's?) view of the Renaissance, and it is somewhat monolithic, he is clearly right in associating the new sense of time with the emergence of the city. A recent informed commentator, Jacques LeGoff, writes,

Perhaps the most important way the urban bourgeoisie spread its culture was the revolution it effected in the mental categories of medieval man. The most spectacular of these revolutions, without a doubt, was the one that concerned the concept and measurement of time.[4]

The Renaissance concept of time began to appear precisely in that period— the end of the thirteenth century and the beginning of the fourteenth—that saw the consolidation of the urban mentality, the growth of commerce and capitalism, the development of the mechanical clock, the cultivation of family feeling expressed in art and literature, new ways of memorialising death and of perpetuating of the self after death through fame, and finally the emergence of the first modern classics in the poetic Renaissance associated with the names of Dante, Petrarch and Boccaccio. Obviously the chronological starting point of this study is Florence of the late thirteenth century. Proud in name and accomplishments this mercantile and banking centre provided a temporary coincidence between its energy and hopefulness, and the higher heroic aspirations of the new writers. The world in which Dante grew up was that of a city flush with victory. Through a series of encounters, the Guelphs had successfully established Florence as a city of burghers, ridding it for ever of the menacing feudal landlords. Past good fortune led to anticipations of future success. The values of its citizens were bourgeois, mercantile values. Yet, however rational and confident, they did not feel their own endeavours to be at odds with their religious heritage. According to Millard Meiss, 'though they remained deeply pious, they . . . believed with increasing assurance that they could enjoy themselves in this world without jeopardising their chances in the next.'[5] And it is in this sort of society, and others like it, that time became a crucial factor of consciousness. To be sure there were antecedents, but as of the fourteenth century, 'le thème se précise, se dramatise.'[6]

Such concern with time places a terrible responsibility on the shoulders of the individual. Whereas medieval man could enjoy a temporal ease and amplitude of being—whether natural (seasonal and agricultural), feudal or religious—to the new Renaissance intellectual time presented threatening possibilities. As Panofsky has pointed out in his essay 'Father Time', Chronos by a mistaken association with Kronos took on some of the destructive implements of Saturn.[7] But even predating such iconographic materials, Dante provided time with shears. He grieves over 'our poor nobility of blood' ('poca nostra nobiltà di sangue') which is doomed to die out unless successive generations make continuing contributions to its merit. He likens nobility to a 'mantle that quickly shrinks, so that if we do not add to it day by day time goes round it with the shears'.[8] Statements such as this make clear that Burckhardt was right when in his classic study of the Renaissance he would usually call Dante as first witness. Despite the complications of his thought and his clear need to locate the basis of action in that which is unchanging, Dante joins later Renaissance thinkers in his consciousness of the destructiveness of time. But as the quotation makes clear such a sense of time is only one part of the argument. Equally important is the possibility as well as need of some human response. Means are available of adding to the distinction of the line or of the self. To do nothing is to waste a bequest. This situation, one which I have called the argument of time, involves an older voice of experience exhorting a younger and usually unheeding figure to wake up, to make use of his time, before it is too late. The danger is that the very energy and brightness of youth could blind it to its inevitable end. In Shakespeare time is described as a tyrant, giving with one hand and taking with the other; its very duality serves to lull the consciousness into feelings of invincibility. For this reason—and Shakespeare's sonnets here are prototypal—the voice of the older man is pressing and urgent. This pattern of the argument of time persists throughout the Renaissance, although there are striking differences between the use to which it is put by fourteenth-century writers like Dante and Petrarch, and by sixteenth-century writers like Spenser or Shakespeare. A study of time in the Renaissance should take these differences into account as well as detect the true, aroused consciousness of time that underlies them.

Since this essay is to be brief, perhaps the most summary way of regarding Renaissance responses to time is to see them as closely connected with the emergence of historical values, those values of continuity effected through identification with a community, or through marriage and its offspring, or through fame. To backtrack slightly, the ideal figure of medieval man was that of a pilgrim, *homo viator*, whose primary focus was stringently directed toward the spiritual goal of his life, with only passing attention given to those things along the way. As Gerhard Ladner has pointed out, medieval man suffered radical alienation from the *temporalia* for the sake of identification with a more perfect and abiding divine order.[9] The process of 'laicisation' discerned in the Renaissance should not be regarded as a direct attack on the divine order of belief (although to weaken it might be one of its ultimate effects), but rather as the attribution of more and more value, even within a religious context, to those earthly goods that promised historical continuity. Rather than alienation from, man achieved identity and security within, the 'horizontal' values of history. At first a kind of mutuality was fostered, where a consonance between those things that are necessary for salvation and those things that are useful for life was allowed, but in moments of crisis—and the fourteenth century, particularly in Florence, seems filled with them—man was compelled to make a radical choice; and the fundamental theocentrism of that period, at least, prevailed.

Given the civic confines of Dante's origins and earlier commitments, what is remarkable about the *Commedia* is the degree to which just those commitments to the commune, to children and to fame, are scrutinised and rejected. While the city remains the dimension of the poem, its earthly repudiation is the *Commedia*'s essential theme. Integral to rejection of the historical values of continuity, of the earthly means offered of triumphing over time is the failure of generation to produce a reliable means of succession. Guido del Duca has much cause to complain in *Purgatorio* xiv, since his native Romagna has been given over to bastardisation ('Oh Romagnuoli tornati in bastardi!'). Summary statement comes from Sordello in *Purgatorio* vii, where his survey of the royal houses of his time and their disappointment in their heirs returns human attention to God and away from the normal processes of succession:

Rade volte risurge per li rami
  l'umana probitate; e questo vole
  quei che la dà, perché da lui si chiami.
Rarely does human worth rise through the branches, and this He wills who
gives it, that it may be sought from him. [121-3]

It is not then in history and continuity that man is able to secure value,
but rather in God's will, the divine comedy of life, and its mysterious and
dramatic rises and falls.

The question that preoccupied Dante as he followed the fortunes of the
Italian houses was 'how from sweet seed can come bitter fruit'. The first
solution absolves nature, and places the fault on human history, particularly
the failures of Church and State (Marco Lombardo's explanation in the
central cantos of the *Purgatorio*). The second sees the disappointments in
line as prompted by God's will and a higher plan (Sordello's account).
Dante will of course never fully relinquish his reliance on historical explanation, so firmly inbred was his early commitment to the possibilities
of civic order and peace as well as his faith in the essential validity of the
Creation. But as his vision in the *Paradiso* deepens and expands, as he
becomes convinced of almost universal patterns of individual, communal
and national decline extending over dynasties and encompassing religious
orders, so overwhelming is the evidence that he seeks out more than
natural explanations. Reaching a third and far more melancholy and re-
signed explanation for man's continuing failure in history, he finally has
recourse to defects in the raw materials of nature itself.

To Charles Martel in *Paradiso* VIII Dante acknowledges (with Aristotle
as his source) that things would even be worse did man not live in a city.
Since the advantages of the city presuppose men of differing talents living
in close proximity in order to share the benefits of diversity, there must
be differences at birth—'so that one is born a Solon, and another Xerxes,
one Melchizedek and another he that flew through the air and lost his son'
(this last a particularly pointed periphrasis for Daedalus). But a problem
results when nature does not make all the members of one family alike; they
can even be brutally different (hence the thematic importance of the great
separations of family members like the Donati, the Montefeltre and others
through the various *cantiche* of the *Commedia*), so that very early conflict
is introduced, competitive rivalry and dissension. But these difficulties,
while open not to historical explanation but only divine and hence properly
situated in the *Paradiso*, are still according to plan, whereby a jealous
Maker visits such trials in order to dislodge man from the comforts and
consolations of history. But the greatness of Dante's speculative intellect
leads him to a more daring suggestion, the idea that the hand of nature,
like that of the artist, can falter and not fully produce in the unformed
material clay the divine idea. A breakdown occurs in the implementation
of the machinery of Creation. 'Hence it comes', Solomon concludes in
Paradiso XIII, 'that trees of one and the same species bear better and worse

fruit and you are born with different talents.' Not only do we have then differing types but imperfect versions of those types, so that some are born equipped to cope and some overready to turn their faces to the wall. Beyond its skill to prevent, the family is presented with turbulent rivalries; and diversity—the central virtue of the city—now becomes the source of conflict that poisons tribal life. Small wonder then that when religious writers from Augustine to Dante and Milton went to indicate the failure of the earthly city and of man in history they had recourse to the Cain–Abel myth as archetype. Discord goes back to the very raw materials of life. These principles of contradictoriness and radical opposition are not transcended by the processes of historical evolution, but are abiding and real. The classic types of Cain–Abel, Esau–Jacob, Romulus–Remus are informed only too well by the present realities of Guelph–Ghibelline, White–Black, Franciscan–Dominican, or Thomist–Averroist. So basic within the very structures of family and birth are these antagonisms, that Dante despairs of their resolution by the efforts of man in time and history.

In fifth-century Athens, Jacqueline de Romilly tells us in her lectures on *Time in Greek Tragedy*, a sense of time and of history emerged together.[10] In the flowering of Elizabethan literature the same mutual concerns developed. But we must recognise that Shakespeare's histories, devoted as they are to the benefits of continuity, are historical not merely in subject but in value as well. We speak readily of the comedies and of the tragedies as having their own dispensations, but are somehow more hesitant to acknowledge that the histories have their particular dimensions. Shakespearian comedy, as Bertrand Evans indicates, depends in its technique upon the manipulation of a hierarchy of discrepant awareness.[11] A similar study for the histories has not been written, but it would have to deal with lateral antagonism, with forces or arguments coming from opposite sides of the stage and meeting in 'mere oppugnancy'. Like the world of Dante's *Commedia*, that of Shakespeare's histories is eminently dramatic, and dialectical, where, however, the principles of opposition are not radical but are finally resolved in a compelling synthesis. Given time—such was the extent of Tudor hopefulness, and hence the tetralogic scope of Shakespeare's histories—things would work out well. In Dante the pessimism of historical tragedy converts into divine comedy; for Shakespeare it was within the historical process that the successful resolutions of comedy were achieved. As Dante's conclusions testify to the evidence of his place and time, so also do Shakespeare's, reflecting not so much Tudor myth as Tudor reality: for more than a century, three generations of Tudors had managed to bring a degree of order and stability which was remarkable in relation to other regimes of those precarious times.

While each of Shakespeare's tetralogies reaches 'historical' conclusions—that is, the reconciliation of opposites—they do this by different ways. From the very outset, the first tetralogy of *1–3 Henry VI* and *Richard III*, covering the Wars of the Roses, is draped with doom. Man does not redeem himself but is rather caught up in a network of consequences. No one can claim justice or mercy for each has violated these principles in the past. The central Yorkist acts of rebellion, therefore, are faulty and fail to evidence any of the positive values necessary for good government. Caught in the great Machiavellian dilemma of charisma and institutions, they have the zest for power and the capacity to seize it, but not the wisdom or values to transform it, once possessed, into a viable political order. The bottom of the pit is reached and the future completely foreclosed with the massacre of the young princes in *Richard III*. Not from within the play, then, but from another source—from one who was not present or too young when all the atrocities were committed, young Henry of Richmond, the future Henry VII—can historical redemption follow.

The second tetralogy differs from the first in that a successful historical figure does emerge from within the action of the play; in Prince Hal we are able to follow the development of a figure whose values and evolution better represent Shakespeare's sense of man's life among men. As a consequence, given these more positive possibilities, time as a principle of reality still permitting right choice is much more active in these plays. In accord with several other related concepts, from *Richard II* on, it has undergone a distinct thematic clarification (perhaps showing the effect of Shakespeare's work on the Sonnets and *Lucrece*).[12] Time, children and the garden (none of them present to the same degree in Shakespeare's chronicle sources) form a network of related themes pointing to the emergence of a vital Shakespearean historical ethic. In each instance Richard II violates this code and is shown, as a consequence, to be unworthy of being king—or representative man—in his world. By seizing Bolingbroke's patrimony he strikes at the very bases of historical life and his own continuing sovereignty; and it is no wonder that in each of these three areas, rebukes of Richard carry implications of self-deposition. The most important for our purposes, and the most telling in the play, is York's warning after he hears that Richard has seized Gaunt's property in order to finance the Irish wars:

> Take Herford's rights away, and take from Time
> His charters and his customary rights;
> Let not to-morrow then ensue to-day;
> Be not thyself; for how art thou a king
> But by fair sequence and succession? [II, i, 195–9][13]

Sequence and succession, linearity and lineality merge in Renaissance temporal concerns. Richard violates both: as he wars against time, so he wars against the great promise of Edward's many sons, and in his way of life and 'unlineal hand' is alien to the historical code of succession.

Given the Shakespearean investment in the processes of history and the need for the lines of growth to respond to the pressures of decay, it is natural that one of the play's more emblematic scenes should take place in a garden, where again Richard's negligence and failures are underscored. The brunt of the imagery suggests a natural inevitability to the process:

> He that hath suffered this disordered spring
> Hath now himself met with the fall of leaf. [III, iv, 48–9]

The 'wasteful King' showed insufficient care and skill:

> Superfluous branches
> We lop away, that bearing boughs may live;
> Had he done so, himself had born the crown,
> Which waste of idle hours had quite thrown down. [63–6]

Richard's sorrows have not dimmed his powers of recognition, and from his own words we get the summary passage concerning the role of time in the Renaissance: 'Ha, ha, keep time!' Richard in prison censures a musician who makes a fault in his playing. With Hamlet-like ratiocination, this remark leads back to himself:

> How sour sweet music is
> When time is broke, and no proportion kept!
> So is it in the music of men's lives.
> And here have I the daintiness of ear
> To check time broke in a disordered string;
> But for the concord of my state and time
> Had not an ear to hear my true time broke.
> I wasted time, and now doth time waste me . . . [v, v, 42–9]

Imprisoned in Pomfret castle Richard is a fantasist, peopling his world with imaginary beings and roles that are impotent to alter the stark reality of the prison walls. His antagonist and competitor, Bolingbroke (now Henry IV) is much more of a realist as the very early challenge scenes indicate, and consequently more capable of coping with the world of time. Norman O. Brown is quite right to assert that 'a sense of time can only exist where there is a submission to reality'.[14] Of the King's Two Bodies Richard had only identified with the permanence of the office, not the reality of the mortal man, and it is upon this body that time works.[15] The role of time then acts to limit the transcendent will, and to promote a more effective will, one that recognises limits, and is able to operate and control events within those limits. It is clear that the sense of time as an urgent pressure was coincidental with the rise of bourgeois society and the middle class. More to the point time figures prominently in the formation of middle class values. It suggests an external world of real limitations against which one must provide if he is to be spared an unsatisfactory reckoning. The absoluteness of any feudal, chivalric or aristocratic will, one not admitting contingencies, is anachronistic and suicidal. If then Shakespeare's England witnessed the great alliance under the Tudors between the throne and the middle class, it is clear why time is so important a force in Shakespeare's second tetralogy.

In the growth of Shakespeare's positive historical ethic in the second tetralogy, the father–son link reaches enormous proportions, uniting the related themes of education and generation. Henry IV embodies the paradox of Machiavelli's new prince, upon whom is placed the terrible burden of effecting historical change, with all the ambiguity and possibility of evil such actions entail, and yet whose success depends upon his ability to bring solid and continuing social institutions out of this murky area of endeavour. In the first tetralogy the parallel actions of the Yorkists were unable to bring about such profound change; generation in their case only resulted in a hellish pejoration, whereby their seizure of power was in no way vindicated, but condemned. The test in the second tetralogy is on the Lancastrians: if Hal turns out to be another Richard II (an accusation his father makes), then the father himself has failed, and the historical process would be, as it was in the first tetralogy, one overridden by a curse. This is the meaning of the father's fear expressed at the end of *Richard II*, 'If any plague hang over us, 'tis he.' Yet the future has only been threatened, not unalterably closed. Despite the exaggerated description of Hal's dissoluteness—made incidentally by Hotspur, and thus setting up the duality of the next play—the father still sees in Hal 'some spark of better hope, which elder years / May happily bring forth.'

In fact, the very presence of the father, summarising the argument of time and exhorting proper action, implies a remnant of freedom still available to Hal. The structure of the play, with its multiple plot levels, and with Hal standing before the three worlds there represented, reminds us that we still live in the realm of choice. The past has not so far constrained him that he is committed to a single line of behaviour. This very multiplicity is a redeeming factor in Hal, and separates his attractiveness from the more simplistic characters like Hotspur, as simple and predictable as his name. In the midst of Renaissance variety and energy, there developed a sense of time, the great and serious force addressed when such multiplicity of possibilities is to be channelled into a single, functioning, historical identity. In these plays time is about its important role of effecting a conversion from what could be called an aesthetic stance to an ethical one. The father, as the spokesman for public seriousness, and the central model of the new secular *paideia*, has an important hand in this transformation. Each of the pivotal scenes in both parts of *Henry IV* contains a crucial exchange between the father and the son. At the heart of the Shakespearean historical world, at the centre of his time-world, is the happy resolution of this generational link. Unlike Augustine or Dante, and summarising more than two centuries of Renaissance growth and speculations, Shakespeare dramatises the real possibilities of value in history.

A deep and far-reaching quarrel seems to exist between the two major responses to time, children and fame, even though these appeared contemporaneously in early Renaissance literature. It is a quarrel that places civic humanist in one camp and stoic humanist in the other. While they agree on the necessity of making response to time, they differ on the more effective means. Some, of course, like Dante or Shakespeare, could entertain both, but there is a strong current, originating with Petrarch (and reflected in Boccaccio's life of Dante), extending to Montaigne and even Bacon ('He that hath wife and children hath given hostages to fortune'), that prefers offspring of the spirit to those of the loins. Late comes the tree that will provide shade for your descendants, Petrarch quotes Virgil approvingly. And in a letter he declares,

> We shall extend our name into the future, God willing, by intellect not marriage, with the aid of books not children and through virtue not wives . . . You will not have a bright and long-lasting name unless you get it for yourself. This is the work of manhood, not women. How much fame would Plato and Aristotle, Homer and Virgil have today if they had thought to possess it through matrimony and offspring?[16]

But even here Dante has some precedence. Showing the strong Roman source and inspiration to the Renaissance idea of fame, Dante has Virgil provide the spur to his sluggishness:

> 'Omai convien ché tu cosi ti spoltre,'
>   disse 'l maestro; 'ché, seggendo in piuma,
>   in fama non si vien, né sotto coltre;
> sanza la qual chi sua vita consuma,
>   cotal vestigio in terra di sé lascia,
>   qual fummo in aere ed in aqua la schiuma.'

> 'Now must thou cast off all sloth,' said the Master, 'for sitting on down or under blankets none comes to fame, and without it he that consumes his life leaves such trace of himself on earth as smoke in air or foam on the water.'
> [*Inferno* XXIV, 46–51]

But fame like children is a rich theme in that it reflects the complex evolution of Dante as well as that of other Renaissance figures. In the *Inferno* the possibility of fame is some consolation to the agonists, although even here the final condition is one of strange silence, the pure perversion of God's original trumpet call. But in the *Purgatorio* requests to have one's name kept alive, so often heard in the *Inferno*, practically disappear as temporary voguish fame is dwarfed by the prospect of thousands of years, and prayerful intercession becomes a more effective bridge between the dead and the living. However, in the *Paradiso* where human affections have been purified as to object and intent, that is, in the theocentric righting of Dante's pilgrimage, even fame returns as part of a divine comedy. Cunizza, the sister of Ezzelino—the destructive firebrand of the north—praises Folco of Marseille for his great fame, and then draws this conclusion:

> vedi se far si dee l'uomo eccelente,
> sí ch'altra vita la prima relinqua.

> consider then if man should not make himself excel so that the first life may leave another after it. [IX, 37–42]

It was Petrarch, through his *Triumphs* and his sonnets, who made the most significant contribution to the iconography and literature of fame. Aspiration towards fame as the high worthy goal of the poet has been as synonymous with his name as has the laurel crown (although in both instances Dante and perhaps others may have been there first). Petrarch's *Coronation Oration* and its companion letter addressed to his sponsor King Robert of Sicily clearly engages the poet in a war with time, from which the great reward would be a fame as sweet-smelling and long-lasting as the laurel leaf itself, whose real and mythical properties (including its immunity from lightning, equated by Petrarch with time) seem to imply diuturnity.[17] The letter is important for the expression it gives to the high endeavour of Renaissance optimism and energy, for its rejection of the arguments of decline, and the monumentality of the models it chooses to follow. The enemies are the abusers of history, those who praise the ancients in order to condemn the moderns; their true motive is envy, and their purpose is the defeat of all literature. But rather than disheartening him, Petrarch states, their despair incites him to become the kind of man they believe only existed in the classical world. Such men are rare and few, he confesses, but there have been some. He then proceeds to the great utterance of his own personal faith, one that is at the heart of Renaissance creativity. 'Who', he asks, 'is to deny us from being among these few? If the scarcity of such spirits frightened everyone, in short time there would not only be few but none at all. Let us endeavour, let us hope, and perhaps it will be given to us to attain what we have sought. Virgil himself has said, "They were able because they believed themselves capable." We, too, believe me, can, if we believe we can.' The active encouragement, the collective imperative *enitamur, speremus*, are typical of the high possibilities envisioned by the men of the Renaissance at its outset. Petrarch's original endeavour to march with ancient greatness will itself be a model for the burgeoning young literatures of France and England, for the *Pléiade* and for the generation of Spenser. The urge to fame is clearly part of the heroic humanist war against time: man is different from the beast, and is not intended for a life of sleeping and feeding. Clearly its active inspiration and ascendant heroism separates this means of warring with time from the more practical and down-to-earth arguments for marriage and the begetting of children.

While to some modern readers this lofty aspiration might appear frenetic, or at least jejune, we must not forget that during the Renaissance, even in the hands of its most fervent advocates, this quest for fame underwent transformations, even repudiations.[18] The very position of Fame in the triumph-structure indicates that it will be overcome, that other forces, time, can surmount it, or that death, outside the structure, can reveal its fragility. In Dante, as we have seen, and in other writers of a religious passion—Petrarch, Spenser, Milton—the study of time uncovers what could be called a 'purgatorial' stage. This is a period of extreme vigilance

and care, or radical shifting of priorities; it is a time of fear and vulner-
ability. In such a time there is little room for largesse, for accommodative-
ness. In fact, the purgatorial stage is a psychic state of necessary exclusiveness, where cruel rejections have to be made. Time becomes extremely
precious, as Forese Donati tells his longtime friend, bidding him adieu,
'che il tempo è caro / in questo regno' (xxiv, 91–2). The father-figure as
stern law-giver here, and casual human affections and associations are
brutally severed by the need for intellectual justice. Cato in the ante-
purgatory has no eyes for his wife Marcia, resident in the limbo of virtuous
pagans, and interrupts the aesthetic rapture of Dante provided by Casella's
song by commanding the souls bent on salvation to run to the mountain
where God will be manifest. This need for undeviating direction and consciousness of time will become the function of the other father-figure,
Virgil, whose own rejection, anticipated by these exchanges of the first two
cantos, reveals how severely theocentric this purgatorial stage can be.

In Petrarch's *Secretum* a similar dramatic confirmation of the purgatorial
mentality occurs. In the crisis of his middle years, the arbitrariness of death
has revealed to Petrarch the insufficiency of literary fame as a principle of
life. His reliance on fame and his devotion to his literary works are instances of a larger historical faith in the processes of life, a faith that somehow he will be allowed to complete all his projects, lead a full life, and then
at the end render his account to God. Augustine, another father-figure and
the necessary antagonist to Petrarch's own more sanguine hopes, cuts
across such historical hopefulness: 'O man, little in yourself and of little
wisdom! Do you then dream that you shall enjoy every pleasure in heaven
and earth, and everything will turn out to be fortunate and prosperous for
you always and everywhere?' Multiple evidence is presented that such
hopes have proved to be fallacious. 'Alas! if (which God may forbid) in the
midst of all your plans and projects you should be cut off—what grief, what
shame, what remorse (then too late!) that you should have grasped at all
and lost all.'[19] Petrarch is here expressing his own fears over what he
thought was a major flaw in his character: his many interests, his ability to
initiate projects and his inability to complete them. He felt he lacked
follow-through, the character strength of simple direction. Petrarch's elder
brother Gherardo was an occasion that informed against him; and his
decision to join a monastery while Petrarch tried to travel along alternate
routes is allegorically told in the famous letter concerning the ascent of
Mont Ventoux (a Renaissance letter, to be sure, but hardly for the reasons
normally proposed). While time is a destructive force against which the
Renaissance poets waged war in their more heroic aspirations, it is also
a principle of reality, serving as a force that channels the Renaissance
energy and love of variety into a functioning unity—the father-figure with
his argument of time is the agent for such a conversion. These are the
patterns within the argument of time that unites the religious purgatorial
stage of Dante and Petrarch with the historical one of Hal and his father

in Shakespeare's dramas while at the same time highlighting some major spiritual differences.

Not all of the objections to the historical means of continuity through children and fame came from religious writers; one of the most persuasive came from Montaigne, that social conservative who turned out to be the most radical thinker of the sixteenth century. After all, humanist and religious poet were alike intent on climbing the mountain, be it Parnassus or Purgatory; and they alike sought to return to the source of perfection and light by a process of rigorous discipline, as they alike looked to the continuity of their existence through a 'second life' after death. It was Montaigne, more so than Petrarch, who qualifies for the title of first modern man; for it was Montaigne who rebelled against Renaissance activism and high endeavour, as he sought to return the men of his day to matters that were present, imperfect, material, earth-bound. A late convert to Stoicism after the death of Etienne de la Boetie, Montaigne tried for some ten years to measure up to the lofty standards of his exemplary friend.[20] But in time he returned to an attitude and a way of life more convenient to his own easy-going temperament and inquiring intellect, irrespective of the ascendant aspirations of fame, or the great hopes of children, or any of the aroused militancy of Renaissance endeavour. Lofty and heroic claims seemed pretentious to Montaigne, ever aware as he was of the 'continual variation of things'. His overwhelming sense of change and vicissitude swept aside human claims for permanence, or even powers of choice, deliberation and control. He saw little reason behind any assertions of human greatness. We cannot even know the world, since the very tools of measurement—human subjectivity and feelings—are in themselves changing and defective, and not apposite to the nature of things. But by what Cassirer has called a 'strange reversal' this profound scepticism and sense of change leads to a redemption of present things in all their defective materiality and passingness.[21] 'La maladie universelle est la santé particulière.' If all are sick then in a certain sense none is sick, since we are compelled to revise our standards of what is attainable health. Disgust with present things, Montaigne complains, had traditionally been thought to be the way of wisdom. In the Renaissance poetic war against time one succeeded by overcoming those distracting present things; indeed, in his defection from Beatrice Dante can admit to having fallen victim to 'presenti cose', and Spenser in his own dedication to high achievement must forswear the tempting allurements of present things. But for Montaigne the present, and present things, are all that man has: 'car enfin c'est nostre estre, c'est nostre tout.' If the men of the Renaissance looked toward the future with its rewards of progeny or fame, Montaigne was not one of them. He took no pleasure in the 'au-dela': 'We are never at home, we are always beyond. Fear, desire, hope, project us toward the future and steal from us the feeling and consideration of what is, to busy us with what will be, when we shall be no longer.' Man is always 'elsewhere' and this diversion of attention introduces a kind of temporal anxiety, a lack of focus on the things at hand. 'Our thoughts are always elsewhere: the hope of a better life stays and supports us, or the hope of our children's worth, or the future glory of our name . . .' And in a passage pregnant with summary and typical sentiment, he concludes:

If I were one of those to whom the world may owe praise, I would ask for payment in advance and hold it quits. Let the praise make haste and pile up all around me, concentrated and massive rather than extended and lasting and let it vanish abruptly together with my consciousness of it, when its sweet sound will reach my ears no longer.[22]

No wonder then that Montaigne enjoys an enormous prestige in the works of literary modernism, whose theme, as Ortega has pointed out, is the rejection of progressivism, that 'attitude of mind in which life for its own sake is a matter of indifference, and only acquires values if it is considered as an instrument or as a basis for the use of a culture operating in the "Beyond" '.[23] The instrumentalisation of life is the great hazard of any triumph of historical values in that things are used as stepping stones to higher goods and virtues rather than as goods in themselves. The connection between Montaigne and modernism on this point is further clarified by the fact that when Leonard Woolf wrote the final instalment of his autobiography, he chose its title from Montaigne, *The Journey Not the Arrival Matters.*

If the Machiavellian need to control time was behind Shakespeare's histories, the Montaignesque impossibility of such control is behind the tragedies. With his insistence on the role of chance, and change and error, Montaigne could well be the Heraclitus to Shakespeare's Sophocles: the body of his thought represents the strongest attack on values of historical continuity, claims to permanence, time-transcendent heroism, or even simple deliberation. If the link of history is generational, in tragedy that link is destroyed; and rather than living in simple continuity with past and present, man endures again—but this time outside the Christian dispensation—an orphanhood in time. Hamlet cries out that there will be no more marriages, as the hopes of the earlier tetralogies are upstaged by the bedroom. The defection of King Lear's daughters call forth amazed questioning: 'Is there any cause in Nature for these hard hearts?' And finally, as in Dante, dissatisfied by any natural explanations, Kent can only blame the stars for the different ways that children take:

> It is the stars,
> The stars above us, govern our conditions;
> Else one self mate and make could not beget
> Such different issues. [IV, iii, 33–6]

If the histories of Shakespeare seem to require the scope of a tetralogy, indicating a kind of historical optimism that, given time, things will finally work out well, the tragedies are contracted down to the single play and the bare moment. Mysterious forces operate that rule out historical sequence and succession. Oppositions in the tragedies are not dialectical: they reveal no higher synthesis, no third force that seems to amalgamate the different forces into a viable unity or to reconcile the hostile parties. There is no sense that the historical process raises itself at the end of tragedy; there is rather a sense of diminishment and valediction. Dualities are not transcended. They are radical; and, like death, are rooted, final and unavoidable.

The parable of the talents, in its complex usage and rich relevance to a study of time in the Renaissance, can help us, looking back, to summarise the twists and turns of the road we have travelled. The argument of time was present in its essential Renaissance patterns in the fourteenth century, but it was largely used by Dante and Petrarch for religious ends. Looked at from another direction, however, the influence of bourgeois time-values can be seen to have had its impact even on religious life, when a Dominican preacher very popular in the fourteenth century, Domenico Cavalca, could use the argument of time and the parable of the talents to castigate his idle, do-nothing clerical brothers, particularly the monastic orders and those clergy who do not render services. [24] The argument of time urges activism and social utility; consequently, the indolent person is more than wasteful —he is, after the parable of the talents, evil, since he scorns the gifts of God (of which time is among the chief) given to man so that he could add to them. One must develop this talent; otherwise he will be inevitably held to a dire reckoning. Of course, one way to accomplish this, and Petrarch is an early leader in this phenomenon, is rigorously to schedule one's activities so that in the language of Gargantua's humanistic mentors, 'il ne perdoit heures quelconques du jour, ains tout son temps consommoit en lettres e honeste scavoir' ('that he lost not any one hour in the day, but employed all his time in learning, and honest knowledge'—*Gargantua*, XXIII). Alberti would recommend beginning-of-the-day lists and end-of-the-day inventories to see if everything scheduled were actually performed.

In discussing time in the Renaissance it is almost impossible to extricate from it the language and methods of economics. The trend in the Renaissance would be toward the intensification of this kind of social pressure as well as a decrease in the tensions produced by theocentrism. Petrarch, for instance, directed his warnings against that terrible final day, the day of judgement, when all achievements will pale before the necessity of rendering a good account of one's spiritual life. Shakespeare, however, is closer to one of Petrarch's classical models, Seneca: not one's soul, but rather to what use he has put his potentialities and expectations will be clarified, and not necessarily at the day of death, but in any moment of crisis. Earthly royalties, capacity, manliness, what one has done and what one has to show for it, are then reckoned. In Petrarch, and to a lesser degree in Dante, the call to account brings one to a singleness, where only religious virtues (as in *Everyman*) can be of assistance. But for Shakespeare—at least in the histories—it is mainly in relation to the goods of life, children and accomplishment, and in association with others and in historical connections with past and future, that man redeems his time. If the young man in Sonnet IV refuses to add to his line, Shakespeare asks, '. . . when Nature calls thee to be gone, / What acceptable audit canst thou leave?' In Sonnet II the young man must needs answer the troubling question: 'where all thy beauty lies, / Where all the treasure of thy lusty days?' With admirable poetic *esprit* Shakespeare adopts the parable of the talents to the themes of the sonnets; if the young man, like him who buried his talent, can only point to his 'deep-sunken eyes', that were 'an all-eating shame, and thriftless praise'. In Shakespeare the culmination of the secularising process of the Renaissance is seen when the parable of the talents, without any religious connotations whatsoever, is used as an argument to further reproduction.

But so rich is the cycle of Renaissance literature that the very triumph of time produces its own reaction. The energy expended to control time creates an anxiety about what will be and from which one seeks release in the insouciant manner of Montaigne; the tendency to relegate immediate things to the status of instruments only serving toward some ultimate, more perfect end, induces a kind of bewilderment and detracts from the significance of things, to which Montaigne's present centredness is a response; or the triumph of time creates a need to rediscover higher-controlling principles as bases for existence, as in Hamlet's shaping divinity and the larger 'let be' of Milton's poetry (each reacting, like Montaigne's responses, to Renaissance activism and the need for control). Milton, particularly, is in conscious reaction against the strong and compelling Renaissance exhortations over time, those comparisons with others, the 'timely-happy', who seem to be making their way in the world. He is more aware of an inner process of ripening that enjoys religious favour, even though deprived of that 'one talent which is death to hide,' and unable to present his 'true account'. In the divine plan there is room for those who stand and wait. His greatest poetry, then, is not only a product of industry and planning, but comes 'easy'—with a sense of wonder—his 'unpremeditated verse'. To be sure, in *Paradise Lost*, his great public poem, Milton shows us the pain of history and the fall into time; in fact, part of Satan's failures stems from his refusal to admit the reality of change, while the first parents—and the name acquires new significance in this poem of history—begin their recovery by such admissions. Still, the truer model in Milton's Christocentric world is the passivity and divine trust exemplified in *Paradise Regained*. But Milton does not stop there. Showing a remarkably rich development that is an attribute of the Renaissance itself, his final voice is heroic:

> Samson hath quit himself
> Like Samson, and heroicly hath finished
> A life heroic. [*Samson Agonistes*, 1709–11]

This essay, following the course of my own current interests, has occasionally moved back and forth between the Renaissance and literary modernism of the twentieth century. Such manoeuvring was practicable because as movements and bodies of literature they are both preoccupied with time, perhaps to an unparalleled extent. Time is a strong and central presence in many of their major works and to discuss these works—it is hardly necessary to name them—is to discuss concepts of time. From the beginning they each began bothering about time. In each literature time is an important sign of character alteration and a means for distinguishing what is new from what is old. But beyond such parallelisms, there are stronger historical connections that justify bringing these two movements together where time is the concern: the time-world that emerged from the Renaissance, based on the values of historical continuity, is the one with which modernism contends, even threatening its very disestablishment. What was new and exciting in the Renaissance had become old and rigidified with damaging personal consequences. That this could have happened can be in part explained by the dual nature of time in the Renaissance, and by what I call the paradox of time in the west. Renaissance time was related to aspects of human heroism and practicality, to the passion and precision of the age. Two quotations from Petrarch can show this to be so. For heroic-minded writers, 'ingentibus animis nichil breve optabile est', and nothing that is brief will finally be desirable for writers from Petrarch to Spenser and Milton. If time can thus elicit fervent heroic aspiration, it also calls for rational control. Energy needs to be channelled and the love of variety brought into some working order. Thus Petrarch can conclude, 'Totum in ipsius temporis dispensatione consistit' ('Everything consists in the ordered disposition of time'). This dual nature of time joins together cosmic argument and practical ends, fervent discovery and the *emploi du temps*, and in their varying ways the major Renaissance writers I have discussed all reflect it. This double face of time could also be called predictive and innovative. A Japanese scholar has declared that 'the sheer need for a more precise control of predictions thus might have helped to give birth to the notion of time as such'.[25] The other side was represented by R. Schlegel in a paper entitled 'Time and entropy': 'In a human culture in which there is active development of new knowledge and new ways of living there is perhaps concomitantly an emphasis on history and the role of time.'[26] In the nineteenth century, or at least in those sectors of society to which the modernist writers were most sensitive, the predictive face of time triumphed over the innovative. In some of the central representatives of the western dynamic this resulted in a feeling of fatigue and enervation, especially in those presumed to be most busy, the type represented by Thomas Buddenbrooks in Mann's first major novel, or by Gerald Crich in Lawrence's *Women in Love*. The latter, the industrial magnate in the pattern of Henry Ford, had set himself to manage an industrial world organised (in the manner of Frederick Winslow Taylor) with the efficiency of a machine: 'a great and perfect machine, a system an activity of pure order, pure mechanical repetition, repetition *ad infinitum,* hence eternal and infinite. . . .' This clockwork efficiency of functioning rather than representing the harmonisation of the energies of the universe, as it does in the superb image concluding Canto x of Dante's *Paradiso,* or the magisterial working of the laws of the universe as it does in the seventeenth century, represents the desolation accruing to the total triumph of time.

The thought [and these are Gudrun's thoughts in *Women in Love*] of the mechanical succession of day following day, *ad infinitum,* was one of the things that made her heart palpitate with a real approach of madness. The terrible bondage of this tick-tack of time, this twitching of the hands of the clock, this eternal repetition of hours and days—oh God it was too awful to contemplate. And there was no escape from it, no escape.

Even sex can bring no release, as the industrial world that Gerald Crich manages with such ruinous success extends to their bodily love, and in the image of the clock, causes revulsion:

He, his body, his life—it was the same ticking, the same twitching across the dial, a horrible, mechanical twitching forward over the face of the hours. What were his kisses, his embraces? She could hear their tick-tack, tick-tack. [27]

In each of these passages a kind of surrealistic horror of empty solitude and total sameness or repetition brings us to the paradox of time, that is, when the predictive face of time almost thoroughly eclipses the innovative a condition is produced that is the exact contrary of the original impulses attending the discovery of time. A world is created where experience is felt to be standardised, and where all things begin to look alike. In a certain sense, according to the notion I am sketching in, the triumph of time paradoxically produces the triumph of space. A reason for this might be in the instrumentalism implied in any progressivistic ethic, and the dangers to which Montaigne first called attention. If things are only instruments to further ends they lose their own substance, and in any time of accelerated tempo, such as our own, it would be very easy for all things to lose their texture and to become unreal. A condition is produced which weighs heavily on the human soul and initiative. If this is the case, the further paradox is introduced that modernism, committed in its early phases to challenging that which the time-sense of the Renaissance had become, could very well be seen as attempting, with somewhat uncertain chances of success, to restore to human experience the combination of qualities and the complex nature of things reflected in the literature of the Renaissance.

NOTES

[1] 'A sociological interpretation of the early Renaissance in Florence', *SAQ,* xxxix (1939), p. 437.
[2] In *Phoenix: The Posthumous Papers of D. H. Lawrence,* ed. Edward McDonald (1936), p. 109.

[3] *The Magic Mountain*, trans. H. T. Lowe-Porter (1927, 1969), p. 243.

[4] 'The town as an agent of civilization', in *The Fontana Economic History of Europe: The Middle Ages*, ed. Carlo M. Cipolla (1972), pp. 86 ff.

[5] *Painting in Florence and Siena after the Black Death* (Princeton, N.J., 1951), p. 15.

[6] LeGoff, 'Le Temps du travail dans la "crise" du XIVe siècle: Du temps médiévale au temps moderne', *Le Moyen Age*, LXIX (1963), p. 597. See also the same author's 'Au moyen àge: temps de l'église et temps du marchand', *Annales, E.S.C.*, XV (1969), p. 425.

[7] Panofsky (see Bibl. *225*), p. 69.

[8] *Paradiso*, XVI. Quotations from the *Commedia* conform to the edition of Natalino Sapegno (Florence, 1957); the English renderings are from J. D. Sinclair's *The Divine Comedy* (1939), 3 vols.

[9] '*Homo Viator*: medieval ideas on alienation and order', *Speculum*, XLII (1967), p. 233.

[10] See Bibl. *333*.

[11] *Shakespeare's Comedies* (Oxford, 1960).

[12] See the unpublished dissertation by R. Thomas Simone, 'Shakespeare and *Lucrece*: a study of the poem in relation to the plays' (Claremont Graduate School, 1972).

[13] Quotations are from *The Riverside Shakespeare*, ed. G. Blakemore Evans, Harry Levin, *et al.* (Boston, Mass., 1974).

[14] *Life Against Death* (1961), p. 95; quoted in Frank Kermode's very relevant study, *The Sense of an Ending* (1967), p. 57.

[15] See Ernst Kantorowicz, *The King's Two Bodies* (Princeton, N.J., 1957).

[16] XIV, 4 in *Epistolae rerum senilium*; in *Francisci Petrarchae opera omnia* (Basle, 1554), vol. 2.

[17] The 'Coronation Oration' is available, trans. Ernest H. Wilkins, *PMLA*, LXVIII (1953), p. 1245. The letter to King Robert of Sicily, not yet translated, is *Fam.* IV, 7; it is contained in *Francesco Petrarca: Prose*, ed. G. Martellotti *et al.* (Milan, 1955).

[18] For a recent, extreme anti-Renaissance view see Russell Fraser, *The Dark Ages and the Age of Gold* (Princeton, N.J., 1973); but see my review in *MLQ*, XXXV (1974), p. 78, as well as Harry Levin's in *CL*, XXV (1973), p. 373.

[19] *Prose* (as above, note 17), p. 200.

[20] For this story see Donald Frame's *Montaigne: A Biography* (1965), pp. 63–84.

[21] See Cassirer's suggestive work *The Individual and the Cosmos in Renaissance Philosophy*, trans. Mario Domandi (1964).

[22] The last three passages are from *The Complete Works of Montaigne*, trans. Donald Frame (Stanford, Ill., 1957), pp. 8, 633, 595–6.

[23] Ortega y Gasset, *The Modern Theme*, trans. James Cleugh (1933), pp. 69–70.

[24] I refer to the *Disciplina degli spirituali*, ed. G. Botari (Rome, 1757).

[25] Quoted from the abstract of Masanao Toda's paper at the Second World Conference for the Study of Time, July 1973. It is hoped that this sentence will reappear in his contribution to the proceedings of that Conference, *Studies in Time*, vol. II (forthcoming).

[26] Printed in *TSP*, p. 29.

[27] These quotations are from *Women in Love* (1960), pp. 220, 456.

Jerome H. Buckley†

The four faces of
Victorian time

All the four faces of that Charioteer
Had their eyes banded . . .
                              Shelley

The Triumph of Life, the great dream-allegory that Shelley left unfinished at the time of his death, depicts the whole human race led captive by the delusions of earthly living. Through all history none have escaped, none 'but the sacred few . . . of Athens or Jerusalem', none, that is, but Christ and Socrates. The vanquished, passing down a broad highway, follow a triumphal car, ill-guided by a four-faced charioteer with all eyes banded. The victor is Life, and the charioteer is apparently Time, turned blindly to all directions, past, present, future, and eternity.

To all modern men life and time seem inseparably linked, and most are content to be led aimlessly through the drift of experience. But the more sensitive ones demand some perspective, a sense of time's course or meaning; like Shelley, who was pre-eminently one of them, they are forever looking before and after, impatient with the mere flux of things and eager to find a way once again of measuring their brief lives under some eternal aspect.

It was in the nineteenth century, especially in Victorian England, that many modern attitudes toward the whole temporal process first emerged. The Victorians, at least as their verse and prose reveal them, were pre-occupied almost obsessively with time and all the devices that measure time's flight. The clock beats out the little lives of men in *In Memoriam*. Adam Bede in an anxious moment listens intently to a clock's 'hard indifferent tick, . . . as if he had some reason for doing so'. Hardy's clock-winder cancels out each sore sad day as he climbs nightly to 'the trackway of Time' in an old church tower, where the clock itself wheezes rheumatically on. Hopkins's watch marches in step with the man, 'Mortal my

† Reprinted from *The Triumph of Time: A Study of the Victorian Concepts of Time, History, Progress, and Decadence* (Cambridge, Mass., 1967), chapter I.

mate, bearing my rock-a-heart / Warm beat with cold beat company.'
The large watch that defines Mr Dombey races with the doctor's watch
across the silence as Mrs Dombey lies dying. A chorus of bells and chimes,
striking the hour in a pawnshop makes time 'instant and momentous' for
Stevenson's Markheim. Rossetti's Blessed Damozel sees 'Time like a
pulse shake fierce / Through all the worlds', while the lovers in his
House of Life cherish a silent noon when time is hushed and 'still as
the hour-glass'. The egoists of Meredith's *Modern Love* have 'fed not on
the advancing hours', and over them, appropriately, 'Time leers between,
above his twiddling thumbs.' Father Time behaves frighteningly as the son
of Jude the Obscure. Time crawls 'like a monstrous snake / Wounded and
slow and very venomous' through the City of Dreadful Night. And the
bird of Time flutters more relentlessly over FitzGerald's *Rubáiyát* than
over Omar Khayyám's. Though sometimes stopped and sometimes accelerated, Time dominates even Wonderland, where the White Rabbit, consulting his pocket-watch, is constantly fearful lest he be late, and the mad
tea-party lingers forever at six o'clock. 'If you knew Time as well as I do,'
the Hatter tells Alice, 'you wouldn't talk about wasting *it*. It's *him*.'

Whatever the gender, time presents its several faces again and again
throughout Victorian literature. Tennyson was concerned early and late
with the relation of the temporal to the eternal; the Mystic of his first
volume[1] hears 'Time flowing in the middle of the night, / And all things
creeping to a day of doom'; and the emperor Akbar, in his last, writes of
the faithful who 'Kneel adoring Him the Timeless in the flame that measures Time.' Arnold made the attrition of private time the main subject
of his poetry; and in his prose he sought constantly to evaluate the public
present, the 'modern element' in life and letters, in the light of the best
that the past had thought and said, yet with an awareness also of 'the
natural and necessary stream of things'[2] leading to the future. Browning all
his life strove both to recreate the dramatic past and to record the infinite
moment. Swinburne on occasion yearned for a life beyond time, a life 'that
casts off time as a robe',[3] but his most characteristic and perhaps his best
poem remains the moving, rhapsodic 'Triumph of Time', which hymns the
irreversible passing of love.

More generous men than Scrooge were haunted by the ghosts of Past,
Present, and Yet to Come. A new generation of historians, both literate
and laborious, enlarged the limits of the human past and speculated on the
possibility of finding patterns of recurrence or meaningful analogies with
their own time. And almost every one of the major novelists attempted at
least one historical novel, somehow relevant in theme to their own age,
while many of the lesser ones peered into a more or less utopian future.
The author of the standard biography frequently designed his work as a
'Life and times'. The philosophic critic made a new appeal to a mysterious
Time-spirit, the *Zeitgeist* determining the availability of modes and ideas.
And the theologian, whether in the conservative *Tracts for the Times* or

the liberal *Essays and Reviews*, asserted the role of the historic Church, the rites and dogmas evolving over long years, in the immediate life of the nineteenth century. Ruskin, who wrote 'Time' in blank verse when he was only seven and *Praeterita*, his warm evocation of the personal past, when he was nearly seventy, paused in mid career—in his letters *Time and Tide*— to reflect that time and life were virtually one: ' "Time is money"; the words tingle in my ears so that I can't go on writing. Is it nothing better, then? If we could thoroughly understand that time was—*itself*,—would it not be more to the purpose? A thing of which loss or gain was absolute loss, and perfect gain.' [4] Carlyle, on the other hand, whose interest in the problem is apparent not only in his histories but even in the titles of some of his other works, 'Signs of the times', *Past and Present, Latter-Day Pamphlets*, could recognise time as the great deluder. Speaking for all Carlyle's contemporaries, though in his own peculiar idiom, Teufelsdröckh complained:

Our whole terrestrial being is based on Time, and built of Time; it is wholly a Movement, a Time-impulse; Time is the author of it, the material of it. . . . O Time-Spirit, how hast thou environed and imprisoned us, and sunk us so deep in thy troublous dim Time-Element, that only in lucid movents can so much as glimpses of our upper Azure Home be revealed to us! [5]

Man's concern with time had, of course, persisted since the beginning of time itself. It had engaged Zeno the Eleatic no less than Heraclitus of Ephesus. The death of time had fired the apocalyptic vision of St John, and the ruins of time had provided a major theme to the Renaissance poets. On a far lower plane, men had long since learned to live by the clock; the Lilliputians had reason to believe that the god Gulliver worshipped was a watch, since he 'called it his oracle, and said that it pointed out the time for every action of his life'.

None the less, as Carlyle suggested, nineteenth-century absorption in the troublous time-element differed both in kind and degree. The notion of public time, or history, as the medium of organic growth and fundamental change, rather than simply additive succession, was essentially new. Objects hitherto apparently stable had begun to lose their old solidity. Drawing on Lyell's geology, Tennyson saw in nature tangible evidence of a fluidity about which ancient philosophy had only speculated:

> The hills are shadows, and they flow
>   From form to form, and nothing stands;
>   They melt like mist, the solid lands,
> Like clouds they shape themselves and go.

The Victorians were entering a modern world, where, according to the angry Wyndham Lewis, 'chairs and tables, mountains and stars, are animated into a magnetic restlessness, and exist on the same vital terms as man. They are as it were the lowest grade, the most sluggish of animals.

All is alive; and, in that sense, all is mental.'[6] When we so dissolve all things in time, said Lewis, we commit ourselves to cultural decadence, for true civilisation has always depended on the classical virtues of permanence and order. Be that as it may, modern man is no more able to repudiate the time dimension than to reject the whole course of modern knowledge.

If the discovery of space has proceeded rapidly ever since the late Renaissance, the discovery of modern scientific time belongs largely to the past one hundred and fifty years, and the theory of relativity has made each dependent on the other. The object now has even less fixity than Tennyson imagined; it has become a range of charged energies in a space-time continuum. The 'new sciences' of the nineteenth century—uniformitarian geology, nebular astronomy, evolutionary biology—were all sciences 'in time', governed by temporal methodologies. The new physics, unlike its older mechanistic counterpart, acquired a time-direction with the second law of thermodynamics and the concept of entropy. And mathematics itself, the key to most scientific advance, turned to time, as it developed new theories of assemblages and continuous functions, theories which first adequately resolved Zeno's ancient paradox: how could the arrow in passage exist if it occupied no fixed point at any given moment of its flight? Meanwhile, the new social studies, emulating the more exact sciences, were all thoroughly grounded in the historical method.[7] Herbert Spencer with his great law of evolution, Karl Marx with his dialectical materialism, Sidney Webb with his faith in the inevitability of gradualism—to name only the three most conspicuous of the many social scientists living and working in Victorian England—each of these responded wholeheartedly to the dominion of public time as the age understood and defined its authority.

Toward the end of his long life Hardy pondered in awkward rhyme the irony of the 'Fourth Dimension':

> So, Time,
>   Royal, sublime,
> Heretofore held to be
> Master and enemy,
> Thief of my Love's adornings,
> Despoiling her to scornings:—
> The sound philosopher
> Now sets him to aver
>   You are nought
>   But a thought
> Without reality.[8]

Though the theorists of relativity were heavily cerebral, this was not precisely the significance of the space-time continuum; for the scientists still dealt in an objective linear time to be measured quantitatively. The poet, on the other hand, was prepared to value only the far more complex private time, a qualitative force to be experienced, not to be measured, intense yet illusory, quite 'without reality' apart from the psychological

life of the individual. Many years before, Tennyson had allowed the erudite
Princess Ida to explain the necessary subjectivity:

> For was, and is, and will be, are but is;
> And all creation is one act at once,
> The birth of light: but we that are not all,
> As parts, can see but parts, now this, now that,
> And live, perforce, from thought to thought, and make
> One act a phantom of succession: thus
> Our weakness somehow shapes the shadow, Time.

And the learned Master Whewell, once Tennyson's tutor at Cambridge,
had apparently distinguished the subjective time from the objective when
he announced that 'time, like space, is not only a form of perception, but of
intuition'.[9]

So conceived, private time is arbitrary, relative in quality to the passing
personal emotion, continuous, yet variable in tempo—now fast, now slow.
Scott weighs 'one crowded hour of glorious life' against 'an age without a
name'; and Byron's Manfred, in unequal argument with the unbyronic
Chamois Hunter, who considers himself by far the older man, insists that
life is to be reckoned by deeds, not years:

> Think'st thou existence doth depend on time?
> It doth; but actions are our epochs: mine
> Have made my days and nights imperishable,
> Endless, and all alike, as sands on the shore. . . .

In a quieter vein, Thackeray comments in *Esmond* that 'at certain periods
of life we live years of emotion in a few weeks—and look back on those
times, as on great gaps between the old life and the new'.[10] Stevenson tells
the fable of a monk who, wandering in the woods, hears a bird sing just 'a
trill or two' and then, on his return, finds himself a stranger at the monastery gates; for he has been absent fifty years, so much has he been beguiled
by 'that time-devouring nightingale', romance.[11] Less fancifully, George
Eliot in *The Mill on the Floss* describes how slowly time moves in the
bedchamber of the stricken Mr Tulliver as he shows signs of gradual
recovery, and meanwhile how rapidly and efficiently the time outside
consummates his financial ruin.

As seen by poet and novelist alike, human time thus defies scientific
analysis and measurement; contracting and expanding at will, mingling
before and after without ordered sequence, it pays little heed to ordinary
logical relations. It is intelligible only as duration, as a constant indivisible
flow in which life will be a continuous, enduring unity of change, and the
consciousness, by memory and desire, will completely merge any given
moment of the present with the whole personal past and future. Such a
formulation, essentially Bergsonian, belongs largely to a post-Victorian
period. But its terms were freely anticipated by various philosophies current in the nineteenth century, especially by those developing Hume's

concept of the self as function rather than substance, a concept which made the psychological life an incessant movement in time. Walter Pater accordingly could argue in the most celebrated of his essays:

This at least of flame-like our life has, that it is but the concurrence, renewed from moment to moment, of forces parting sooner or later on their ways. . . . Analysis . . . assures us that those impressions of the individual mind to which, for each one of us, experience dwindles down, are in perpetual flight. . . . It is with this movement, with the passage and dissolution of impressions, images, sensations, that analysis leaves off—that continual vanishing away, that strange, perpetual weaving and unweaving of ourselves.[12]

Yet the endless movement was not limited to the insubstantial self. The Victorian, looking outwards, could see his whole age in perpetual motion. And in this respect at least, personal time was to a high degree consonant with public time, with the actual objective history of the period. So widespread and so rapid were the changes wrought by the nineteenth century in the material conditions of living that no one, however much he might wish to dwell in the spirit, could altogether escape a sense of almost physical exhilaration or bewilderment rushing in upon him. The hero of 'Locksley Hall' thrilled to the steamship and the railway as to 'thoughts that shake mankind'. And there were to be other and scarcely less impressive inventions: the telegraph, the telephone, the sewing machine, the photograph, the gas burner, the electric light, the typewriter, the motor car. Increasingly rapid communication and transportation made the farthest corners of the globe seem less and less remote. The spread of the factory system virtually destroyed the British agricultural economy and transformed the way of life of toiling millions. The population multiplied at a staggering rate, and London grew eight- or ten-fold within the century, until it had become by far the world's largest and most prosperous city.

'Take it all in all,' wrote Frederic Harrison early in the 1880s, 'the merely material, physical, mechanical change in human life in the hundred years, from the days of Watt and Arkwright to our own, is greater than occurred in the thousand years that preceded, perhaps even in the two thousand years or twenty thousand years.'[13] Moreover, the speed of the change seemed to be constantly increasing. At the time of the Second Reform Bill, which widened the base of the franchise and so prepared the way for a broader democracy, Carlyle complained that his political fears were being realised with frightening rapidity:

the series of events comes swifter and swifter, at a strange rate; and hastens unexpectedly,—'velocity increasing' (if you will consider, for this too is as when the little stone has been loosened, which sets the whole mountain-side in motion) 'as the *square* of the time':—so that the wisest Prophecy finds it was quite wrong as to date; and, patiently, or even indolently waiting, is astonished to see itself fulfilled, not in centuries as anticipated, but in decades and years.[14]

Carlyle's use of the ratio anticipated Henry Adams's 'rule of phase', according to which each successive phase would last for the square root of the number of years in the preceding. And his dread of the acceleration foreshadowed the mood of Adams as he stood awe-struck at the Paris Exposition in 1900 before the great silent dynamo, the symbol to him of an incalculable 'ultimate energy'.[15]

In literature however, and in the cultural life generally, the actual changes and the rate of their occurrence count for less than the emotional and intellectual responses positive, or negative, that they engender or that the very concept of change elicits. Though often opposed to particular innovations, many Victorians accepted the idea itself almost as an article of faith. If Carlyle came to resist a dreadful velocity, he was from the beginning convinced that progressive adaptation and sudden or gradual conversion, the constant rebirth of the spirit symbolised by the phoenix, were essential both to the individual man and to the whole society. For the latter, remarked Teufelsdröckh, 'is not dead; that Carcass, which you call dead Society, is but her mortal coil which she has shuffled off, to assume a nobler; she herself, through perpetual metamorphoses, in fairer and fairer development, has to live till Time also merge in Eternity.'[16] Ruskin, who deplored many modern developments, found Changefulness a central attribute of the creative achievement he most admired:

It is [he wrote] that strange *disquietude* of the Gothic spirit that is its greatness; that restlessness of the dreaming mind, that wanders hither and thither among the niches, and flickers feverishly around the pinnacles, and frets and fades in labyrinthine knots and shadows along wall and roof, and yet is not satisfied, nor shall be satisfied. The Greek could stay in his triglyph furrow, and be at peace; but the work of the Gothic heart is fretwork still, and it can neither rest in, nor from, its labour, but must pass on, sleeplessly, until its love of change shall be pacified for ever in the change that must come alike on them that wake and them that sleep.[17]

And Hopkins saw in all changeful, incandescent, dappled things a witness to the God 'whose beauty is past change', and in the daily renewed freshness of nature a testimony that the Holy Ghost still broods over the whole bent world.

Other poets were less sanguine in their view of change, especially in so far as new modes and attitudes seemed to threaten the great traditions of art and society. Clough urged that the innovators 'yet consider it again' before sweeping aside the legacy of two thousand years.[18] Tennyson as a young man declared 'Raw Haste' to be 'half-sister to Delay', and years later as Laureate he warned against the 'fierce or careless looseners of the faith' and the new literary naturalists, dealers in 'poisonous honey stolen from France'.[19] But it was Arnold above all who indicted his contemporaries for rushing precipitantly into action. 'Let us think', he counselled, 'of quietly enlarging our stock of true and fresh ideas, and not, as soon as we get an idea or half an idea, be running out with it into the street, and trying to make it rule there.' Yet he had at no time much confidence that his advice would be heeded. In 'The scholar-gypsy' he had diagnosed 'this strange disease of modern life, / With its sick hurry', and bidden the legendary wanderer to remain immortal by escaping infection,

> For what wears out the life of mortal men?
>   'Tis that from change to change, their being rolls;
> 'Tis that repeated shocks, again, again,
>   Exhaust the energy of strongest souls
>   And numb the elastic powers.

Goethe, to be sure, he felt, had kept his luminous vitality in an age of violence and confusion, but only because his youth had been passed 'in a tranquil world'. The Victorians, on the other hand, he contended, had been conditioned from childhood to a general instability:

> . . .we brought forth and reared in hours
> of change, alarm, surprise—
> What shelter to grow ripe is ours?
> What leisure to grow wise?

Arnold was troubled by the vision of universal change governing all human affairs of past, present, and foreseeable future, and on occasion was even driven to imagine with dismay 'Far regions of eternal change' beyond human experience altogether.[20]

The sense of history, of a perpetually changeful public time, was central to the intellectual life of the nineteenth century. But there were as many expressions on each of time's faces as there were thoughtful observers. 'Each age', sang Arthur O'Shaughnessy, 'is a dream that is dying, / Or one that is coming to birth';[21] and the assessment of rise or fall depended in large part on the predisposition of the dreamer. Whatever the historian's effort to achieve objectivity, public change could seldom for long be contemplated with a calm detachment; it called for evaluation as advance or decline, change for the better or change for the worse. The great polar ideas of the Victorian period were accordingly the idea of progress and the idea of decadence, the twin aspects of an all-encompassing history. Poised at their high moment in time, the Victorians surveyed their world, its past and its future, with alternate hope and fear. Distant echoes of their positive faith and stronger reverberations of their misgiving reach down the years to our own age. We listen as to those who have already, like Stein in Conrad's *Lord Jim*, braved 'the destructive element', which is time itself; and like Marlow, his auditor, we weigh the possibilities of our response: 'The whisper of his conviction seemed to open before me a vast and uncertain expanse, as of a crepuscular horizon on a plain at dawn—or was it, perchance, at the coming of night? One had not the courage to decide. . . .'

NOTES

1 'The mystic' appeared in Tennyson's first independent book, *Poems, Chiefly Lyrical* (1830) and was thereafter suppressed. But an even earlier piece, 'Time: an ode,' had been included among Alfred's contributions to *Poems by Two Brothers* (1827).

2 Arnold, 'Literature and science' (1882), *Discourses in America* (1885), p. 135.

3 Swinburne, 'The lake of Gaube' (1899).

4 Ruskin, *Time and Tide, Works*, Library Edition, 39 vols. (1903–12), xvii, pp. 395–6.

5 Carlyle, *Sartor Resartus, Works*, Centenary Edition, 30 vols. (1896), i, pp. 103–4; cf. *Heroes and Hero-Worship, Works*, v, p. 8.

6 Wyndham Lewis, *Time and Western Man* (Boston, Mass., 1957), p. 433; cited also in Geoffrey Wagner's *Wyndham Lewis: A Portrait of the Artist as the Enemy* (1957), p. 163.

7 Cf. Hans Meyerhoff's observation that in the nineteenth century 'all the sciences of man—biology, anthropology, psychology, even economics and politics—became 'historical'' sciences in the sense that they recognized and employed a historical, genetic, or evolutionary method'. *Time in Literature* (Berkeley, Calif., 1955), p. 97.

8 Hardy, ' "So, time," ' a continuation of 'The absolute explains' (which argues that science has made a mockery of time), dated 1922, *Collected Poems* (1928), p. 723. On the modern scientific view of time, see Hans Reichenbach, *The Direction of Time* (Berkeley, Calif., 1956).

9 William Whewell, *History of Scientific Ideas*, 2 vols. (1858), I, 133.

10 Thackeray, *Esmond*, book ii, chapter i.

11 Stevenson, 'The lantern-bearers', *Works*, 25 vols. (1912), xvi, p. 207.

12 Pater, 'Conclusion', *The Renaissance* (1873).

13 Harrison, 'A few words about the nineteenth century', *The Choice of Books* (1907), p. 424; the essay first appeared in *Fortnightly Review*, April 1882.

14 Carlyle, 'Shooting Niagara', *Works*, xxx, pp. 2–3.

15 See *The Education of Henry Adams* (Boston, Mass., 1918), p. 380, and for Adams' 'rule of phase', see his *Degradation of the Democratic Dogma* (1919; 1949).

16 *Sartor Resartus, Works*, i, p. 188. I have discussed the concern of Carlyle and others with spiritual rebirth in 'The pattern of conversion,' *The Victorian Temper* (Cambridge, Mass., 1951), pp. 87–108.

17 Ruskin, 'The nature of Gothic', *The Stones of Venice, Works*, x, p. 214.

18 Clough, 'Ah! Yet consider it again' (1851), the first stanza of which runs: ' "Old things need not be therefore true."/O brother men, nor yet the new;/Ah! still awhile the old thought retain,/And yet consider it again!'

19 Tennyson, 'Love thou thy land' (1834) and 'To the Queen' (1872), epilogue to *Idylls of the King*.

20 Arnold, quotations from 'The function of criticism at the present time' (1864), 'The scholar-gypsy' (1853), 'Stanzas in memory of the author of "Obermann" ' (1849), 'Resignation' (1849).

21 Arthur O'Shaughnessy, 'Ode' (1874), where poets are pictured as above time in all ages, prophesying things to come, 'the dreamers of dreams', yet 'the movers and shakers/Of the world forever.'

II

'Time present':
general premises

A. A. Mendilow†

The time–obsession of the twentieth century

Never perhaps have our feelings about time changed so radically and assumed such importance in our eyes as in this century. Spengler maintains that this interest in time is peculiar to our whole civilisation, and that 'We men of the western Culture are, with our historical sense, an exception and not a rule' in the series of cultural cycles of human-kind. For such as accept this theory, the twentieth century is merely formulating more consciously a philosophy implicit in the whole outlook of the west, which presents a view of life 'not as things-become, but as things-becoming'; our culture sees 'the world-as-history in contradistinction to the world-as-nature', and has thereby acquired 'the sense of the logic of time, additional to the logic of space'.[1]

The antiquarianism of the seventeenth and eighteenth centuries tended to constitute a cult somewhat self-consciously adopted by a number of individuals. The interest shown by these scholars in the past was largely confined to bringing its long-buried relics to light as curiosities or at best as things of some value in their day. The living, human implications of the past were left for the new historians like Burke, and the new poets like Wordsworth to develop. The new approach was made less in a spirit of abstract scholarship than as the expression of their growing need to revivify the past, to incorporate it as an organic part into their world-scheme and scale of values. Through them a new temporal sensibility was infused into the climate of opinion of their day; or, alternatively, around them this new sensibility held in solution in society crystallised itself. Hence the frequent recurring of the Romantics to the childhood of the race, of the nation, of the human being. The noble savage was still present actually or potentially in civilised people, the child was father of the man, the medieval world lived on in the modern. In more modern terms, they realised that the subconscious, 'the twilight realms of consciousness'[2] as Coleridge called it, where the primitive instincts and infantile experiences maintain their subterranean existence, plays an important role in determining the directions along

† Reprinted from *Time and the Novel* (1952), chapter i.

which the human personality pursues its way. This explains the significance attached by Wordsworth to dreams in the *Prelude*, an interest far removed from the dream-convention of medieval poetry; it underlies Shelley's recognition of the permanent validity of the myth and Scott's evocation of historical atmosphere.

One of the qualities that marks the romantic from the classical attitude derives from this difference in the feeling for time. Classicism had developed the spatial sense and conceived even literature in terms of the plastic arts; it saw the past as an accreting cumulation of independent events and states which were complete in themselves and could be laid, as it were, side by side, fixed in a uniform medium for the curious to survey; of value chiefly as providing precedents for present or future claims, of interest only as a lesson to guide future action, to point a moral or adorn a tale. The Romantics, on the other hand, saw significance rather in the creative temper that went to the forming of one state out of another. They tended to look on human nature and human development in terms of the organic unity underlying the process of history, the growth of the individual, and the constantly self-adjusting equilibrium that determines the pattern of behaviour of men in groups. For them, civilisation was an evolving biological unit. The tradition which that civilisation embodied was a contemporary condition of all living existence, always and inevitably part of the Now. Their duty as artists and thinkers was to modify the trend of their times by throwing their weight into some or other of the innumerable component forces which, meeting at a point, affected the strength and direction of the resultant that is society.

Some of the earlier antiquaries like Stukely had already pointed the way to the deeper understanding of the customs of the past which Blake seized upon so eagerly as forms into which he might read the problems of his own day. Shelley had already equated the myth-making of the savage with the play of the child and the creative urge of the artist.[3] From the nineteenth century onwards, sociology and anthropology began with ever-increasing conviction to see contemporary significance in the patterns of earlier or more primitive behaviour and belief. These sociological and other investigations were subsumed in the new arts or pseudo-sciences of psychoanalysis and analytic psychology, while the theory of the unconscious with its emphasis on regression intensified the awareness that past states were incorporated into an all-inclusive presence and presentness. The individual and the group had not merely passed through certain phases of development; these phases were all present at the same time in the unconscious, individual or collective, and were constantly modifying conscious behaviour. The theories associated chiefly with Bergson, Freud and Einstein in the fields of philosophy, psychology and science gave a new turn to modern thought. The first two in particular have exerted a direct and powerful influence on the whole trend of modern fiction.

Whether, and if so to what extent, this interest in time is inherent in our civilisation, and whether it has led to or has resulted from recent revolutionary theories in science, from novel trends in philosophic thought, and from new conceptions of the nature and functions of art are questions that are open to dispute. Philosophy, science and art, while they influence the way people think and feel, are themselves affected by the way people live. It would seem to be not unlikely, therefore, that what is widely referred to as 'the time-obsession of the twentieth century' is conditioned by the increasing pace of living, by the widespread sense of the transience of all forms of modern life, and more particularly perhaps, by the rapidity of social and economic change. These factors have taken from people that feeling of stasis in society, that assurance of permanence that appears to have marked more confident and more slowly changing periods.

Disintegration is menacing every form of life: 'Things fall apart, the centre cannot hold.'[4]

Everywhere the old configurations have broken up. Out of the flux that has superseded the comparative fixity of the past, new forms rapidly crystallise, only to dissolve once more into the flow of things. Society no longer offers the appearance of an entity, but is instead conceived as an aggregate of conflicting forces. People in olden times believed that there was a symmetrical framework that held a stable universe together in a regular series of relationships. A cosmic order embraced God, man and beast, the vegetable and mineral kingdoms, in a hierarchy of degrees and orders that extended to the humours, the elements and the stars in their courses. The exact correspondence of sphere with sphere and the balance of the elements in each against those in every other brought macrocosm and microcosm into a satisfying world pattern. This harmonious unity based on divinely imposed principles filled up the perfect closed circle of the universe and offered an assured resting place for the questing soul. Donne had lamented the disintegration of that medieval scheme,[5] and his revival of popularity in our day is perhaps symptomatic of a similar sense of indeterminacy and aimlessness.

There was once a general veneration for authority that derived from the partly unconscious identification of father, King and God, of the family, the state and the Kingdom of God. 'Les Rois', wrote an eighteenth-century historian, 'sont nos Dieux visibles . . . et même nos pères en quelque sorte.'[6] But this propensity to submission and unquestioning acceptance in all realms of thought and action can no longer be taken for granted. The protestation of the Protestants against the 'Papa', the Pope, and 'the dissidence of dissent' have been carried many steps further. And as the discrepancies between the two attitudes of acceptance and rejection become steadily wider in more and more fields, they themselves serve as causes for further conflicts and for the division of the world, the nation, society, the family unit, even science, into hostile camps. The old integration of religion, science, philosophy and art no longer holds. The universe has proliferated into a multiverse and no one can as Bacon did take all knowledge for his province. The enormous accretions to knowledge and the trends to specialisation have sent off the different branches of intellectual and artistic activity into widely divergent directions. We are bewildered and frustrated in our attempts to synthesise for ourselves a new harmonious and stable pattern of living and thinking. Once more a 'new Philosophy calls all in doubt', and a disrupted world shows itself 'all in pieces, all coherence gone'. We are confronted with an 'open *Gestalt*' and cannot guess how and whether the configuration will be completed.

This loss of assuredness has contributed to an urgency that relates closely to the modern absorption in time. The frantic quest for something to replace the old certainties has ended for a few in a new religious revival leading back to a pure faith that denies the right to question; others it has led to a mysticism that negates all questioning. Marxists have been directed by it to attempt the realisation of a dialectic which, they believe, will resolve all antinomies under one universal principle of society. The scientist searches for it by probing deeper into the infinite and infinitesimal. But here too both quantum physics and astronomy come up against the problem of time. Even the comforting concreteness and fixity of material things is denied us. Dr Johnson could refute Berkeley to his own satisfaction by stubbing his foot against a stone, but since his time, the stone has dissolved into an arabesque of time-experiences and become a set of abstractions drawn from the measurable aspects of patterns of movement. For the novelists, the static symmetry of the old self-contained plot can no longer be imposed on the dynamic formlessness of life which they feel as a variable flowing rather than as an unchanging being. Heidegger's principle of indeterminacy has its equivalent in their technique of the stream of consciousness; they try to 'record the atoms as they fall upon the mind';[7] by breaking up the categories of language and syntax, they strive to express their sense of life as a sequence of non-causal impressions in which direction can be predicted only of the larger units, never of the smaller components. Some of them, looking for fragments to shore up against their ruins have set against the bewildering chaos of their personal vision the clarity of an earlier pattern, hoping thereby to find emerging a wider rhythm that will correspond to the archetypal motifs that were caught and held in the finest writings of older civilisations. Behind the fragmentation of *Ulysses*, Joyce has posed step by step the complex movement of a story that shades into myth. The waste land of our modern life is itself one of the regions through which we must find our way in the Quest for the Grail, and is related to the mythic alternation of light and dark, of life and death.

The economic and demographic changes of recent times have placed woman in a new position in society; this, in conjunction with the instability of modern society has created a parallel instability in human relationships. The time-significance of this for fiction is very clear. It has forced serious novelists to abandon the old convention of the novel ending with a wedding. Once marriage did imply a considerable degree of certainty, and a final solution of the main problems of sexual relations. One could safely assume that after the wedding, these relations would pursue their usual and therefore fictionally unexciting course. But the realisation has now been forced upon reader and writer alike that very often the ceremony begins rather than ends the series of highly varied and highly individual emotional adventures. As an ending it is no longer satisfying; the problems are only beginning. The structural beginnings and endings provide the limiting framework that determines the form of every time-art; but they can no longer be truthfully consonant with the thematic beginnings and endings. And this again brings to the fore the time-problem of modern literature that has changed the novel into what is virtually a new form.

Urged to look for some solution, to find some way out of the *impasse* of modern living, filled with a general sense of major transition from one cultural or social cycle to another not as yet fully determinable, we become more appetitive. While the final goals tempt us on and yet evade us, we can on the other hand look to the realisation of more and more immediate desires which modern technical achievement renders possible more and more quickly. Hence the bounds of our future perspectives shift continually and new horizons open in rapid succession to our eager gaze. In Whitehead's words,

in the past the time-span of important change was considerably longer than that of a single human life. Thus mankind was trained to adapt itself to fixed conditions. Today this time-span is considerably shorter than that of human life, and accordingly our training must prepare individuals to face a novelty of conditions.[8]

It is this appetitiveness that makes Time so emphatic for us on the level of everyday life. The more we have or can get, the more we want; the quicker we can get what we want, the more we become aware of change and movement (not necessarily synonymous with improvement or progress), that is, of Time.

The keynote of modern existence is speed, which is the relation of distance to time. It is significant that 'speed' originally had the meaning of 'success'; western civilisation measures success today by the increase in the rate of movement towards some spatial point or towards some goal we set up before us. Achievement is estimated in terms of the length of time taken to accomplish our purposes, for time is money, and in a changing universe we have no time to waste or to lose. While the factories turn out thousands of new appliances to save time, the entertainment industry spends millions on amusements to kill time. Life seems to be resolving itself into a feverish scramble for the last drinks before the inexorable barmaid calls out her fatal 'Time, gentlemen, time' and shuts up shop for good.

In a sense our age has seen the conquest of space by time. The four quarters of the world have drawn together, and to reach them has become not merely quicker and easier but more common. There has been an enormous increase in the number of people who travel, whether to find relief from boredom or sanctuary from persecution or death, whether as members of armies or as refugees from them. The consequent sense of change and the fullness of incident in relation to duration are closely connected with the sense of Time. The increase so marked in recent decades in the life-span of the civilised man who has not fallen victim to civilised wars, taken with the quicker tempo of living and more frequent changes in the manner of living, means that there is an increase in the number of events per life-time both in density and spread. This closer texture in the process of living has clearly a direct relationship to the awareness of Time.

Our Universe has changed. We are no longer confined comfortably in Time between the limits of the creation and the Day of Judgement. Our temporal horizons have withdrawn, they no longer remain a matter of faith. Every man carries his own time-system about with him. Again and again, writers and thinkers remind us that 'The demand that time shall be taken seriously is one of the fundamental notes of modernism.'[9]

NOTES

[1] Spengler, *Decline of the West* (1918–23), introduction; trans. C. F. Atkinson (1926–9).
[2] *Biographia Literaria* (1817), chapter XXII.
[3] Shelley, *The Defence of Poetry* (1821).
[4] W. B. Yeats, 'The second coming', in *Michael Robartes and the Dancer* (1921).
[5] e.g. *The First Anniversary*.
[6] L'abbé Lenglet du Fresnoy, *De l' Usage des Romans* (1734), I, p. 142.
[7] Virginia Woolf, 'Modern fiction', in *The Common Reader* (1925), p. 190: see above, p. 2.
[8] *Adventures of Ideas* (1933), Pelican ed., p. 94.
[9] Wilbur Urban, *The Intelligible World* (1929), p. 238.

Charles Mauron (translated by T. S. Eliot)†

On reading Einstein

I hope that it is no longer fashionable to talk about Einstein. It was a rather sad, even rather shameful comedy, during some years past, to see, swarming about the lecturers and the popularisers, a crowd which I shall not call 'incompetent'—for incompetence is not ignominious—but in reality hardly even interested in the Truth, the revelation of which it called for so vociferously. People who would never have taken notice of the reflections of Poincaré, of the work of Lorentz or of the experiments of Michelson, if by chance they had heard of them: why should they suddenly pretend to give so much importance to a theory directly derivative from these enquiries? Now that this hubbub of snobism has died away, I imagine that there are still a few individuals here and there who read Einstein and his sources and his commentators, simply because they are directly interested: the scientists, as a guide or as a reference in their own work; others, to appease that strange craving which we call the need to understand. They read Einstein, as they would read God knows what, when that odd hunger grips them: and when they meet, they talk contentedly of the velocity of light or of non-Euclidian space, with the simplicity of peasants discussing the weather. In that spirit I wish to set down here, as briefly as possible, a few ideas which have struck me in re-reading the little book into which the famous German scientist has condensed his essential conceptions.

I have no ingenuous presumption here to establish the truth or falsity of the theories of relativity: that must be left to experiment and to physicists. What I wish to remark are the modifications of our theories of knowledge, and consequently of all our philosophy which these theories, provisionally taken to be true, must bring.

Can man understand the world? To this vague and extensive question, religions and philosophies have replied with an equally extensive and vague series of attempts in every direction. And as it happens in every branch of enquiry—the behaviourists know this well enough—the first result of these has been not to resolve the question but to define it. What does 'under-

† Reprinted from *The Criterion*, ed. T. S. Eliot, x (1930–1), pp. 23–31.

stand' mean? Little by little, two types of opinion have been formed: the first holds that any profound knowledge of any reality, implies an intimate fusion of the mind with that reality: we only understand a thing in becoming it, in living it. In this way St Theresa believed that she knew God; in this way a Bergsonian believes that he knows at the same time his self and the world. The second type of opinion, on the contrary, holds that this mystical knowledge is meaningless, that to try to reach a reality in itself is vain, inasmuch as our mind can conceive clearly nothing but relations and systems of relations. Those who adhere to the first view recognise truth by internal evidence, by the intensity of their sensation, and, in fine, by the degree of their ecstasy: these are the mystics. Those who maintain the second demonstrate the truth of their theories by their powers of prediction: for them, to understand is to be able to predict; knowledge which fails or which merely is incapable of predicting, must be set aside: such are the men of science.

These two types of knowledge coexist still; it is even possible that they will always coexist. Their fields are indeed so distinct that theoretically neither can overstep the other. But in the practice of thought—and in practice absolutely—the case is quite altered; these transgressions are incessant. They swarm in our everyday thinking, which as a matter of fact is nothing but a muddle of the two kinds of knowledge. When I say, 'I know this path well', that evidently means that I know ahead all its twists and turnings, that, if asked, I can even sketch a map more or less in detail. But after sketching this map, I am well aware that I have not told all I know about that path: within myself I have my own impression, personal, useless, and incommunicable; and this impression also went to the making of my phrase, 'I know this path well.' Possibly I deplore this confusion of language, but in the end that is a small matter. Languages are not perfect, and what matters is to understand each other; and if I can keep these two modes of knowledge side by side but distinct, everything will go smoothly. But there are men who, because they have had a strong impression of the path, fancy that they will never lose their way; or who because they are keenly aware of the original character of a particular human soul, fancy that they can foresee all its reactions; such people confuse, often to their cost, the two modes of knowledge. The Bergsonians commit the same confusion, in my opinion, when they affirm that scientific psychology is *a priori* impossible, because the reality of the soul is *quality*; they merely confess the incapacity for prediction of their own kind of knowledge. It is as if a tree should say: 'it is impossible to measure my growth because I feel it to be purely qualitative.' But the men who look at the tree from the outside measure it without worrying about its quality.

We seem to be a long way from Einstein, and yet we are very close to the subject; for it is confusion of this sort upon which limited relativity throws light. But before tackling this matter, let me venture the following remark: this profound sense that we have of the reality of things is rooted in the first kind of knowledge, which I have called *mystical*. A man of science who has dreams at night and who spends his days in writing out chemical equations will tell you that the equations are *true* and the dreams illusory, and on that point he will be wholly in accord with his doctrine; but he will admit nevertheless, that the dreams give him a much stronger impression of reality than the formulae; for the latter are relations which may well serve for prediction, but have not, like dreams or any other sensation, the flavour of immediate experience. From this remark we may draw the conclusion: that wherever this sense of an absolute reality intervenes, this sense of a reality 'in itself', we must suspect the presence—intrusion, sometimes—of mystical knowledge.

Now, it is difficult to believe what a place, not only in everyday life, but in science itself of the purest and most abstract kind, is occupied by this mystical knowledge and the sense of absolute reality that goes with it. It is only in recent times that a physicist dares to say that he knows the laws of 'something' which he calls electrical current, but that he is and always will be completely ignorant as to what an electrical current is 'in itself'—because the question has no meaning for him. That is, however, the only correct attitude; only, it is very difficult to adopt in a thoroughgoing manner, because it shocks most of our everyday habits (and in particular those habits of language in which a substantive designates an object, a thing, as what we call 'real'). In this correction of attitude there is a whole adaptation, a whole evolution of the mind which is nothing else than that great attempt, remarked at the beginning of this note, which is pursued from age to age, toward fixing the meaning of the word 'know'. Qualitative knowledge, for its part, is incapable of evolution, because it is individual and in consequence untransmissible; and for the same reason, it cannot be contradicted, because it cannot be verified. But scientific knowledge can be refined from generation to generation; one of these processes of refining has been precisely that of eliminating as much as possible of qualitative knowledge to the point of no longer admitting it except as the foundation, in the crude fact, the crude sensation, the point of departure and the point of arrival of its speculations. So that its reply to the question set, 'can man understand the world?' is today more or less the following: 'Our sensations are given. By systematising them, and by only admitting the relations which they verify, we may hope that the system of relations thus constructed will be a kind of reflection, or image of that which gives them to us. As to knowing *that* which gives them to us, that is impossible. If immediate knowledge had any value, it would be able to predict.'

A proposition like this is far from being explicit on every scientific head. Not all the things-in-themselves are dead. How many may still be discovered, for example, in biology? *Life* in particular, that 'reality' of which we have so lively a conviction: can it not be resolved, in the end, into a system of physico-chemical relations? But I wander from the subject. What I wanted to say here is that before the first relativists (Lorentz, Poincaré) two vast things-in-themselves dwelt quietly at the very heart of abstract science: Space and Time. We have a very distinct sense of our personal time; we distinguish quite well, qualitatively, our present from our past and our future. From that we inferred quite naturally that there was a time-in-itself, an immense unique time in which all the phenomena of the universe took place in succession and in a determined order. This time, the time of common sense, was also the time of the physicists. When it was said that this phenomenon took place after that, or at the same time as that, no one questioned that this phrase had a universal and absolute sense. In the same way, no one doubted the existence of a limitless Space, the same for all observers. 'The trajectory of a body in space' was an expression which appeared precise. All that was admitted rather than demonstrated, but the conviction was none the less profound for being implicit, because it was the conviction of our daily life, in which everyone always believes in things-in-themselves. When Pascal wrote 'the eternal silence of these infinite spaces terrifies me', he was thinking of this Time and this Space; and of them we are thinking every time we try to imagine the Universe. I am not going to set forth the relativist theories; I shall only remind you of a few of their consequences. Space and Time cannot be defined absolutely but only in relation to a particular observer, so that there is an infinity of possibles. To be more precise: the shape of objects, the movement of clocks, vary according to the velocity with which they are endowed—not their velocity-in-itself, but their velocity in relation to a particular observer. And as a body may have one velocity in relation to one observer and another velocity in relation to another observer, so it will have for them two different shapes. And as there is no one observer-elect, we shall never know what may be the 'real' shape of the body. Similarly, two events which are simultaneous for one observer, will not be so for another, and we shall never know whether they were 'really' simultaneous or not. In short, Time and Space in themselves vanish, to yield place to a multitude of particular systems which have no meaning except in relation to each other.

It has been said that this shocks our *bon sens moyen* of which Descartes thought so highly. No; it shocks nothing but our habits—unless we agree that our common sense is simply our habits, which is quite possible. Was not our common sense once shocked by a man in the antipodes walking 'upside down'? I should be gratified if, after the reading of this note, the relativist theories seemed a little less strange. They are in the direct line of scientific evolution: they are only a stage in the vast process of elimination of things-in-themselves which is going on every day in human minds. And why do we want to eliminate them? To remain loyal to the primitive definition: that knowledge is prediction. Thanks to the precision of our means of measurement, the idea of an absolute Space and Time, acceptable in the day by day world, had come into contradiction with experiment. Science could no longer predict: thanks to the salutary operation of relativity it has been able to take up business again. We are a bit shaken up by this; but the fact is that the organisations of the intellect have a sensibility as do those of life; the extraction of a 'thing-in-itself' is always painful.

Qualitative and mystical knowledge has also its relativity, as has long been known, a relativity which we must take care not to confuse with the other. It is what the individual expresses when he states: 'my sensation is not yours, nor is my taste yours, nor my universe.' It is a probable truth about which there is nothing to say. The God revealed in ecstasy to St Theresa was probably not the God who revealed Himself to St Francis. Shall we say that the true God was what these two had in common? That would be to argue like a scientist who sets about at once to establish relations. The only correct attitude for a mystic is this: 'You will not know God (or the world, or one thing or another) by comparing two ecstasies from without, but only by having yourself your own ecstasy.' The attitude of a relativist in science is the contrary; when Einstein dissipates absolute Space and Time, to replace them with an infinity of particular little systems, we must understand that at the same time he is giving us the formulae which allow us to pass from one system to another. Shall we revert now to the reply which we put in the mouth of the scientist when asked the question, 'can man understand the world?' If we set the question, can 'man understand Space and Time?' he would, thanks to the relativist theories, give us a parallel reply: 'Our measurements are given to us. By systematising them and allowing only such relations as they verify, we may hope that the system of relations thus constructed will be a kind of image of that which gives them to us.' Otherwise: if there exist somewhere an absolute space and time, the only scientific image which we can have of them is at present to be found in the relativist formulae of transition.

And that is logical; it is, if I may say so, in the scientific tradition which is only a huge sublimation of the world into mathematical equations. Shall we say that these equations have nothing real left to them? Doubtless, they have no longer that bouquet of reality which belongs to immediate experience. But how should they have lost all reality, when they predict phenomena, that is to say our sensations? They are even the only conceivable reality.

In every event, returning to the precise purpose of this note, it seems to me difficult (and after the theory of relativity more difficult than it was before) to build up a theory of knowledge, and consequently any philosophy whatever, without taking account of the following essential fact: the progressive necessary elimination of every reality-in-itself is the obligation of any knowledge conscientious to remain in harmony with experience. This affirmation does not ruin immediate mystical knowledge; it merely digs more abruptly than ever the trench which separates it from scientific knowledge, and prohibits all confusion.

I would add one more observation. To the question, 'can man understand the world?' is balanced another: 'can man know himself?' The mystic replies, 'Yes, in awareness of his own life.' Any other answer belongs to the other domain, and tends more or less blindly towards a scientific psychology. But I want to show how science in its current developments comes to reveal automatically to us an image of our intellect. A student of the relativist theory cannot fail to be struck by the bounds which it suddenly sets to our thought; for example, a velocity greater than that of light becomes inconceivable, and the universe (space-time) is probably finite. These limitations are not the first; long before Einstein, we know that temperature cannot fall below 273 degrees Centigrade, and that certain equations representing phenomena quite real between exact numerical limits lose their meaning outside these limits. But we must admit that more lately these examples have multiplied. And then we are led to enquire the exact meaning of this expression: 'such and such an equation loses its meaning beyond that limit'—'that limit cannot be exceeded.' The mind has no trouble with a descent to 274 degrees or in outstripping the velocity of light. But experiment does not follow the mind, and our expression above evidently signifies that with these figures we take leave of the zone of observable facts. To come out where? Evidently to enter the domain of the mathematicians, the kingdom of the mind. It is as if the zone of abstract speculation was more comprehensive than that of real facts, and as if, by the very progress of science, we were brought to define more and more precisely this strange frontier. That can be understood by conceding that we can think several possibles, and that out of these possibles the universe has only realised one. But we can imagine also that many possibles are realised in the universe which we shall never be able to think; and this second proposition would slightly rectify the fatuity of the first. However that may be, we have been spectators, for the last century, of a veritable outburst of unrealised worlds, of two or n dimensions, or otherwise non-Euclidean. None is completely unrelated to our sensible experience, but all exceed it. And after a little time the mind feels itself at ease among them; it is at home. Around this solid system of relations which constitutes science and which is doubtless a reflection of the real are displayed these mathematical fantasies, airy plumes which are reflections of possibles. The ensemble is not without beauty. I am convinced for my part that we shall acquire in these researches clearer notions of what our intellect really is. A scientific psychology will have to take the closest notice of these constructions in the unreal. But furthermore, who knows whether our successors will not see matter as invested with a new magic, more correct but stranger than the phantasmagoria of old? 'There are more things in heaven and earth, Horatio . . .'

Sigfried Giedion†

The new space conception: space–time

Social, economic, and functional influences play a vital part in all human activities, from the sciences to the arts. But there are other factors which also have to be taken into account—our feelings and emotions. These factors are often dismissed as trivial, but actually their effect upon men's actions is immense. A good share of the misfortunes of the past century came out of its belief that industry and techniques had only a functional import, with no emotional content. The arts were exiled to an isolated realm of their own, completely insulated from everyday realities. As a result, life lost unity and balance; science and industry made steady advances, but in the now detached realm of feeling there was nothing but a vacillation from one extreme to the other.

The scope and strength of the emotions are both greater than we sometimes suppose. Emotion or feeling enters into all our affairs—speculation is never completely 'pure', just as action is never entirely practical. And, of course, we are far from having free choice in this matter of feeling. Large tracts of our emotional life are determined by circumstances over which we have no control: by the fact that we happen to be men, of such or such a kind, living at this or that period. Thus a thoroughly integrated culture produces a marked unity of feeling among its representatives. For example, a recognisable common spirit runs through the whole baroque period. It makes itself felt in activities as distinct from each other as painting and philosophy or architecture and mathematics. This is not particularly surprising. Techniques, sciences, the arts—all these are carried on by men who have grown up together in the same period, exposed to its characteristic influences. The feelings which it is the special concern of the artist to express are also at work within the engineer and the mathematician. This emotional background shared by such otherwise divergent pursuits is what we must try to discover.

† Reprinted from *Space, Time and Architecture: The Growth of a New Tradition*, 5th rev. ed. (Cambridge, Mass., 1967), pp. 430–48. The original text is profusely illustrated.

I   DO WE NEED ARTISTS?

Some people question whether any pervasive unity of feeling is possible in a period like ours. They regard science and industry as inimical to art and feeling: where the former prosper, the latter decline. Or they see science taking over the arts, opening up new means of self-expression which make us independent of them. There is some basis for views like these. Do we, then, really need artists any longer?

In any civilisation, feeling continues to filter through every activity and situation. An environment whose chief aspects remain opaque to feeling is as unsatisfying as one which resists practical or intellectual control. But just this sort of emotional frustration has prevailed for a long time past. An official art has turned its back upon the contemporary world and given up the attempt to interpret it emotionally. The feelings which that world elicits have remained formless, have never met with those objects which are at once their symbols and their satisfaction.

Such symbols, however, are vital necessities. Feelings build up within us and form systems; they cannot be discharged through instantaneous animal outcries or grimaces. We need to discover harmonies between our own inner states and our surroundings. And no level of development can be maintained if it remains detached from our emotional life. The whole machinery runs down.

This is the reason why the most familiar and ordinary things have importance for the genuinely creative artists of our generation. Painters like Picasso, Juan Gris, the lyricist of cubism, and Le Corbusier have devoted themselves to the common objects of daily use: bowls, pipes, bottles, glasses, guitars. Natural materials have received the same attention: stones hollowed out by the sea, roots, bits of bark—even weather-bleached bones. Anonymous and unpretentious things like these scarcely figure at all in our normal consciousness, but they attain their true stature and significance under the artist's hand. They become revealed as *objets à réaction poétiques*, to borrow Le Corbusier's phrase. Or, to put it somewhat differently, new parts of the world are made accessible to feeling.

The opening up of such new realms of feeling has always been the artist's chief mission. A great deal of our world would lack all emotional significance if it were not for his work. As recently as the eighteenth century, mountain scenery was felt to exhibit nothing except a formless and alarming confusion. Winckelmann, the discoverer of Greek art, could not bear to look out the windows of his carriage when he crossed the Alps into Italy, around 1760. He found the jumbled granite masses of the St Gotthard so frightful that he pulled down the blinds and sat back to await the smooth outlines of the Italian countryside. A century later, Ruskin was seeking out the mountains of Chamonix as a refuge from an industrial world that made no kind of aesthetic sense. Ships, bridges, iron constructions—the new artistic potentialities of his period, in short—these were the things Ruskin pulled down the blinds on. Right now there are great areas of our experience which are still waiting to be claimed by feeling. Thus we are no longer limited to seeing objects from the distances normal for earth-bound animals. The bird's-eye view has opened up to us whole new aspects of the world. Such new modes of perception carry with them new feelings which the artist must formulate.

The artist, in fact, functions a great deal like an inventor or a scientific discoverer: all three seek new relations between man and his world. In the artist's case these relations are emotional instead of practical or cognitive. The creative artist does not want to copy his surroundings, on the one hand, or to make us see them through his eyes, on the other. He is a specialist who shows us in his work as if in a mirror something we have not realised for ourselves: the state of our own souls. He finds the outer symbols for the feelings which really possess us but which for us are only chaotic and—therefore—disquieting, obsessive stirrings. This is why we still need artists, however diffcult it may be for them to hold their place in the modern world.

But if the artist is so necessary to us, how is it that he seems to have lost contact with all but a small number of his contemporaries? Ordinary people make it almost a point of pride to insist that, so far as they are concerned, his vocabulary of forms is totally incomprehensible.

This is often said to be a consequence of the revolt against naturalism. Actually, however, it dates from quite another event: the *proclamation de la liberté du travail* of 17 March 1791, which dissolved the guild system. The abolition of all legal restraints upon the choice of a trade was the starting point for the tremendous growth of modern industry and the isolation of the artist.

Cut off from the crafts, the artist was faced with the serious problem of competing with the factory system for his living. One solution was to set himself up in the luxury trades, to cater, quite unashamed, to the lowest common denominator of public taste. Art-to-public-order flooded the world, filled the *salons*, and won the gold medals of all the academies. With no serious aims and no standards of its own, the most such an art could hope for was a financial success, and this it often achieved. The most favoured of these cultivated drudges—a Meissonier, for example—sometimes saw their canvasses sold at a thousand francs the square inch.

As far as the public and the critics were concerned, this was art—and this the work the artist was meant to do. The half-dozen painters who carried on the artist's real work of invention and research were absolutely ignored. The constituent facts in the painting of our period were developed against the will of the public and almost in secret. And this from the beginning to the end of the century, from Ingres to Cézanne.

The same situation held for architecture. Here too the advances were made surreptitiously, in the department of construction. The architect and the painter were faced with the same long struggle against *trompe l'œil*. Both had to combat entrenched styles by returning to the pure means of expression. For some four decades painter after painter makes the effort to reconquer the plane surface. We have seen how the same struggle arose in architecture as a consequence of the demand for morality. Painters very different in type but sharing a common isolation from the public worked steadily toward a new conception of space. And no one can understand contemporary architecture, become aware of the feelings hidden behind it, unless he has grasped the spirit animating this painting.

The fact that modern painting bewilders the public is not strange: for a full century the public ignored all the developments which led up to it. It would be very surprising if the public had been able to read at sight an artistic language elaborated while its attention was elsewhere, absorbed by the pseudo art of the *salons*.

II   THE RESEARCH INTO SPACE: CUBISM

In many places, about 1910, a consciousness that the painter's means of expression had lost contact with modern life was beginning to emerge. But it was in Paris, with cubism, that these efforts first attained a visible result. The method of presenting spatial relationships which the cubists developed led up to the form-giving principles of the new space conception.[1]

The half-century previous to the rise of cubism had seen painting flourish almost nowhere outside of France. It was the high culture of painting that grew up in France during this period which formed the fostering soil for our contemporary art. Young people of talent—whether Spanish like Picasso, or Swiss like Le Corbusier—found their inspiration in Paris, in the union of their powers with the artistic tradition of that city. The vitality of French culture served to the advantage of the whole world. Among the general public, however there was no sympathetic response to this achievement. It was from a form of art which the public despised that nineteenth-century painting drew its positive strength. Cubism, growing up in this soil, absorbed all its vigour.

Picasso has been called the inventor of cubism, but cubism is not the invention of any individual. It is rather the expression of a collective and almost unconscious attitude. A painter who participated in the movement says of its beginnings: 'There was no invention. Still more, there could not be one. Soon it was twitching in everybody's fingers. There was a presentiment of what should come, and experiments were made. We avoided one another; a discovery was on the point of being made, and each of us distrusted his neighbors. We were standing at the end of a decadent epoch.'

From the Renaissance to the first decade of the present century perspective had been one of the most important constituent facts in painting. It had remained a constant element through all changes of style. The four-century-old habit of seeing the outer world in the Renaissance manner—that is, in terms of three dimensions—rooted itself so deeply in the human mind that no other form of perception could be imagined. This in spite of the fact that the art of different previous cultures had been two-dimensional. When earlier periods established perspective as a constituent fact they were always able to find new expressions for it. In the nineteenth century perspective was misused. This led to its dissolution.

The three-dimensional space of the Renaissance is the space of Euclidean geometry. But about 1830 a new sort of geometry was created, one which differed from that of Euclid in employing more than three dimensions. Such geometrics have continued to be developed, until now a stage has been reached where mathematicians deal with figures and dimensions that cannot be grasped by the imagination.

These considerations interest us only in so far as they affect the sense of space. Like the scientist, the artist has come to recognise that classic conceptions of space and volumes are limited and one-sided. In particular, it has become plain that the aesthetic qualities of space are not limited to its infinity for sight, as in the gardens of Versailles. The essence of space as it is conceived today is its many-sidedness, the infinite potentiality for relations within it. Exhaustive description of an area from one point of reference is, accordingly, impossible; its character changes with the point from which it is viewed. In order to grasp the true nature of space the observer must project himself through it. The stairways in the upper levels of the Eiffel Tower are among the earliest architectural expression of the continuous interpenetration of outer and inner space.

Space in modern physics is conceived of as relative to a moving point of reference, not as the absolute and static entity of the baroque system of Newton. And in modern art, for the first time since the Renaissance, a new conception of space leads to a self-conscious enlargement of our ways of perceiving space. It was in cubism that this was most fully achieved.

The cubists did not seek to reproduce the appearance of objects from one vantage point; they went round them, tried to lay hold of their internal constitution. They sought to extend the scale of feeling, just as contemporary science extends its descriptions to cover new levels of material phenomena.

Cubism breaks with Renaissance perspective. It views objects relatively: that is, from several points of view, no one of which has exclusive authority. And in so dissecting objects it sees them simultaneously from all sides— from above and below, from inside and outside. It goes around and into its objects. Thus, to the three dimensions of the Renaissance which have held good as constituent facts throughout so many centuries, there is added a fourth one—time. The poet Guillaume Apollinaire was the first to recognise and express this change, around 1911. The same year saw the first cubist exhibition in the Salon des Indépendants. Considering the history of the principles from which they broke, it can well be understood that the paintings should have been thought a menace to the public peace, and have become the subject of remarks in the Chamber of Deputies.

The presentation of objects from several points of view introduces a principle which is intimately bound up with modern life—simultaneity. It is a temporal coincidence that Einstein should have begun his famous work, *Elektrodynamik bewegter Körper*, in 1905 with a careful definition of simultaneity.

'Abstract art' is as misleading a term for the different movements which depart from the spatial approach as 'cubism' is for the beginnings of the contemporary image. It is not the 'abstract', it is not the 'cubical', which are significant in their content. What is decisive is the invention of a new approach, of a new spatial representation, and the means by which it is attained.

This new representation of space was accomplished step by step, much as laboratory research gradually arrives at its conclusions through long experimentation; and yet, as always with real art and great science, the results came up out of the subconscious suddenly.

The cubists dissect the object, try to lay hold of its inner composition. They seek to extend the scale of optical vision as contemporary science extends the law of matter. Therefore contemporary spatial approach has to get away from the single point of reference. During the first period (shortly before 1910) this dissection of objects was accomplished, as Alfred Barr expresses it, by breaking up 'the surfaces of the natural forms into angular facets'. Concentration was entirely upon research into a new representation of space—thus the extreme scarcity of colours in this early period. The pictures are grey-toned or earthen, like the grisaille of the Renaissance or the photographs of the nineteenth century. Fragments of lines hover over the surface, often forming open angles which become the gathering places of darker tones. These angles and lines began to grow, to be extended, and suddenly out of them developed one of the constituent facts of space-time representation—the plane.

The advancing and retreating planes of cubism, interpenetrating, hovering, often transparent, without anything to fix them in realistic position, are in fundamental contrast to the lines of perspective, which converge to a single focal point.

Hitherto planes in themselves, without naturalistic features, had lacked emotional content. Now they came to the fore as an artistic means, employed in various and very different ways, at times representing fragments of identifiable objects, at others such things as bottles or pipes flattened out so that interior and exterior could be seen simultaneously, at still others completely irrational forms equivalent only to psychic responses.

Around 1912 new elements entered; the planes were accentuated, assumed strength and dominance, and were given an additional appeal—to the tactile sense—by means of new materials (scraps of paper, sawdust, glass, sand, etc.). And when, though always meagrely, colour was employed, it was often corrugated and roughened in order to strengthen the pigment. In such *collages* fragments of newspapers, fabrics, or handwriting, sometimes even single words, achieved the force of new symbols.

The process continued, from the greyish background of the first period through the *collage*, to the reappearance of colour, which gradually became stronger and more varied, until its brilliant culmination in Picasso's and Braque's still-lifes toward and at the beginning of the twenties. In this period, perhaps cubism's happiest, colour was used in pure strength. At the same time curvilinear forms were introduced, taken from such everyday objects as bowls and guitars, or simply invented. Colour no longer had the exclusive function of naturalistic reproduction. Used in a spatial pattern, it was often divorced from any object, asserting a right to existence in itself.

Cubism originated among artists belonging to the oldest cultures of the western world, the French and the Spanish. More and more clearly it appears that this new conception of space was nourished by the elements of bygone periods. Its symbols were not rational, were not to be utilised directly in architecture and the applied arts, but they did give force and direction to artistic imagination in other fields. Following upon the first efforts of the cubists, there came, as has already been said, an awakening in various countries. In France appeared Le Corbusier and Ozenfant; in Russia, Malewitsch; in Hungary, Moholy-Nagy; in Holland, Mondrian and van Doesburg. Common to them was an attempt to rationalise cubism or, as they felt was necessary, to correct its aberrations. The procedure was sometimes very different in different groups, but all moved toward rationalisation and into architecture.

When Ozenfant and Jeanneret (Le Corbusier) came together as young painters in 1917, they called their painting *Purisme*. In comparison with the movements preceding it (constructivism in Russia or neo-plasticism in Holland), purism, coming out of French soil, was the closest of all to the aim of cubism and, at the same time, to architecture.

Two years after the exhibition of the cubists in the Salon des Indépendants, there appeared in Russia an abstract-art movement, fostered by Kasimir Malewitsch, which completely eliminated the object. It was a flight from and a protest against the naturalistic object, with painting reduced to a few signs of symbolic intensity. What its paintings achieve are fundamentally only pure interrelationships. Flatly extended rectangles and strips float in continuous interrelation in space for which there is no true human measure.

Interrelation, hovering, and penetration form the basis of Malewitsch's half-plastic architectural studies, which he calls 'architectonen'. These objects are not intended for a particular purpose but are to be understood simply as spatial research. Interrelations are created between these prisms, slabs, and surfaces when they penetrate or dislodge each other. They come close in spirit to the so-called megastructures of around 1960.

Neo-plasticism, an expression used by the Dutch painter Mondrian, signifies that three-dimensional volume is reduced to the new element of plasticity, the plane. Mondrian sacrifices every contact with illusionistic reproduction, going back to the fundamental elements of pure colour, of planes, their equipoise and interrelations.

The small circle of young artists who gathered around Theo van Doesburg and his periodical, *Stijl*, after 1917 progressed much more radically than the French painters and architects. Van Doesburg and Mondrian sought 'pure art' not in any way deflected by external motives. With them everything rests on the distribution and juxtaposition of planes of pure colour: blue, red, yellow. To these are added black and various tones of white, all being placed in a network of panels.

The Belgian Vantongerloo, who also belongs to this circle, demonstrated with the prisms, slabs, and hollows of his plastic of 1918 that contemporary sculpture, like painting, was not to be limited to a single point of view.

Van Doesburg, the moving spirit of the circle, was painter, man of letters, and architect. Although he executed few buildings, he cannot be omitted from the history of architecture, since, like Malewitsch, he possessed the gift of recognising the new extension of the space sense and the ability to present and explain it as though by laboratory experiments.

One of van Doesburg's drawings in which an attempt is made to present 'the elementary forms of architecture' (lines, surface, volume, space, time) may very well have appeared to many at that time as so much disjointed nonsense. The present-day observer, who has the advantage of being able to look back upon intervening developments, has a very different attitude toward these mutually penetrating flat surfaces. He sees how the enormous amount of contemporary architecture which has since appeared acknowledges this vision of space.

In 1923 van Doesburg, together with van Eesteren who later became a town planner of Amsterdam, produced a house that is bolder than any other building executed during the period. The breaking-up of the compact mass of the house, the accessibility of the roof, the horizontal rows of windows—in fact, all the features that were later to be realised in numerous examples were indicated in it. If a *collage* by Georges Braque, produced ten years earlier, consisting of different papers, scraps of newspaper and fragments of planes, is placed alongside a reproduction of this house, no words are necessary to indicate the identity of artistic expression. An architectonic study of Malewitsch might be likened to it equally well. The effect is as if the blind surfaces of the Malewitsch sculpture had suddenly received sight. It is obvious that in the second decade of this century the same spirit emerged in different forms, in different spheres, and in totally different countries.

III   THE RESEARCH INTO MOVEMENT: FUTURISM

In the first decade of this century the physical sciences were profoundly shaken by an inner change, the most revolutionary perhaps since Aristotle and the Pythagoreans. It concerned, above all, the notion of *time*. Previously time had been regarded in one of two ways: either realistically, as something going on and existing without an observer, independent of the existence of other objects and without any necessary relation to other phenomena; or subjectively, as something having no existence apart from an observer and present only in sense experience. Now came another and new way of regarding time, one involving implications of the greatest significance, the consequences of which cannot today be minimised or ignored.

In 1908 Hermann Minkowski, the great mathematician, speaking before the Naturforschenden Gesellschaft, proclaimed for the first time with full certainty and precision this fundamental change of conception. 'Henceforth,' he said, 'space alone or time alone is doomed to fade into a mere shadow; only a kind of union of both will preserve their existence.'

Concurrently the arts were concerned with the same problem. Artistic movements with inherent constituent facts, such as cubism and futurism, tried to enlarge our optical vision by introducing the new unit of space-time into the language of art. It is one of the indications of a common culture that the same problems should have arisen simultaneously and independently in both the methods of thinking and the methods of feeling.

During the Renaissance the common artistic perception, perspective, was expressed by one group of artists primarily through lines, and by another primarily through colours. So in our own day the common background of space-time has been explored by the cubists through spatial representation and by the futurists through research into movement.

For Jakob Burckhardt there reigned in Italy 'the quiet of the tomb'. The futurists were a reaction against this quietness; they felt ashamed that Italy had become simply a refuge for those seeking to escape from the demands and realities of the present. They called upon art to come forth from the twilit caves of the museums, to assert itself in the fullness of modern thought and feeling, to speak out in authentic terms of the moment. *Life* was their cry—explosive life, movement, action, heroism—in every phase of human life, in politics, in war, in art: the discovery of new beauties and a new sensibility through the forces of our period. Not without right did they claim to be 'the first Italian youth in centuries'.[2]

So, from the beginning, they plunged into the full struggle, and carried their cause militantly to the public. The poet Marinetti, whose apartment in Rome even to this day bears the escutcheon of the 'Movimento futurista', proclaimed in the Parisian *Figaro* of 20 February 1909, 'We affirm that the splendour of the world has been enriched by a new beauty: the beauty of speed.' And later, in 1912, in the 'Second technical manifesto of futurist painting', the futurists developed their principal discovery, that 'objects in motion multiply and distort themselves, just as do vibrations, which indeed they are, in passing through space'. The most exciting of their paintings realise this artistic principle.

The productions of futurist painting, sculpture, and architecture are based on the representation of movement and its correlates: interpenetration and simultaneity. One of the futurists' best minds and without any doubt their best sculptor, Umberto Boccioni, who died much too early, in 1916, has most clearly defined their purposes. In an effort to penetrate more deeply into the very essence of painting, he sought terms for his art, terms which, now obscurely felt, now shining clear and immediate in his increasing creative experience, anticipated those that later appeared in the atomic theory. 'We should start', he said, 'from the central nucleus of the object wanting to create itself, in order to discover those new forms which connect the object invisibly with the infinite of the apparent plasticity and the infinite of the inner plasticity.'

Boccioni tried in these words to circumscribe the sense of a new plasticity which conceives objects (as they are in reality) in a state of movement. This was reflected directly in his sculpture, *Bottle Evolving in Space*, 1911–12, with its intersecting spatial planes. One of the few sculptural masterpieces of the epoch, this sculpture expresses the inherent significance of an object of daily use by treating it with new artistic invention. Sometimes, as in this instance, cubistic and futuristic works are closely bound together on a common basis of the same spatial conception.

The French painter, Marcel Duchamp, who belonged neither to the futurists nor to the cubists, painted at the same time (1912) his *Nude Descending the Staircase*, in which the movement is dissected mathematically and yet fully surrounded by the multi-significance of irrational art.

Usually the futurists present movement as such, as subject matter (*Elasticity*, Boccioni, 1911; *Dynamisme musculaire, Simultanéité*, Carrà, 1912; *Speed*, Balla, 1913), or show objects and bodies in motion (Gino Severini's study of the dance as a movement in mass, *The Dance Pan-Pan*, 1911; *Walking Dog*, Balla, 1913; *Rattling Cab*, Carra, 1913).[3]

In both futurism and cubism this enlargement of the optical was achieved before 1914, before the first world war. The cubists were the more passive and less vocal. Not fighters in the futuristic sense, more purely research men in their work, they kept to their ateliers, preparing quietly and without fanfare the symbols of our artistic language. Braque and Picasso wrote no ponderous tomes expounding their theories. Even the name 'cubist' was a label fixed upon them by outsiders. They did not try to paint 'movement' itself, or the dynamism of muscles, or the automobile, but through their still lifes of things of daily use sought to find artistic means for our spatial conceptions. This is the reason cubism found extension into so many ramifications. This is why laboratory painters, who had no thought beyond their own artistic problems, could also give an impulse to the expression of the new spatial conceptions in architecture.

To try to introduce the principle of movement directly into architecture did not touch the fundamental problem. In his projects for his 'Città Nuova', in his skyscraper apartment houses connected with subways, elevators, and traffic lanes at different levels, Antonio Sant' Elia tried to introduce the futuristic love of movement as an artistic element in the contemporary city. Sant' Elia's 'Città Nuova', as well as Malewitsch's sculptural studies of the same period, expressed trends that were first implemented in the 1960s when movement in cities came to be recognised as a problem of urban form and obliged different levels to be created for pedestrians and vehicles. We do not know if Sant' Elia's talent would have developed. He died in 1916, at a time when his contemporary, Le Corbusier, was still far from self-realisation. Although Sant' Elia's prophetic vision did not direct the way architecture then followed, it did present a new viewpoint in a period when everyone was looking for a signpost. In his manifesto of 14 July 1914, which he published in connection with the exhibition of his schemes in Milan, he demanded an architecture imbued with the utmost elasticity and lightness, utilising all the newly developed elements of construction from iron and ferroconcrete to composite materials made by chemical processes, including textile fibre and paper. Behind these technical demands loomed his artistic aim: mobility and change. What he wanted to realise he condensed into the few words: 'Every generation its own house!'

There are times when the man of the laboratory is compelled to go forth into the street to fight for his work. On occasion this may be advisable. But normally he endangers his work by so doing. The futurists were perhaps too much bound up in trying to apply their ideas to all kinds of human activities; the result was that their movement—which our period cannot ignore—had a comparatively short span of volcanic productivity. It was unfortunate in that some of its ablest exponents died too early and that others lapsed into regrettable routine work, bequeathing nothing to the future except the few years of their youth.

Futurism did not have the opportunity of the cubist movement: to accumulate, through all the many-sided stages of modern development, the results of artistic research, until they should appear united and in full power in a single great work—*Guernica*.

NOTES

[1] We shall treat contemporary movements in art here only so far as their methods are directly related to the space conceptions of our period, and in order to understand the common background of art, architecture, and construction. For an understanding of these movements the elaborate catalogues of the Museum of Modern Art, New York, are very useful. See Alfred H. Barr, Jr, *Cubism and Abstract Art* (1936), and Robert Rosenblaum, *Cubism and Twentieth Century Art* (1960). For a short survey with emphasis on historical relations, see J. J. Sweeney, *Plastic Redirections of the Twentieth Century* (Chicago, Ill., 1935); for the relation of contemporary art to education, industrial design, and daily life, see L. Moholy-Nagy, *The New Vision* (1938). The close relation of contemporary sculpture to primitive art, on the one hand, and, on the other, to an enlargement of our outlook into nature is stressed in C. Giedion-Welcker, *Contemporary Sculpture* (1955).

[2] For the literary intentions of futurism cf. the article of its founder, F. T. Marinetti, in *Enciclopedia italiana*, vol. XVI, 1932. See also above, p. 12.

[3] For illustrations of this first and most important futuristic development cf. Boccioni, *Pittura, scultura futuriste* (Milan, 1914), a volume of over 400 pages, with bibliography of exhibitions, manifestos, etc.

Alan Stanbrook†

The time and space
of Alain Resnais

Alain Resnais' first feature film was shown at Cannes in the year that brought Truffaut's *Les Quatre Cents Coups* and Marcel Camus' *Black Orpheus*. What more natural, then, than that Resnais' name should have been linked with the group of young directors making their initial productions and subsequently termed the *nouvelle vague*? But Resnais has as little in common with Truffaut as the latter has with Camus. He had already established himself with art films and documentaries and had made a masterpiece in *Night and Fog* in 1956, long before Chabrol had privately produced *Le Beau Serge*. *Hiroshima mon amour* was the logical development of an existing talent and it was pure chance that it was made concurrently with the first products of the *nouvelle vague*.

To understand Resnais we must forget the misleading expression 'new wave' and consider him as the supreme individualist that he is. Resnais is a true intellectual. He has made the most significant contribution to film language since the early Orson Welles and evolved a highly idiosyncratic and inimitable style. We need only consider *Une Aussi Longue Absence* by Henri Colpi, Resnais' editor, or Albicocco's *Girl with the Golden Eyes* to realise how far short his imitators fall. Though he has collaborated with the best literary minds of his generation—Paul Eluard, Raymond Queneau, Marguerite Duras and Alain Robbe-Grillet—and the most imaginative composers—Maurice Jarre, Hanns Eisler and Giovanni Fusco—his films do not emerge as illustrations, more or less bolstered by an apt musical score. They are transformed into original creations in which the perfect symmetry of each aspect of production attests the controlling hand of a single, dedicated personality. And the thematic continuity of his work, taken from scripts by very different writers indicates an artist with an extremely personal vision. The same intelligence and attitude inform a commissioned documentary on the French national library as the labyrinthine complexities of *Marienbad*.

Memory and the corrosive effects of time are the twin poles of Resnais'

† Reprinted from *Films and Filming*, x, No. 4 (January 1964), pp. 35–8.

world. On the one hand there is the desire and obligation to remember the past if mistakes are not to be repeated ('Eighty thousand wounded in nine seconds. It will happen again') and working against this is the inevitable decay in once burning thoughts and feelings brought about by time and distance ('I shall forget you. Look, I am forgetting you already'). The positive and the negative, the forces of light and dark, of life and death are engaged in mortal conflict in Resnais' work and out of this conflict comes a compulsion for truth and the need to perpetuate it before it is too late. The director is an artist committed to a cosmic view in a sense far transcending the ivory tower aestheticism deplored by some critics.

Of course Rensais is concerned with style. He is exploring territory that has never been opened up in the cinema before and it requires a treatment radically different from all known models. Eisenstein had to invent a language to express the needs of the new Soviet state; Resnais has had to do the same for the cinematic space-time continuum. He has used the two basic elements of film technique—the cut and camera movement—in a revolutionary way. Hitherto the cut has been used almost exclusively for variety or emphasis (occasionally for ironic association as in *October* or *A propos de Nice*) and the track or pan in order to follow a subject in motion. In both cases they refer to present time and can be traced back respectively to Eisenstein and Murnau. Resnais looks back farther than this to *Intolerance*, where Griffith applied parallel editing to the presentation of four stories widely separated in time in order to accentuate an underlying unity. It is only an extension of this to use the track to similar effect in *Night and Fog*. The establishment of connections between past and present is Resnais' counterblow to the corruption of time, even as the operation of the involuntary memory was for Marcel Proust.

Proust was obsessed by similar problems to those of Resnais and it is scarcely surprising that the latter should have had recourse to *A la recherche du temps perdu* for the clarification of his vision. Some acquaintance with Proust's work is, if not essential, a considerable help in the elucidation of Resnais' unfamiliar world. For Proust all the past is stored in the subconscious and is available to the sleeping man, but only a fraction filters through to the waking state. This can be recalled by a conscious effort but the rest lies buried in sleep, to be recaptured only fragmentarily by such accidental experiences as the working of the organic or involuntary memory, that faculty whereby past feelings and incidents are recovered through bodily attitudes or objects associated with them. The scene in *Hiroshima mon amour* in which the position of the Japanese lover's hand spontaneously evokes an image of the death of the German soldier many years ago is a powerful example of this. Resnais makes the association more difficult for us to follow than Proust: the latter has words to explain it but the film director has only the juxtaposition of two shots to create the same effect. Without a knowledge of Proustian method the unprepared spectator might well find such a bridge difficult to make. It is such involuntary memories which revive the past in all its intensity far more than any conscious attempt to preserve it and Resnais' heroine would regretfully echo Proust's statement: 'There comes a day when we suffer no longer from a grief we had thought inconsolable.'

For Proust and Resnais alike any given moment is an amalgam of past and present. We *are* our past whether we recognise it or not and Proust's remark: 'We cannot recall our memories of the past thirty years but we are steeped in them utterly' finds its equivalent in :'Your name is Hiroshima. And your name is Nevers—Nevers in France.'

Many of Resnais' art films and documentaries have never been shown in Britain. His *Gauguin* remains an unknown quantity, his industrial documentary *Le Chant du Styrène*, with a script by Raymond Queneau, has mysteriously never been imported and the controversial *Les Statues meurent aussi*, with its impassioned anti-colonial propaganda disguised as a study of African art, still languishes under French governmental proscription. We have, however, seen *Van Gogh*, for which he collected an Oscar, in a dubbed American version.

The art film is a severely constricting *genre*. The director is limited to whatever is already on canvas and his task lies in bringing out the salient points in other people's compositions. He has certain abilities denied to the spectator at an art gallery: he can isolate a particular area of the canvas for consideration by moving his camera and he can juxtapose two pictures for comparison by cutting. Resnais made full use of these advantages in his art films and they proved an invaluable training ground for his developing talent.

By the time we reach *Guernica*, however, it is apparent that Resnais is exceeding his duty to clarify and explain. He is beginning to use existing material, in this case a work of art in its own right, as the point of his own. *Gurnerica* is a *tour de force* in which the director uses a single painting by Picasso to make a heartfelt protest against brutality and oppression. The tragic violence, already present in Picasso's picture, is deepened by repetition and the jagged cutting style.

Resnais cuts from one fragment of the picture to another, revealing now a naked light bulb, now a startled horse, now a woman running in terror, the whole accompanied by a disquieting, clangorous musical score. Out of a limited canvas area Resnais extracts countless images of horror and ugliness, joining them now in one order, now in another to produce a composite impression even greater than that of the picture seen in its entirety. To heighten the effect, he intercuts fragments of the picture with composed shots of machine gun bullets raking a wall. Picasso's abstract portrait of the Guernica incident is thus made concrete by the addition of live material and also by the reading of Paul Eluard's poem, compellingly delivered by the formidable Maria Casarès in that tragic intonation normally reserved for Racine. *Guernica* serves as a reminder of an act of brutality that time has almost erased from men's minds. 'Eighty thousand wounded in nine seconds'—the distance between *Guernica* and *Hiroshima mon amour* is shorter than one might think.

In a sense *Guernica* is a prelude to the mature expression of a similar concept in *Night and Fog*. Out of this half-hour picture barely five minutes constitute material shot especially for the film. The remainder is sifted from footage shot by the Germans themselves or by the Allies at the time of the liberation. Yet the careful integration of those five minutes into the pre-existing shots results in a work that is stamped with Resnais' personality from beginning to end.

Resnais, of course, is assisted immeasurably by the contributions of Hans Eisler, who has supplied him with one of the most authentically moving scores in film history, and of Jean Cayrol, whose script commentary, spoken in a grey, matter-of-fact monotone, achieves a poetic understatement comparable with the best of Hemingway. But the co-ordination of the separate talents working on the film into a single tragic harmony is the director's achievement and the ultimate credit for this, the most remarkable of documentaries, must lie with Resnais himself.

Night and Fog is an essay in forgetfulness, in the fallibility of the human memory. It opens with Ghislain Cloquet's beautifully photographed colour shots of a typical afternoon landscape. It is peaceful and soothing and the commentary echoes the mood: 'Même un paysage tranquille . . .' And then the camera begins a long lateral track to reveal buildings, outhouses and barbed wires. This landscape, bathing in the sun, was the location of one of Hitler's concentration camps a mere ten years before. The tracking speed is gradually increased until with the words 'camp de concentration' we are abruptly transported in time to 1933 to the origins of Nazism. The colour vanishes to be replaced by the stark monochrome of Nazi newsreels. The alternation between colour and black and white is no mere director's whim but a dramatic illustration of our weak memories. Who would have thought that so calm a setting concealed such an appalling history? And who would have thought in 1945 that ten years would be sufficient to make men forget?

It is Resnais' intention to show just how much we have forgotten: human skin used as parchment, useless scientific experiments, systematic sterilisation, severed heads—the very words are bad enough but Resnais' images underline the full force of the atrocities. The British censor has trimmed some of these scenes for public exhibition. We do not see the bulldozer heaving mounds of corpses into a lime pit (though, curiously enough, we do in *Judgement at Nuremberg*) or the remnants of bodies withdrawn from the crematoriums but nothing is more indicting than the seemingly endless pan across acres of human hair.

These scenes are intercut with further shots of the same locale a decade later, where tourists have their pictures taken in the same gas chambers whose ceilings are scored with the marks of the victims' nails. Weeds push up amid the railway tracks along which innumerable Jews were transported to destruction in the night and the fog. Resnais' relentless tracking camera, moving along rows of latrines where prisoners spent nights racked by dysentery, makes us embarrassingly aware of our own comfort and idle consciences.

The film builds inexorably to its emotional climax in which culpability is denied by all concerned. 'I am not responsible, says the Kapo. I am not responsible, says the officer. I am not responsible! Then who is responsible?' Here Cayrol breaks his dispassionate reporting to provide the film's one moment of direct emotional colouring. In contrast to the rest of the script, the word 'who' is emphasised and Eisler times the most emotive passage in his score to coincide with it.

The last scene, reverting to the colour camera, proceeds to answer the rhetorical question. We are all responsible and the forces which produced Nazism are still latent in our midst, waiting for the moment when a relaxed conscience will permit them to rise again. And significantly Resnais' film, which began with a sunny shot of a peaceful landscape, closes with a black, menacing image of the crematorium waiting patiently for its prey.

After the inspired didacticism of *Night and Fog* nothing would seem more unlikely than a study of the French national library. Yet *Toute la memoire du monde* is every whit as personal a film. The title gives the clue. The Bibliothèque Nationale is not merely a storehouse for books; it is the granary of the human mind. Everything that man has discovered in science and letters is here preserved in permanent written form and is available for consultation. Man's history is irrevocably written on the leaves of these books and protected from effacement by regular injections of insecticide. It is all here, all the memory in the world, and Resnais contrives to invest these dusty tomes with a personality while the librarians in constant attendance are dehumanised.

Stylistically the film is one of the most audacious in the documentary field: it consists from start to finish of a series of carefully matched tracking shots. The camera is never still as Resnais conducts us on a mesmeric tour to the very heart of memory—books. Each movement is dovetailed into the next to create a forceful drive and there is more wit, suspense and mystery in this seemingly dry-as-dust subject than in many a thriller. The film opens with a practical joke as the camera pans from a pile of books to reveal a strange, three-eyed, metallic monster. Is it a cousin of Robby The Robot? No; it is a camera skeleton, as we realise when a microphone suddenly dips into the frame and the commentary begins. These are to be our guides on the tour.

With introductions completed, we set out with our guides through the library's security system of bars, grills, lifts, and locks, and Resnais uses parallel or contrasting tracking speeds and directions and the visual pattern set up by the constantly shifting miasma of light and shade to create a hypnotic rhythm, not unlike atonal music.

The second half of the film is the return journey. We have reached the book vault, where every published volume is numbered and catalogued, and seen the methods of preservation. Now someone has filled in a demand slip and we follow the book back from its shelf, up through the labyrinth of passages and lifts to the reading room, where it will serve mankind by revealing its captive knowledge. Earlier in the film a witty superimposition had made a vast simulacrum of the human brain contract to the size of the dome of the library filmed from an aeroplane. To conclude the film Resnais constructs a complex pan and track up to the roof of the dome seen from the inside—a veritable temple of the mind, housing the lessons of the ages to the end of time.

Hiroshima mon amour is at once the sequel and the consummation of the particular attitude towards Time revealed in the director's previous work. Nothing quite so complex and exacting had been seen since *Citizen Kane*, and with his first feature Resnais proved to be the most advanced artist currently working in the cinema. The film is again about time and memory, about man's tragic propensity for forgetfulness, even of an event like the annihilation of Hiroshima. To illustrate this theme, Resnais and his scenarist, Marguerite Duras, tell three stories concurrently—a love story set in Hiroshima today, another set in Nevers involving the same woman, and the story of the atomic holocaust itself. All three are intertwined in an allusive pattern so that a concept of universal relevance emerges through a particular and apparently trivial narrative.

The stories are told fragmentarily and it is only subsequently that the full import of certain scenes and even isolated shots is realised. The heroine is a French film star on location in Hiroshima, who falls in love with a Japanese student. Both have incidents in their past, the memory of which they had once thought of permanent durability. His is the destruction of his home town in a single, cataclysmic explosion; hers a romance with a German soldier, killed in Nevers on the last day of the war, for which her head was shaven as a punishment for collaboration. The chance alliance between these two people brings home to them the transience of human emotion and the film concludes with the realisation that the legendary healing powers of Time can sometimes be a killer in disguise. We forget the true meaning of Hiroshima as easily as we forget our past loves. 'Just as in love this illusion exists, this illusion of never being able to forget. I had the same illusion in Hiroshima—that I will never forget. Just as in love.' And the title itself, with its arbitrary coupling of two dissociated concepts, reveals a hidden unity between events on a personal and a universal plane.

Hiroshima mon amour is one of the most formally dense and rewarding films ever made. It opens with gigantic close-ups of parts of two bodies locked in close embrace. Their skin is covered with a tracery of whorls resembling the photographs of atomic survivors. It is in fact the effect of sand clinging to their wet bodies but we are not aware of this. An association of ideas has been established, a connection between the creative act of love and the destructive force of the atomic bomb. And then, before we can even see who these lovers are, the first, disturbing words are heard. 'You saw nothing at Hiroshima, nothing. I saw everything, everything.' It is not the way one expects a film to begin and one is totally unprepared for the way it develops.

Pursuing Robert Bresson's experiments, Resnais allows his script writer full rein to launch into a purely literary prologue. A strange, hypnotic duologue is maintained between these two lovers, whom we have still not seen, and who are discussing things far removed from their physical actions. With the aid of shots from Japanese films and documentary material filmed in Hiroshima today, a complete image, visual and aural, is built up of the most tragic city in human history.

Words are used only partly for their literal meaning, far more for the poetic effects that can be obtained by reiteration: 'The seven branches of the estuary in the delta of the river Ota empty and fill at the usual time, at exactly the usual time with a water fresh and teeming with fish, grey or blue according to the time and the seasons. Along the muddy banks people no longer watch the slow rising of the tide in the seven branches of the estuary in the delta of the river Ota.' We are mesmerised as much by the insistent incantation as by the appalling images Resnais directs us to see. This is a true sound film in which the sound track is an integral part of the whole poetic and philosophical structure.

Throughout the film no indication is given when a flashback is to be used. There are no helpful dissolves but direct cuts to past experience. From a sequence in Hiroshima we are dramatically reverted to Nevers to see the same woman fifteen years younger cycling to meet her lover. We follow her in a panning shot of prodigious length, shot through a telescopic lens, as she rides through a bleak wood vastly different from the world of Hiroshima. Again, in the midst of a scene in Hiroshima there is a single, brief shot of a foreboding pavilion. In the context it is meaningless but later we learn that it was from this pavilion that the bullet was fired that killed the German lover.

The flashbacks are both untelegraphed and interjected with total disregard for chronology. The logical presentation would be the shaving of the girl's head and then her incarceration in the cellar. In the film, however, the order is reversed and we see her in the cellar already shaven before we see the actual shaving operation. The screen order of events is not a chronological but a subjective one as various parts of the past are revived in the heroine's mind in conversation with the Japanese.

Hiroshima mon amour is, then, very much an editor's film and Resnais' own apprenticeship in the cutting rooms stands him in good stead in the implementation of Astruc's theory of the *Caméra-stylo*. As Henri Colpi points out, the scene in which the heroine crosses a square, enters the hotel, climbs the stairs and arrives at her room on the second floor in a series of jump cuts, with the omission of thirty yards of space and a whole flight of stairs, is an exact visual transcription of the novelist's shorthand: 'she crossed the square, went into the hotel, climbed the stairs to her floor.' Remarkable, too, to think that the sequence towards the end of the film, in which the heroine's nocturnal perambulation through the streets of Hiroshima is intercut with a fragmented tracking shot through the streets of Nevers, was in fact shot by two different cameramen in different parts of the globe. The tracking speeds are identical, creating an impression of spatial and temporal unity. She is walking in Hiroshima as her mind wanders through Nevers fifteen years before.

The cutting is as precisely calculated as anything since Eisenstein. Long stretches of stillness are punctuated by staccato cuts as in the scene in which the Japense slaps the girl's face or in the sudden recapitulation of the German soldier's violent death. Time and space are swept aside to provide a reality in which past and present, France and Japan co-exist in a single continuity. This is also achieved by a manipulation of the sound track. Sounds from Hiroshima—Japanese music, frogs croaking, a barge hooting —overscore the Nevers scenes, with which they have no visual association while Giovanni Fusco creates musical motifs for the Nevers and Hiroshima sequences and then proceeds to apply them to the very opposite of the scenes they govern. *Hiroshima mon amour* is a synthesis, in which hope for the future is contained in the memory of the past. As the heroine re- marks: 'What else can one make in Hiroshima but a film about peace?' and Resnais has done just this, fusing particular and general into a fully autono- mous work of art.

Initially Resnais' latest picture presents a problem of sheer comprehension unmatched by any of his previous works. We are familiar with the device of superimposing the credit titles on the opening sequence of a film. *L'Année dernière à Marienbad* asserts its individuality from the start by commencing the sound track as the credits unroll. As they follow each other against a grey, nondescript background, a disembodied voice is heard behind an oppressive organ score. The words at first are indistinct and we are puzzled, irritated even. Gradually, however, they become audible and we realise that they are an incantation, like the opening of *Hiroshima mon amour*. Phrases echo and repeat each other while the camera tracks endlessly down long, deserted corridors and past elaborate, baroque ornamentation. We are effectively disorientated in time and space and wonder whether this is not, after all, a documentary about architecture.

Suddenly the human element supervenes: a brief glimpse of two statu- esque footmen and the camera skirts past a playbill to bring us into a private auditorium, where a theatrical performance is in progress. The words of the unseen narrator are taken up by the actor: the actress replies 'And now I

am yours'; the curtain falls and Resnais cuts to the audience applauding to
bring the first sequence to its perplexing close. We doubt the reality of
what we have seen. Has the whole monologue been part of a dramatic
speech and are those measured tracking shots merely the visualisation of
the actor's words? Are the words or the pictures true, or both, or neither?
It is not Resnais' intention that we should be able to answer these points
at this stage. It is sufficient that we have been induced to question the
validity of our senses in determining the nature of external reality. This
bold sequence is an overture to the film, evoking the atmosphere of what is
to follow and stating some of the themes that are later to be developed.

The quantities of literature written about this film, both by critics and
by Resnais and his scenarist, Alain Robbe-Grillet, have made it seem a
perverse and incomprehensible riddle. Certainly it is rigorously intellectual
but once it has been seen the various explanations fit marvellously into
place and the film becomes clear as day.

There is a later sequence, involving a status of a man and woman, which
perhaps provides the key to the interpretation of the whole work. The
narrator, whom Robbe-Grillet designates simply as X, exclaims: 'I talked
about the statue. I told you that the man wanted to keep the young woman
from venturing any further: he had noticed something—no doubt a danger
—and he stopped his companion with a gesture of his hand. You answered
that it was actually the woman who seemed to have seen something, but
something marvellous, on the contrary—in front of them, which she was
pointing to. But this was not incompatible.' As these words are delivered,
Resnais' camera edges around the statue, which seems to change its whole
meaning with every new angle, so that reality becomes dependent, not on
absolute truth, but one's momentary perspective. The same judgement
could be applied to the entire film, which has not one meaning but many,
all of which are equally viable.

However disputed the interpretation, the story is extremely simple. At a
country mansion at Marienbad, or is it Frederiksbad, or somewhere else?
a man (X) reminds a woman (A) that he met her last year at Marienbad.
She denies it but is eventually persuaded and leaves the man (M) who may,
or may not, be her husband, for this man who may, or may not, have been
her lover last year at Marienbad, or was it Frederiksbad.

This extraordinary story, which includes flashback references to what
may, or may not have happened and flashforward allusions to what may, or
may not occur, encourages limitless explanation. A mere recital of the facts
recreates the impression of nonsensicality that strikes every reader before
he sees the film. If clarity is achieved only by seeing the work, it reinforces
its claim to be cinema rather than a literary conceit of Robbe-Grillet.

It can be considered most simply as a mythological love story, repeating
itself from age to age: Prince Charming who carries off the Princess from
her petrified surroundings. Significantly all the characters except X and A

parade through the film like zombies and are frozen in sculptural poses. The attempt to awaken the 'sleeping beauty' has been made again and again in different settings and times. The game of cards, at which X perpetually loses to M, epitomises the ritual element of the quest, which continually fails until finally Beauty remembers, is awakened and the spell is broken for ever. The End—nothing further can happen.

Secondly, it may be taken as an examination of Time as an irregular circle, which repeats itself but never quite in the same way, as in Priestley's play *I Have Been Here Before*. Hence the casual reference to the name Anderson, which reappears sometimes as Ackerson, somes as Patterson. In one Time the husband shoots the wife and concludes one version of the story; in another the lover is killed when a balustrade collapses. Both events are cut into the present with no explanation. If these things happened, why are X and A still here this year at Marienbad? Simply because Time is an irregular circle that never quite retraces its original track.

Again, the theme is reality as an infinite number of possibilities: at every moment we make a choice between two lines of conduct, yet there may be another world, parallel to this, in which we made the opposite decision—a world which is equally true, and not only one world but infinite worlds and infinite permutations of choices. Resnais symbolises this concept in the drawerful of identical photographs, which the heroine sets out in the pattern of the card game mentioned above. This is not a real image but an evocative one, implying the existence of multiple realities from which we can extract those that please us, even as X takes his pick from the neatly arranged rows of cards.

We can even deduce that we are onlookers at a mental asylum to which M has brought his ailing wife. There is a certain similarity to *The Cabinet of Dr Caligari*, where the reality that we have implicitly accepted is finally revealed as the delusion of a madman. The authors, too, have acknowledged the possible analogy of a Breton myth in which Death comes to claim a maiden but allows her a year's respite until the ultimate reckoning. In short, all these interpretations are valid and the film could well be subtitled *What You Will*. A search for truth and the discovery that no single viewpoint can encompass it all—what could be more relevant for our ideology-ridden times?

Resnais directs this complex subject magisterially, with the cinematic vocabulary of an artist at the height of his powers. He cuts back and forth in time present, past, future and imagined with lightning rapidity. The characters' dress changes every few seconds to imply the possibility of parallel realities. In one startling scene we see successive shots of A sitting down on a bed, alternately on the left and right and each time in a different pose. The effect is not gratuitous but a succinct visual representation of the cyclic nature of time.

If one seeks an antecedent for the daring experiments it is to Cocteau's *Orphée* that one must look, where Maria Casarès' costume changes miraculously from black to white and the narration of Cégeste's death is accompanied by a direct, interpolated cut to the event itself. But Resnais is not eclectic: the influence has been fully digested and shaped to serve his own ends, which do not correspond with Cocteau's. And if the director has models for certain aspects of his style, others betoken his complete originality. To evoke the terror of remembering past, possible or future occasions when the lover may have been killed, Resnais overexposes the image to a glaring, disquieting white. Again, his skilful isolation of the principals from the amorphous body of spectators at the theatrical presentation is a model of persuasive camera movement and cutting. In accordance with the grail-like quest of the plot, we even see X addressing his words to a variety of women before he finally communicates with the right person.

Resnais' Proustian preoccupation with time and memory seems here to have reached its zenith in the most original film since *Citizen Kane* and one recalls Lawrence Durrell's dictum: 'it would be worth trying an experiment to see if we cannot discover a morphological form one might appropriately call "classical"—for our time. Even if the result proved to be a "science-fiction" in the true sense.'

III

Time in modern literature

Dorothy van Ghent†   ANTICIPATIONS IN LAURENCE STERNE
(1713–68)

On *Tristram Shandy*

What is the 'action' of *Tristram Shandy*?

Presumably, like that of *Moll Flanders*, it is episodic and biographical: *The Life and Opinions of Tristram Shandy, Gent.* is the full title. But Tristram is not born until a third of the way through the book; not christened until fifty pages later; the story is more than half over when we are told that he has reached the age of five (scarcely yet the age for 'opinions'); is two-thirds finished by the time he is put into breeches; suddenly he appears as a gentleman on his travels in France; and the novel ends with an episode that concerns not Tristram but Uncle Toby. Those sporadic flickers of narrative in which Tristram is seen in chronological circumstances as the hero-in-action (if the hero *can* be thought of as 'acting' while he is being born, christened, circumcised, etc.) evidently do not serve the same purposes of narrative that we have observed elsewhere; they are, rather, if they are anything, an intentional mockery of 'action'. Chronological and plot continuity are, then, not definitively organisational to *Tristram Shandy*. The fact appears extraordinary when we stop for a moment to consider how naturally, habitually, almost stubbornly we tend to think of all experience as somehow automatically dished up to us, like a moulded pudding, in the form of chronology; and how this tendency to see experience as actions related to each other undisturbedly by the stages of the clock and the calendar leads us to expect of fiction that it will provide a similar unidirectional action, or series of episodes, taking place chronologically. *Tristram Shandy* pays lip service to this expectation—at least by allowing Tristram to be born before he is christened, to be put into breeches before he goes travelling in France—but so mockingly as to make us aware that Sterne is engaged in deliberate demolition of chronological sequences and (inasmuch as our notions of 'time' and 'action' are inextricably related) deliberate destruction of the common notion of 'action'.

But this is the negative aspect only, and if this were all there were to Sterne's concern with the relationship of time and human experience, then

† Reprinted from *The English Novel: Form and Function*, 2nd ed. (1961), pp. 84–93.

the squiggly lines he draws in chapter XL of book VI, to show the haphazard
progress of his novel, could represent just that: its hapharzardness, its lack
of form. But it is anything but haphazard or formless. Obeying formal
laws of its own, it is as skilfully and delicately constructed as *Tom Jones*.
Having ruled out plot chronology as a model of the way experience
presents itself, Sterne offers another model: that of the operations of consciousness, where time is exploded, where any time-past may be time-
present, or several times-past be concurrently present at once, and where
clock-time appears only intermittently as a felt factor. As Cervantes' *Don
Quixote* offers fiction as many models as it can use, so it offers this one also.
In the episode of Quixote's descent into the cave of Montesinos, the time
of dream and of poetry, the ancient heroic time of Montesinos and Duran-
darte, the modern time of coinage and 'new dimity petticoats', and the
actual hours of the descent and of the Knight's sleep, merge as one time.

But what is to give unity to this model, if we cannot plot the hero's
adventures on clock and calendar in order to know when they begin and
when they are ended, and if we do not, after all, have a hero to 'act' in the
ordinary sense—no time, no hero, no action? The unity of any novel may
be described on several different levels; we may speak of the unifying
function of theme, or of plot, or of symbolism, or of other elements that
appear to have superior importance. The total structural unity of a work
does not yield itself to a simple description, but only to a quite lengthy
analysis of the complex interrelationships of all major elements. At the
most conspicuous level, the unity of *Tristram Shandy* is the unity of
Tristram's—the narrator's—consciousness.[1] This is a representative kind
of unity, psychologically true to the way in which experience appears to
all of us to have its most rudimentary unity; for though clock-time may
seem to cut experience into units, these are arbitrary units that melt into
each other unrecognisably in the individual's self-feeling; and though we
may assume that a shared experience has a certain definite, common form
and description for those who share it, yet we know that, as the experience
is absorbed into and transformed by the individual consciousness, it is
something very different for different people, and that its form and 'one-
ness' or unity are felt most concretely only as the experience is stamped
by the character of the individual consciousness.

Sterne's project in *Tristram Shandy* was not to have a parallel, in the
work of a major novelist, until Proust wrote *Remembrance of Things Past*.
Though each of these books, so far separated in time and in local culture,
carries the highly singular and special flavour of its historical circumstances
and of the original genius of its author, they have a strong kinship in subject and plan and quality. Each makes of the narrator's consciousness its
subject matter; the artistry of each lies in the 'objectifying' of this 'subjective' material in its own right and for its own sake, so that the 'subjective' becomes an object to be manipulated and designed and given aesthetic
form according to laws inherent in it; and each creates an Alice-in-

Wonderland world that is unique and inimitable because the individual consciousness is itself unique and inimitable. We do not think of Proust's work as the story of Swann, or the story of Charlus or of Albertine or even of Marcel, or as the history of a transmogrification of social classes and manners in the late nineteenth century, although it is these and other stories and histories as well. Nor do we think of *Tristram Shandy* as a series of character sketches of Uncle Toby and Walter Shandy and Corporal Trim and the Widow Wadman, or as a mosaic of sentimental and slapstick anecdotes, some gracefully pathetic, some uproarious, and all peppered with off-colour puns and double entendres, the indulgences of a neurotic clergyman—although it is these and still other things. We think of *Tristram Shandy* as we do of Proust's *Remembrance of Things Past*, as a mind in which the local world has been steeped and dissolved and fantastically re-formed, so that it issues brand new. Still more definitive of the potentialities of Sterne's method, as these have been realised in a great modern work, is James Joyce's *Finnegans Wake*, where the hero's dream swallows and recomposes all time in its belly of mirrors, and where the possibilities lying in Sterne's creative play with linguistic associations—his use of language as a dynamic system in itself, a magic system for the 'raising' of new perceptions as a magician's formulas 'raise' spirits—are enormously developed. Joyce himself points out the parallel. In the second paragraph of *Finnegans Wake*, 'Sir Tristram, violer d'amores', arrives from over the sea 'to wielderfight his penisolate war' in 'Laurens County', and we know—among these puns—that we are not in wholly unfamiliar territory.

Sterne's project, like Proust's, was to analyse and represent in his novel the creative process; and that Sterne should be the first practitioner of what is called the technique of the 'stream of consciousness' in fictional writing is consonant with the kind of subject he set himself. Our fictional centre of gravity is not a happening or confluence of happenings nor a character or concourse of characters under emotional or moral or social aspects of interest; it consists rather in the endlessly fertile rhythms of a consciousness, as those rhythms explore the comic ironies of a quest for order among the humdrum freaks of birth and paternity and place and time and language. In reading *Tristram Shandy*, we are never allowed to forget that the activity of creation, as an activity of forming perceptions and manoeuvring them into an expressive order, *is itself the subject*; the technique does not allow us to forget it—for let alone the harum-scarum tricks with printer's ink, the narrator plunges at us in apostrophes, flirts his addresses at us with 'Dear Sir' or 'Dear Madam', explodes into the middle of a disquisition or a scene in defiance of time, space, and logic. Uncle Toby, pursuing the theory of projectiles in the pages of half a dozen military authors, becomes involved among parabolas, parameters, semi-parameters, conic sections, and angles of incidence, and Tristram suddenly cries out,

O my uncle;—fly—fly, fly from it as from a serpent . . . Alas! 'twill exasperate thy
symptoms,—check thy perspirations—evaporate thy spirits—waste thy animal
strength,—dry up thy radical moisture, bring thee into a costive habit of body,—
impair thy health,—and hasten all the infirmities of thy old age.—O my uncle! my
uncle Toby.

Toby's investigations were conducted some years before Tristram was
born. It is as if Tristram were sitting in a moving-picture theatre, watching
on a most candid and intimate screen the performance of his progenitors.
He is struck suddenly with admiration or consternation, stands up, waves
his arms, applauds, boos, wrings his hands, sheds tears, sits down and
tickles the lady next to him.

We have said that the unity of *Tristram Shandy* is the unity of the
narrator's consciousness; but—without the discipline afforded by chronology, or an objective 'narrative line', or a moral thesis of some sort—what
is the principle of selection by which the contents of that consciousness
present themselves? Sterne himself puts the question, in chapter xxiii,
book iii, and as he states it, it is the problem of every novelist. In setting
up a certain body of human experience novelistically, what should come
first? what last? what should follow or precede what? Should not everything
appear at once and in fusion, inasmuch as this is the way the author's
consciousness grasps it in its fullest truth? But the novel itself is an artifact
subjected to time law; words follow words and pages follow pages in
temporal sequence, necessarily imposing temporal sequence upon the
material; everything cannot be said at once, although this disability may
seem to injure the wholeness and instantaneousness of the material as the
author grasps it. Trim announces to Toby that Dr Slop is in the kitchen
making a bridge. ''Tis very obliging in him', Toby says, mistaking the
bridge (meant for Tristram's nose) for a drawbridge. The author must
elucidate Toby's error; but when? Right now, at the moment Toby makes
the remark? But the goings-on upstairs in Mrs Shandy's bed-chamber are
of the greatest consequence now. Later, then, among the anecdotes of
Toby's amours with Widow Wadman? Or in the middle of Toby's cam-
paigns on the bowling green? All of these circumstances press upon the
author at once, and are, in the atemporal time of consciousness, contem-
poraneous. By what principle of selection is he to subject them to the time
demands of the novel?

O ye powers! [Sterne cries] . . . which enable mortal man to tell a story worth
the hearing—that kindly shew him, where he is to begin it—and where he is to
end it—what he is to put into it—and what he is to leave out— . . .
    I beg and beseech you . . . that wherever in any part of your dominions it so
falls out, that three several roads meet in one point, as they have done just here—
that at least you set up a guide-post in the centre of them, in mere charity, to
direct an uncertain devil which of the three he is to take.

Sterne's uncertainty is not really uncertainty at all. His cry of authorial distress is one of the many false scents he lays down humoristically in order to give to his work the appearance of artlessness and primitive spontaneity. At the same time it points up the paradox of all novel writing, the paradox of which Sterne is very much aware: the antagonism between the time sequences which the novel imposes, and the instantaneous wholeness of the image of complex human experience which the novel attempts to present. Sterne has his guidepost in the philosopher John Locke, and it is according to Locke's theory of the human understanding that he finds his way down all the several roads that are continuously meeting in one point in *Tristram Shandy*, or, conversely, we might say that it is with the guidance of Locke that he contrives continually to get his roads crossed. To the French Academician, M. Suard, he said in conversation that 'those who knew the philosopher (Locke) well enough to recognise his presence and his influence would find them or sense them on every page, in every line'. Locke had attempted to explain the genesis of ideas from sensation. Simple sensations produce simple ideas of those sensations; associated sensations produce associated ideas of sensations, a process which becomes immensely complicated with the accretion of other associations of this kind. Besides the capacity of the mind to form ideas from sensations, it has the capacity of reflection. By reflection upon ideas acquired from sensation, it is able to juggle these into new positions and relationships, forming what we call 'abstract ideas'. Thus the whole body of logical and inferential 'knowledge' is built up, through association, from the simple primary base of sensation. There are two aspects of this theory which are of chief importance in Sterne. The one is the Sensational aspect, the other the Associative.[2] From the notion of sensation as the prime source of knowledge and as the primitive character of experience, arises that doctrine of 'sensibility' or 'sentimentality' which Sterne made famous: the doctrine that value lies in *feeling* as such. With this we shall concern ourselves later. But at this point let us see how the associative aspect of Locke's theory appears structurally in *Tristram Shandy*.

Tristram starts out on the first page with a disquisition on the need of parents to mind what they are about when they are in the act of begetting. Why? Because their associated sensations at that moment have a determining effect upon the nature and destiny of Homunculus. (Note the manner in which, from the beginning, Sterne's imagination *concretises* an abstraction, here an abstraction of eugenical theory: the 'humours and dispositions which were then uppermost' in father and mother suffuse, somewhat like a glandular tincture, the 'animal spirits' passed on to the son; and the metaphor hinted by 'animal spirits' becomes instantly a picture of unharnessed horses—'Away they go cluttering like hey-go mad.' This concreteness of imagination is one of the secrets of Sterne's surprises and of the Alice-in-Wonderland character of his world, for it is with a shock of astonished and delighted recognition that we suddenly see the pedantic abstraction taken literally and transformed into physiology, complete with hooves, feathers, or haberdashery.) What went wrong with Tristram's begetting was that Mrs Shandy, accustomed to associate the winding of the clock with the marital act (like Pavlov's dogs and the dinner bell), missed the association appropriate at the moment, and in speaking of it to Mr Shandy distracted his attention and prepared for poor Homunculus nine long months of disordered nerves and melancholy dreams (a pre-Freudian comment on the assumed bliss of this period). Mrs Shandy's remark about the clock leads then, by association of ideas, to a portrait of Uncle Toby, typically 'wiping away a tear'; for it was to Uncle Toby that Tristram owed the information as to the circumstances of his begetting. Again, by association with conception, origin, *the egg*, we are led, in chapter IV, into a discussion of Horace's dictum on the technique of beginning a literary work: 'as Horace says, *ab Ovo*'—which Horace did not say but which nevertheless serves as an apology for Sterne's own technique; thence to an explanation of Mrs Shandy's unfortunate association between the winding of the clock and the marital act—'Which strange combination of ideas', says Sterne,

the sagacious Locke, who certainly understood the nature of these things better than most men, affirms to have produced more wry actions than all other sources of prejudice whatsoever.

The explanation leads to a determination of the date of Tristram's geniture and the manner of his birth, which involves a digression into the history of the midwife, which in turn involves a digression into the history of the parson Yorick, who was responsible for establishing the midwife in her vocation; and the history of Yorick necessitates first a description of his horse (before we can get back to the midwife), but the parson's horse recalls Rosinante, and that steed, that belonged to a famous gentleman with a hobby, sets Sterne off on the subject of hobbyhorses in general, which leads to . . . (When are we going to learn the circumstances of Tristram's birth?) Sterne's comment on 'the sagacious Locke', who understood the 'strange combination of ideas' to which men's brains are liable, indicates the method here. It is precisely in the *strangeness* of the combinations or associations that Sterne finds the contour of his subject, the logic of its grotesquerie and the logic of its gaiety. At the same time, he is in perfect control of the 'combination', as we are slyly reminded again and again; for we *do* come back to the midwife.

Nor, in terms of total structure, is Mrs Shandy's remark about the clock, in the first chapter, quite as irresponsible as it would seem. In this odd world, where the methodical Mr Shandy winds up the house clock, together with 'some other little family concernments', on the first Sunday night of every month, time is of the utmost importance and the utmost unimportance.[3] Let us illustrate. In chapter IX of book IV, Toby and Mr Shandy are conversing as they walk down the stairs, and from chapter IX to chapter XIV we are kept, presumably, in the chronological span that covers their descent from the top to the bottom of the staircase. Chapter IX starts out,

What a chapter of chances, said my father, turning himself about upon the first landing, as he and my uncle Toby were going downstairs—

and there follows a page of conversation between the two. With the beginning of chapter X, Sterne reminds us that we are still on the staircase and have not got to the bottom yet.

Is it not a shame to make two chapters of what passed in going down one pair of stairs? for we are got no farther yet than to the first landing, and there are fifteen more steps down to the bottom; and for aught I know, as my father and my uncle Toby are in a talking humour, there may be as many chapters as steps: . . .

a remark which obviously calls for the insertion here of his 'chapter upon chapters' (which he has promised us, along with his chapter on noses, his chapter on knots, and his chapter on whiskers). Hence, by chapter XI, we are not yet to the bottom of the staircase.

We shall bring all things to rights, said my father, setting his foot upon the first step from the landing.—

With chapter XII, Susannah appears below.

—And how does your mistress? cried my father, taking the same step over again from the landing, and calling to Susannah, whom he saw passing by the foot of the stairs with a huge pin-cushion in her hand—how does your mistress? As well, said Susannah, tripping by, but without looking up, as can be expected.—What a fool am I! said my father, drawing his leg back again—let things be as they will, brother Toby, 'tis ever the precise answer—And how is the child, pray?—No answer. And where is Dr Slop? added my father, raising his voice aloud, and looking over the ballusters—Susannah was out of hearing.

Of all the riddles of a married life, said my father, crossing the landing in order to set his back against the wall, whilst he propounded it to my uncle Toby—of all the puzzling riddles . . .

In chapter XIII, Sterne desperately appeals to the critic to step in and get Uncle Toby and Mr Shandy off the stairs for him. Obviously, what has been presented to us in this bit of fantasy is the incongruity between the clock-time which it will take to get the two conversationalists down the stairs, and the atemporal time—the 'timeless time'—of the imagination, where the words of Toby and Mr Shandy echo in their plenitude, where their stances and gestures are traced in precise images (as a leg is lifted or a foot withdrawn from the step), and where also the resonances of related subjects (such as chance and chapters and critics) intertwine freely with the conversation of Toby and Walter; and we are made aware of the paradox of which Sterne is so actively aware, and which he uses as a selective principle and as a structural control: the paradox of man's existence both in time and out of time—his existence in the time of the clock, and his existence in the apparent timelessness of consciousness (what has been called, by philosophers, 'duration', to distinguish it from clock-marked time).

But the time fantasy still piles up, in chapter XIII, in odder contours. We are reminded, by Walter's interchange with Susannah, that this is the day of Tristram's birth; and the autobiographer Tristram is also suddenly reminded of the passage of time.

I am this month one whole year older than I was this time twelve-month; and having got, as you perceive, almost into the middle of my fourth volume—and no farther than to my first day's life—'tis demonstrative that I have three hundred and sixty-four days more life to write just now, than when I first set out; so that instead of advancing, as a common writer, in my work with what I have been doing at it—on the contrary, I am just thrown so many volumes back . . . at this rate I should just live 364 times faster than I should write—

—a piece of mathematical calculation which the reader will follow as he is able. Again, what the fantasy suggests is the paradoxical temporal and yet atemporal status of consciousness, whose experience is at once past experience, as marked by the passage of time, and present experience, inasmuch as it is present within the mind. These considerations of time may seem abstract as set down here, but we are simplifying Sterne's performance so that we may see it structurally and schematically. Actually, what his acute time sense provides is never the dullness and ponderousness of abstraction, but the utmost concreteness of visualisation—a concreteness which we have seen in the descent of Toby and Mr Shandy down the stairs. Mr Shandy turns himself about upon the first landing; he sets his foot upon the first step from the landing; he takes the same step over again; he draws his leg back; he looks over the balusters at Susannah; he crosses the landing to set his back against the wall. These are Sterne's typical time markings, showing the procession of stance and gesture, but they are also the delicately observed details of the dramatic picture, which bring it alive and concrete before our eyes, and which make of the characters of *Tristram Shandy* creatures who, once known, remain unfadingly, joyously vivid to our imagination.

It is because of Sterne's acute awareness of time passage and of the conundrums of the time sense, that he is also so acutely aware of the concrete moment; or, conversely, we could say that it is because of his awareness of the preciousness of the concrete moment, that he is so acutely aware of time, which destroys the moment. In the eighth chapter of the final book, he engages in an apostrophe.

Time wastes too fast: every letter I trace tells me with what rapidity Life follows my pen; the days and hours of it, more precious, my dear Jenny! than the rubies about thy neck, are flying over our heads like light clouds of a windy day, never to return more—everything presses on—whilst thou art twisting that lock,—see! it grows grey; and every time I kiss thy hand to bid adieu, and every absence which follows it, are preludes to that eternal separation which we are shortly to make.—

Which calls for a new 'chapter' of a single line—'Now, for what the world thinks of that ejaculation—I would not give a groat.' The philosopher, with his head full of time, is never allowed to sail off into abstraction; he keeps his eye on Jenny twisting her lock. For it is not in abstract speculation about time, but precisely in Jenny's gesture as she twists her lock, that the time sense finds profoundest significance; or in Trim's gesture as he drops his hat on the kitchen floor to illustrate to Obadiah and Susannah the catastrophe of mortal passage into oblivion.

NOTES

¹ This point of view with regard to the unity of the work is developed by Benjamin H. Lehman in his essay on Sterne, 'Of time, personality and the author', in *Studies in Comedy* (Berkeley, Calif., 1940).
² These observations are based on Herbert Read's discussion of Sterne in *The Sense of Glory* (1930).
³ Sterne's treatment of time, as discussed here, is central to the essay by B. H. Lehman, cited above.

Conrad's word–world
of time

Conrad's fiction mimics the fictions of human experience—the futuristic fantasies created out of the canards of habitual thought and language. The scenarios of these inadvertent impostures inhere in the artificial structuring of western time. The past, present, and future exist only in words (by analogy with the three-tense system). They possess no integral connection with the legitimate awareness of time (the consciousness of latering) or with the sovereign continuum of cosmic duration. Always objectified in relation to events, they evolve into mental signposts, spatialised locations in an imaginary geography of time. Efficient in pragmatic hoodwinking, they transmute into metaphors of bewitching credence, like the 'B.C.' and 'A.D.' that historians adduce in chopping up time into logical blocks of nonsense. Of course, Conrad never philosophises about these matters in his own voice (if such there be); his prefaces, like those of James, rarely illuminate the strategies of execution in the work under discussion. Only his narrators (Marlow, for instance, in 'Youth', 'The heart of darkness', and *Lord Jim*) expound the rational lunacies of this cultural fixation on time, and, concurrently, his heroes act out its postulates under the same spell of reflex prompting.

The autotelic impressions of time reciprocate Conrad's structuring of his fiction, at least in the longer short stories and the novels. This practice takes form in the staggering, telescoping, and paralleling of random sequences of action. As they ultimately converge in the present of the main narration, they comprehend all of the influences that shape an individual destiny. In the irrational unfolding of these patterns of cause and effect lies the reality of human existence—its utter incomprehensibility. Once forced to bear the crushing burden of this truth, programmatically always linked with the collapse of a planned future, the hero subjectifies his despair in the trauma of a plunge into the abyss, that is, the loss of a historical identity. With no goal left to pursue, his sense of tripartite time disintegrates, precipitating a total dissociation of consciousness. So his conditioned modes of thought, hopelessly estranged from the rhythms of nature (the unalterable cycle of making, unmaking, and remaking that looms in the background of Conrad's tales), beget the nightmares of illusion. Though individualised in focus, the motif of disintegrating time also functions to debunk the catchword repertoire of progress. Passed off as proverbial wisdom, these specious assumptions of evolutional development colour the economic, political, and ideological atmosphere of Conrad's plots. On the one hand, they shape the aspirations of his protagonists; on the other, they promote the destructive dislocations in their careers. In the international struggle for power and wealth they serve the cause of expediency, and hence they victimise literalistic interpreters of the beliefs (all of Conrad's infatuates). In elaborating the erratic course of European civilisation on the march towards a paradisaical secularism, Conrad consistently converts this monomania into a *reductio ad absurdum* of Christianity and science. Thus the quests for Utopia, for New Jerusalem, and for evolutionary perfection simply mark different stages in the slippery verbalisation of futuristic bliss—man enslaved to inherited systems of reference and figuration.

Almayer's Folly launches Conrad's exploration of the cultural obsession to impose a tripartite order of time upon the irremediable disorder of his terrestrial environment. Almost simplistically, he employs the complicating episodes of the novel to catalogue the external forces that consolidate the rule of chaos: the dislocations of war, the disruptions of nature, and the unpredictable behaviour of human beings. As he declares in the 'Author's note', the refusal to contain personal aspirations within the bounds of unforeseeable contingencies breeds inevitable disillusionment:

For, their land—like ours—lies under the inscrutable eyes of the Most High. Their hearts—like ours—must endure the load of the gifts from Heaven: the curse of facts and the blessing of illusions, the bitterness of our wisdom and the deceptive consolation of our folly.[1]

The manipulated clash of sentimental religiosity and cynical observation anticipates Conrads' rhetorical irony in his representation of Almayer's experience of his temporal ecstasies and adversities. However much purporting to deal specifically with the predicament of backward races, the remarks capture the ambivalent attitudes of Europeanised man towards his historical fate. Depending upon its character, it reflects either divine superintendence or phenomenal disorder. At the outset of the novel the recapitulation of the hero's youthful past centres on his fantasies of success that eventually fail to materialise upon the mysterious disappearance of his benefactor, another romantic materialist:

[I]n the far future gleamed like a fairy palace the big mansion in Amsterdam, that earthly paradise of his dreams, where made king amongst men by old Lingard's money, he would pass the evening of his days in inexpressible splendour. [XI, 10]

The same contradictory mixture of profane and sacred reference colours

the imagery of the reverie. Spanning the gamut of prosaic recollection, it
intermingles the vision of apocalyptic Christianity and the wish-fulfilment
of the fairy tale with the popular magazine story. Though seemingly
restricted to Almayer and his particular situation, the grab bag of aspirations rehearses the yearning of every member of the middle class, young
and old, male and female. Whether in Borneo or England, the standardised
imagination of western culture feeds on the same fictions of egotistical
glory. For in the present of the narration, twenty-five years later, Almayer
still clings to the same aspiration, albeit stripped of the golden promises
of a New Jerusalem:

He absorbed himself in his dream of wealth and power away from this coast where
he had dwelt for so many years, forgetting the bitterness of toil and strife in the
vision of a great and splendid reward. They would live in Europe, he and his
daughter. They would be rich and respected. Nobody would think of her mixed
blood in the presence of her great beauty and of his immense wealth. [xi, 3]

In establishing this rigid perspective on a world subject to continual
change, Conrad isolates the deficiencies of blind futurism. Unlike, say,
Heidegger, he refuses to impose the stamp of authentic existence upon the
inauthentic automatisms of habit. Expectations promiscuously floated out
on the crosscurrents of history pull apart the unity of consciousness. With
no genuine sense of newness, life ceases to be anchored in its natural
environment. It subsists artificially in a cortical nursery, estranged from
reality.

   As Conrad applies this outlook to Almayer, the latter's rejection of the
present creates a yawning gulf at the centre of his being. Hence when his
adult dream of wealth and respectability dissipates under the impact of
chance events (a violent tropical storm and a series of unexpected be-
trayals), he plunges into a subjective abyss:

It seemed to him that for many years he had been falling into a deep precipice.
Day after day, month after month, year after year, he had been falling, falling,
falling; it was a smooth, round, black thing, and the black walls had been rushing
upwards with wearisome rapidity. [xi, 99]

In effect, Almayer loses his hold on time. Bereft of a future and defrauded
of the present, he owns only an empty past that he detests. Thus he faces
the task of discovering a consoling formula of temporality to replace the
one that experience disavows. An impossibility under the circumstances,
Almayer chooses to become an opium addict, unconsciously intent upon
lifting the curse of petrified associations.

   In *An Outcast of the Islands* Conrad writes a mock romance to ridicule
the fashionable option for neutralising the follies of futurism. Borrowing a
motif from the escape literature of the middle class, he builds the climax of
his story around an episode of sexual passion that succeeds in obliterating
the anxieties of tripartite time. Recognising the authority of hack authors
on this subject, he follows these oracles of arrested imagination in formulat-

ing his anatomy of vicarious thrills. Until he suddenly introduces his sardonic reversal of customary expectations, the plot recapitulates all the stale conventions of melodrama: disgrace, exile, desperation, girl (beautiful and uninhibited), ecstasy. But before this interlude occurs, Conrad sketches out the hero's involvement in time. Though almost offensively obtrusive, this effort, paradoxically, rarely commands the attention of critics, even as in *Almayer's Folly*. Obviously, familiar outlooks occasion no curiosity, and, as in so much great fiction, the art of irony goes begging. Not unlike Almayer in his greed for material well-being but more daringly enterprising, the hero Willems devises a scheme to acquire a fortune through illegal transactions in opium and gunpowder:

Willems walked on homeward weaving the splendid web of his future. The road to greatness lay plainly before his eyes, straight and shining, without any obstacle that he could see. He had stepped off the path of honesty, as he understood it, but he would soon regain it, never to leave it any more! It was a very small matter. [xiv, 11]

Stuck in the opening lines of the novel as an invitation to perception, this background information allies Willems' attitudes with the general delusion that underlies the code of action in western society (a topic picked up in another context in *Victory*). As an exemplification of the belief that man can make his personal history and, by extension, the history of his civilisation, the impelling desire evolves into an *idée fixe*. As a consequence the means of achieving the goal, a betrayal of confidence, excites no moral considerations. Conrad orchestrates the implications of this ruthlessness in the economic and political manoeuvrings of contesting European and regional factions that parallel and overlap Willems' activities. Like Kurtz in the jungles of Africa, Willems acts out his existence in accordance with a predetermined script—the aggressive self-assertion that characterises the progress of modern civilisation.

And Conrad fiendishly employs this mark of cultural branding to resolve Willems' entanglement in lust. Contradicting the idyllic unfolding of the affair, he locates a cancer in the rose of untrammelled sexuality:

And while she was near there was nothing in the whole world—for that idle man—but her look and her smile. Nothing in the past, nothing in the future; and in the present only the luminous fact of her existence. But in the sudden darkness of her going he would be left weak and helpless, as though despoiled violently of all that was himself. He who had lived all his life with no preoccupation but that of his own career, contemptuously indifferent to all feminine influence, full of scorn for men that would submit to it, if ever so little; he, so strong, so superior even in his errors, realized at last that his very individuality was snatched from within himself by the hand of a woman. [xiv, 76–7]

Here Conrad evokes the only state of love that the western world will ever know, given the abstract dimensions of time that shape the male sensibility. Regardless of her race and colour, Willems' paramour represents the woman of fantasy who appeals wholly to the body—to pure sensuality. Judged in thought, she becomes hateful and dangerous, *la belle dame sans merci* who unmans the masculine conception of manhood. And this persona degrades the enslaving pleasures of sexuality, preferring the power to dominate other men, even as instanced in Willems' patronising treatment of his wife's relatives and of his less successful bar-room companions. His infatuation with Aïssa threatens the function of his habits of temporalisation. The past of his dishonest prosperity supplies the basis of his self-love in the present, encouraging the hope that his predatory cunning will provide the opportunity in the future to recoup what he lost. Unfortunately, Aïssa understands none of these things. She lives solely on the affection that she gives and receives. As a consequence Willems begins to loathe her dream of abiding love, of a duration without the pricking sense of masculine superiority:

He was robbed of everything; robbed of his passion, of his liberty, of forgetfulness, of consolation. She, wild with delight, whispered on rapidly, of love, of light, of peace, of long years. . . . He looked drearily above her head down into the deeper gloom of the courtyard. And, all at once, it seemed to him that he was peering into a sombre hollow, into a deep black hole full of decay and of whitened bones; into an immense and inevitable grave full of corruption where sooner or later he must, unavoidably, fall. [xiv, 339]

Stripped of its exotic setting, the narrative exhibits Willems as a stereotype of bourgeois individuality. He robots the sentiments of his counterparts all over the world, affirming the perverting authority of the Protestant ethic, of the sainthood of money and respectability. As Conrad indicates in the reference to Willems' continence, the rigours of materialistic futurism produce the phenomenon of capitalistic Puritanism. Success displaces salvation. As a result of habituation, the one abstraction assimilates the ascetic imperatives of the other, leaving the body and mind split in their pursuit of eudemonic bliss. Conrad's fiction, early and late, supports this position. He never creates a protagonist with the capacity to love a woman only for her womanhood. All legal unions turn out to be marriages of convenience. The flirtations, courtships, and liaisons end up in absurd distortions of erotic passion. As in *Almayer's Folly, An Outcast of the Islands, Victory, Lord Jim, Nostromo,* and *The Secret Agent,* the woman occupies the status of a handmaid to ambition or of a sop for the frustration of an ambition, naught else.

The three internal narratives of Marlow continue to explore the quest for the socially enshrined goals of convention. Involuting his representation of linear succession, Conrad converts the retrospects into ironic refutations of the external events that they embrace. His gambit of a story within a story disintegrates the function of integration that Marlow assigns to the partitive dimensions of time under the metaphors of youth and old age. The past, the present, and the future dissolve into words, into a glossolalia that looks back to the divine monkeyshines of Babel. Consider the semantics of the present in the stories; one applies to the framing occasion, the other to Marlow's unfolding tale. A deliberate artifice, the structuring reduces the present as a term into the idiom of mumbo-jumbo. Then take the past. In 'Youth' it assumes a quantitative form, a slice out of the span of years that supposedly defines a specific phase of age. In *Lord Jim* it becomes a mosaic of dissociations, a haphazard record of Marlow's compilations of opinion and of his debauches of subjectivity in creating the myth of Lord Jim. In the process the present and the future exist only in the illusion-making of artistic rhetoric. The hero pursues his intentionalised repute over the quicksands of atemporality, evolving into the eternal youth of Marlow's fancy. When he dies, he still lives in his creator's mind as the personifying agent of an ideal ethnic future. In this confusion of reference time degenerates into a teasing figure of speech, objectifying the Humpty-Dumpty world of Marlow's consciousness. The pattern of Marlow's recollections in 'The heart of darkness' completely inverts the primary narrator's conception of the stream of history:

Hunters for gold or pursuers of fame, they all had gone out on that stream, bearing the sword, and often the torch, messengers of the might within the land, bearers of a spark from the sacred fire. What greatness had not floated on the ebb of that river into the mystery of an unknown earth! . . . The dreams of men, the seed of commonwealths, the germs of empires. [xvi, 47]

The evolutionary determinism of this poetic outburst dissolves into an insane reversal later on. Conrad reinvokes 'the torch' and 'the sacred fire' to commemorate the devolution of progress into Kurtz's barbarism. And as Marlow spins a cocoon of mentalised time (primordial, mythic, religious, and historical) around the subject of his yarn, the past, present, and future plummet pell-mell into a vortex of meaninglessness. Time simply eludes description in the grammars of the western tongue.

Hardly by coincidence, Conrad turns this defect into a disease of language in the three stories. Employing the device of a choric rallying cry or slogan in each instance, he couples these psittacisms with the printing press, that is, with the subliminal atrocities of its relentless assaults on human gullibility. Broached at the outset of 'Youth', the influence of journalism emerges as the monitor of his thought:

[M]eantime I read for the first time *Sartor Resartus* and Burnaby's *Ride to Khiva*. I didn't understand much of the first then; but I remember I preferred the soldier to the philosopher at the time; a preference which life has only confirmed. One was a man, and the other was either more—or less. [xvi, 7]

In the witty juxtaposition of Carlyle's *tour de force* on the clothes of self-deception and of Burnaby's adventures in the clothes of melodrama, Conrad establishes the fashions of received opinion that cloak Marlow's outlook on life, specifically his verbigerative invocation of the motto on the stern

of the lost ship, '*Judea*, London. Do or Die' [xvi, 12, 18, 20]. But more
to the point, Conrad employs the trite phrase to sketch out the conversion
of spiritual zeal into materialistic avarice. The apposition of Judea (hardly
an accidental etymological reverberation of Judas) and London (the largest
seaport in the world) consecrates the liaison between religion and capitalism, and the auditors of Marlow's tale, an accountant, a lawyer (a Tory
and High Churchman), and a financier, supply the audit of this unholy
contract. The pun on the crucifixion plots the overthrow of the prophets of
eternity by the profits of finitude (earthly success at any cost). But Conrad's tricks of syntax and rhetoric convey a still more shocking revelation.
Through centuries of habituation by the futuristic dogmas of theology, the
patterns of human thought take on the stamp of this abstract notion of time,
defrauding life of its vibrant immediacy. Marlow provides his record of this
fate in 'Youth'.

In *Lord Jim* he frames the career of the hero in the same web of dearth.
Though he constantly undermines the validity of belief in intentionalised
time by citing the tyrannical role of contingency in historical affairs, he
never heeds the implications of such statements. Like all of Conrad's
Europeanised characters, he mistakes the projection for the actuality of
experience. He spaces Lord Jim's successes and failures in a subjective
framework that belies the passage of time. This tendency skulks behind an
empty catch phrase that he employs to rationalise the absurdity of his
hero's pursuit of honour: 'He is [or "was"] one of us' (xxi, 43, 78, 224,
331, 361, 416). As in 'Youth', Conrad resorts to the absolving repetition
in order to disclose the mesmerising effect of the watchword. Addressed
to the social and nationalistic biases of Marlow's auditors (or, by extension, the readers of the novel), the unifying bond of undefined sentiment
at last develops into a justification for moral anarchy. It becomes an end
that nullifies the importance of means:

The seventeenth-century traders went there for pepper, because the passion for
pepper seemed to burn like a flame of love in the breast of Dutch and English
adventurers about the time of James the First. Where wouldn't they go for pepper!
For a bag of pepper they would cut each other's throats without hesitation, and
would forswear their souls, of which they were so careful otherwise: the bizarre
obstinacy of that desire made them defy death in a thousand shapes; the unknown
seas, the loathsome and strange diseases; wounds, captivity, hunger, pestilence,
and despair. It made them great! By heavens! it made them heroic; and it made
them pathetic, too, in their craving for trade with the inflexible death levying its
toll on young and old. [xxi, 226]

According to this pedigree Lord Jim incarnates the spirit of the buccaneers
in the disguise of a romantic pilgrim. Conrad ironically observes no distinction between Lord Jim and any other self-seeker in the story. But this
involves no condemnation. Every white man in one way or another
pursues the same goal in life, adhering to an ethic that society evasively
describes in circumlocutions like 'He is one of us.' A nihilism built into the

aladdinising power of the illusions of language, it presages the ultimate conquest of western civilisation by empty rhetoric. At least so Conrad condenses this notion in the title of *Victory* and in its terminal word, 'nothing' (xv, 412).

'The heart of darkness', the last of Marlow's reminiscences, brings this view full-turn in Conrad's translation of a philosophy of history into a code of action. Assigning the primary narrator the role of a dupe, Conrad systematically betrays his lack of perception in his dissociations of language. When he catalogues the auditors of the tale, 'The Director of Companies' (xvi, 45), 'The Lawyer' (xvi, 46), 'The Accountant' (xvi, 46), he inadvertently exhibits the displacement of person by function or vocation, Conrad's anaphoric capitalisation indicating the play of hugger-mugger inversion. For these epithets of the present link up with those of the past to bear witness to the almost total externalisation of Western identity as a consequence of the glorification of the makers of history:

Hunters for gold or pursuers of fame, they all had gone out on that stream, bearing the sword, and often the torch, messengers of the might within the land, bearers of a spark from the sacred fire. What greatness had not floated on the ebb of that river into the mystery of an unknown earth! . . . The dreams of men, the seed of commonwealths, the germs of empires. [xvi, 47]

Not men but high sounding denominations, all with an appeal to pride and pursuit, elicit this eulogy. Of course, the spellbound listeners fall into the genealogy of personified ambition. They garner their self-images from the cultural storehouse of titular vanity. From the vantage point of the unforeseeable future made by the adventurers, they listen to Marlow's tale of the future sought by Kurtz without any awareness of the verbal fragility of time. Plagiarising the past of history, biography, and autobiography, the untrustworthy vehicles of ventriloquised values for an understanding of the logic of human existence, they screen the meaning of Marlow's tale through the very system of references that he invalidates in attempting to make sense of his past in the present. Whether he succeeds or not depends upon how a reader decides to interpret his vision of human subjectivity:

. . . No, it is impossible; it is impossible to convey the life-sensation of any given epoch of one's existence—that which makes its truth, its meaning—its subtle and penetrating essence. It is impossible. We live, as we dream—alone. . . . [xvi, 82]

If Conrad supplies any rubric for the hermetics of 'truth' and 'essence', it lies in the necromantic power of words—the galvanising energy of the activities of Kurtz and all the other bewitched pursuers of fame and fortune. Regardless of the predicament of any given European character in Conrad's fiction, he hallucinates reality in the world of rhetoric that chance and conditioning accidentally fashion.

The pun in the title of *The Secret Agent* affirms the source of this existential dream. The ubiquitous subverter of tomorrow's plans of man remains the cut-throat spectre of abstract time. Grounding his plot on the factual status conferred on a reified idea (a logical fallacy), Conrad ties the fate of the nominal revolutionary Verloc (an intriguing surname) to an unsuccessful attempt to dynamite the Greenwich Meridian Observatory. A Conradian jest with its own radical implications, the momument to a mathematical abstraction of time and space monitors the clock that super-intends all purposeful activity in the civilised world. The station exists only as a cerebral actuality, by a common agreement among nations to accept the reality of the imaginary line of zero meridian. Or as Conrad in-terprets the accord by cross-reference to Stevie's compulsive geometric art, the nothingness of time, created by the insidious conspiracy between thought and language, envelops human existence:

[T]he innocent Stevie, seated very good and quiet at a deal table, drawing circles, circles, circles; innumerable circles, concentric, eccentric; a coruscating whirl of circles that by their tangled multitude of repeated curves, uniformity of form, and confusion of intersecting lines suggested a rendering of cosmic chaos, the sym-bolism of a mad art attempting the inconceivable. [XIII, 45]

As the madness of reason and unreason so converge and unite, the reality of the unreal no longer hides behind the looking glass of factuality. As usual when he deals with such pragmatic truths (lies), Conrad resorts to the double and/or opposite to shadow forth his *reductio ad absurdum*. Accordingly, the lethargic Verloc acts out the conventions of mechanised temporality, the neurotic Stevie the conventions of mechanised atempor-ality. The unforeseen collision of these objective and subjective states of consciousness topples the structure of futurism in the novel. An ideologue with a mind shaped by the doctrines of anarchism, 'His voice, famous for years at open-air meetings and at workmen's assemblies in large halls, had contributed . . . to his reputation of a good and trustworthy comrade' (XIII, 23), Verloc falls victim to the autosuggestion of his formulaic principles. He takes for granted that the preconceived deed harbingers the future. Predictably, Stevie who carries the time bomb to demolish the observatory ignores caution in the handling of the assignment, ending up blown to pieces—destroyed by his brother-in-law's infatuation with the clock of nothingness.

Though no less a farce of ideological puppetry than *The Secret Agent*, *Nostromo* veils its forboding social implications in the swagger of heroism. Ostensibly the founding of the Occidental Republic (a microcosm of demo-plutocracy, probably of America) redeems the irrationality of events and the cross-purposes of human conduct from the stigma of a collective in-sanity. Using his bewildering chronological discontinuities as a metaphor for the actual unfolding of the historical process, Conrad contrasts the chaos with the roted order of its interpreters. As the primary mouthpiece of this inherited outlook, he chooses Captain Mitchell, the self-important apologist for the economic and political opportunism of the foreign silver interests.

In an abrupt retrospect hatched out of the immediate circumstances of the revolution, Conrad records the smugness of the reflex judgements that dictate the intellectual patterning of western annals. Devoting almost all of chapter XIII to an overt elucidation of this cant of thought, he converts Mitchell's boasting into a burlesque of patriotic pride by alternating his eulogies of the present opulence with his descriptions of the carnage of the past that produced the miracle of progress. Wholly oblivious to the massacre of the contesting factions of the revolution, all in one way or another tools of the mining company, Mitchell exalts wealth above morality and things above humans:

'. . . I saw the first and last charge of Pedrito's horsemen upon Barrio's troops, who had just taken the Harbour Gate. They could not stand the new rifles brought out by that poor Decoud. It was a murderous fire. In a moment the street became blocked with a mass of dead men and horses. They never came on again.'

And all day Captain Mitchell would talk like this to his more or less willing victim—

'The Plaza. I call it magnificent. Twice the area of Trafalgar Square.'

From the very centre, in the blazing sunshine, he pointed out the buildings—

'The Intendencia, now President's Palace—Cabildo, where the Lower Chamber of Parliament sits. You notice the new houses on that side of the Plaza? Compañia Anzani, a great general store, like those coöperative things at home. [IX, 476]

Conrad's rhetorical itemisation of these perverted attitudes transforms Mitchell into an impassive bookkeeper of the horrors of history-making. The counterpart of the bureaucrats at the African stations in 'The heart of darkness', he anticipates the function of the attendants in the European concentration camps of the twentieth century. Conrad's handling of this unconscious black comedy also stresses the role of the press in establishing the vulgate of glozing, as evidenced in Mitchell's invocation of the authority of the printed word: ' "[T]he Treasure House of the World," as *The Times* man calls Sulaco in his book, was saved intact for civilization—for a great future, sir' (IX, 483). Though he sets up Mitchell as the chief scapegoat of the shibboleths of imperialism, no character, major or minor, escapes this fate. All of them, including the swashbuckling Nostromo, succumb in one way or another to the siren babble of silver dreams.

'The future' in *Nostromo* (and all of Conrad's fiction) ultimately takes on a meaning wider than that of a mere catch-phrase. It becomes a fetish expression, the engram of cultural identity. Never coinciding precisely in substance from person to person, the memory trace evolves into a nonsense vehicle of communication, the catalyst of subjective feelings that at once bind and separate the perceptions of different individuals. As Conrad modulates Mitchell's peroration, he exposes the superficiality and stupidity of the latter's conception of a magnificent future. In contrast with the vision of Holroyd, the American capitalist who subsidises the expansion of the silver enterprise of the Goulds, it diminishes into a total failure of the imagination:

'But there's no hurry. Time itself has got to wait on the greatest country in the whole of God's Universe. We shall be giving the word for everything: industry, trade, law, journalism, art, politics, and religion, from Cape Horn clear over to Smith's Sound, and beyond, too, if anything worth taking hold of turns up at the North Pole. And then we shall have the leisure to take in hand the outlying islands and continents of the earth. We shall run the world's business whether the world likes it or not. The world can't help it—and neither can we, I guess.' [ix, 77]

A spokesman for the delusion of manifest destiny, Holroyd conceives of linear progress in terms of a universal social revolution, a gradual extension of control over the minds and bodies of all peoples through the power of the word. Sheer megalomania, this ambition, as in so many modern dictatorships, achieves realisation, regardless of the degree, under the impetus of followers like Mitchell whose petty illusions preclude understanding of what they see, hear, and read. In the collision of these discrepant associations Conrad shows that the miracle of language unfortunately clouds the reality of time and existence. But as an artist, even if only for his own amusement, he undertakes to illustrate that the logodaedaly of fiction also takes the form of simultaneous revelation and concealment. He introduces this lucid configuration of his rhetoric in a casual remark by the wife of Viola (the proponent of 'Liberty and Garibaldi', ix, 16):

'That is all he cares for. To be first somewhere—somehow—to be first with these English. They will be showing him to everybody. "This is our Nostromo!" ' She laughed ominously. 'What a name! What is that? Nostromo? He would take a name that is properly no word from them.' [ix, 23]

The deliverer of Sulaco receives this nonsense name in baptismal blunder by Mitchell. Hearing him proclaimed *nostre uomo* (our man) by worshipping natives, the superintendent attaches his mispronunciation of the epithet to the former sailor. Though aware of the mistake, Nostromo never protests. Instead he proceeds to imbue the anonym with the same potencies as his admirers, European and otherwise, until he completely desecrates the inspiration behind his given and family name—Gian' Battista (John the Baptist). Blasphemy or jest, this rhetorical huggermugger throws suspicion on the talismanic power that historians find in mythic figures—in fictions of messianic enthusiasm. Viewed in the context of Holroyd's (perhaps a crazy pun on 'holy rood') vision of the salvational word, Conrad's sly dissociations and associations comment gratingly on the gift of tongues.

More than any of his other novels, *Nostromo* relies on the discordances of language to convey its meaning. A cynosure on the first page of the novel, 'the Republic of Costaguana' (ix, 3), provides the compatible setting for the corruption that marks the action. The coast of manure perfectly describes the treachery and deceit that underlie the flowering of the fortune of Charles Gould, with the surname echoing 'gold' and the robber baron Jay Gould. The delusions of grandeur of the silver entrepreneur, all sanc-

tioned by Holroyd, take on a mock sanctity from the name of the vessels
in the fleet of the company, 'Minerva' (IX, 12), 'the *Ceres*, or *Juno*, or
Pallas' (IX, 486). But here Conrad the trickster has the final word, for the
last addition to the fleet claims the name of 'Hermes' (IX, 505). And for
the founding of the new republic he reserves still another slash of irony:
'[T]he United States cruiser, *Powhattan*, was the first to salute the
Occidental flag—white, with a wreath of green laurel in the middle encir-
cling a yellow amarilla flower' (IX, 487). Named after an Indian chief who
terrorised both his own people and the English of the first permanent
colonial settlement for many years, the ship and the ship of state meet and
merge under a banner of symbols that clothe the lies of history. The concluding scene in *Nostromo* picks up the absurdity of the ultimate word of
betrayal:

> 'It is I who loved you,' she whispered, with a face as set and white as marble in
> the moonlight. 'I! Only I! She will forget thee, killed miserably for her pretty face.
> I cannot understand. I cannot understand. But I shall never forget thee. Never!'
> She stood silent and still, collecting her strength to throw all her fidelity, her
> pain, bewilderment, and despair into one great cry.
> 'Never! Gian' Battista!'

Invoked with equal passion and anguish, with equal ignorance and artlessness, by fool and sage, 'love' remains the most bemusing monosyllable in
the vocabulary of western culture. For Conrad it sums up the contradictory
meanings that reside in all the abstractions that trammel the process of
thought. Something, everything, and nothing, it destroys the meaning of
subject and object. It usurps all identities. And so Linda's 'never' indicates.
In its juxtaposition with 'John the Baptist', Conrad proclaims his judgement on time and eternity. Both embody a verbal illusion.

Put in a total perspective, the horology of Conrad's fiction recreates the
cultural habits of conceptualised horoscopy. Always reified into intentionalised action on a projected map of the future, this conditioned mode of
thought traces out in the inscrutable heavens the shadow lines of wishful
dreams. As Conrad's persistent focus on the catch-phrases in the public
domain argues, western man lives in an abstract logocracy far more
imprisoning than the autarchy of phenomenal reality. Under the tyranny of
rehearsed responses, he cultivates the delusion that his finitude violates
macrocosmic duration. Out of this egotistic misconception he fashions his
consoling theories of tragic experience. Actually he fails to recognise that
his gift of consciousness is only an accidental product of nature's anarchic
impulses. No wonder, then, that the absurd rules the word-world of
Conrad's fiction.

NOTE

[1] *The Complete Works*, Kent Edition (Garden City, N.Y., 1926), XI, p. x. Hereafter all paren-
thetical volume and page references will be to this twenty-six volume collection.

R. W. Stallman†   HENRY JAMES (1843–1916)

'The Sacred Rage':
the time-theme
in *The Ambassadors*

> Look, he's winding up the watch of his wit.
> By and by it will strike.
> *The Tempest*: ii, i, 12–13

I

The Ambassadors begins on a question: The beginning phrase of the novel is a question, 'Strether's first question', and that first question starts off the sequence of questions and answers or theories that comprise the nimble-minded substance of the narrative. It initiates the recurrent motif of Strether's uncertainties, misgivings, doubts: 'I'm always considering something else; something else, I mean, than the thing of the moment. The obsession of the other thing is the terror' (chapter i).[1] He prejudges events by theorising about them. His characteristic utterance is the question mark. Little Bilham's answer is: 'Don't put any question; wait, rather—it will be much more fun—to judge for yourself' (x). Maria Gostrey gives him the same advice: 'Don't make up your mind' (ix). She teaches him the message of Europe: How to Live. 'You must take things as they come' (xii). And then some months later, in Gloriani's garden, Strether—like a schoolboy echoing his teachers—instructs Bilham that here in Europe 'people can be in general pretty well trusted, of course—with the clock of their freedom ticking as loud as it seems to do here—to keep an eye on the fleeting hour' (xi). It is ironical that Strether, who has missed out on life by not letting himself go, should deliver to Bilham, who has lived his youth up to the hilt, the Jamesian burden: 'Live all you can; it's a mistake not to' (xi).

Live! Yes, but how? There are two ways 'to keep an eye on the fleeting hour'. Strether's way is by keeping his eye on the clock and simultaneously considering something else than 'the thing of the moment'. He is addicted to watching the clock lest he miss the train, and that is why he has always missed it. Chad Newsome's way of keeping his eye on the fleeting hour is,

† Condensed by the author for the present volume from *The Houses that James Built and Other Literary Studies* (East Lansing, Mich., 1961), pp. 34–51.

on the contrary, by fully enjoying it. Chad represents unclocked Time Now. Chad and Miss Gostrey and Chad's mistress, Mme de Vionnet, represent the Sense of the Present; and as such they bring Strether around to a recognition of what it means to live in Time Now, how to take things as they come. In the final scene of *The Ambassadors*, Strether in the garden retreat of Miss Gostrey's house leans back in his chair 'with his eyes on a small ripe round melon'. He is at last taking things as they come, his curiosity mixed with indifference. Although the melon suggests Strether himself, 'ripe' and 'charming', he nevertheless senses that however ripe and charming he may be, he's not in real harmony with what surrounds him. 'You *are*', he tells Miss Gostrey. 'I take it too hard. You *don't*. It makes—that's what it comes to in the end—a fool of me.' Even when he lets himself go in moods for perceiving things rather than reflecting upon them, as now in his observation of the melon, even then while momentarily he succeeds at enjoyment of things he fails; for what intrudes always is the opposite pole of his consciousness.

Strether is obsessed by 'something else', namely by Mrs Newsome's theory—he is 'booked by her vision' (xxix). To be booked by Mrs Newsome is to be held back by the Sense of the Past. That is why Mrs Newsome is an invalid. The Ambassadors signify those who would draw us away from giving over to the fleeting hour. It is ironical then, that she embodies the Law of the Calculable. ' "It's the sacred rage," Strether had had further time to say; and this sacred rage was to become, between them [between Miss Gostrey and Strether], for convenient comprehension, the description of one of his periodical necessities' (ii).

The Sacred Rage means, for one thing, the rage for life moulded into the ordered forms of conventional morality and conventional time (life ticking according to the clock). It is the rage for what's fixed and scheduled and calculable. It is another name for the conscience of Woollett, Milrose, and Boston, where morality and time—boxed in by puritanical conceptions and fixed ideas—are regulated according to fixed formulas for controlling them. Woollett, that 'prison house' of tradition, manifests a monopoly on how everything must be run: on time.

Timing the return of Chad Newsome, that is Strether's mission. He must bring Chad back from Paris to Woollett *on time*: 'He can come into the business now—he can't come later' (iv). It's a manufacturing business; but what the Newsomes manufacture in Woollett, Strether never discloses —'It's vulgar.' 'Unmentionable?' 'Oh no, we constantly talk of it; we are quite familiar and brazen about it' (iv). At the theatre with Miss Gostrey (iv) what blocks Strether's enjoyment of the play is the recollected image of the Newsome factory; it overlays the stage precisely when Miss Gostrey reverts to that key topic, the Woollett question: 'And what is the article produced?' Strether procrastinates: 'I'll tell you next time.' As at the beginning of the novel, so at the very end Miss Gostrey reminds him 'of his having never yet named to her the article produced at Woollett.

"Do you remember our talking of it in London—that night at the play?" '
What follows immediately this Woollett riddle is the image of a clock:

He found on the spot the image of his recent history; he was like one of the figures
of the old clock at Berne. They came out, on one side, at their hour, jigged their
little course in the public eye, and went in on the other side. He too had jigged his
little course—him too a modest retreat awaited. [xxxvi][2]

As his enjoyment of the play was blocked by the image of Newsome's
roaring factory, the 'hum of vain things' drowning Miss Gostrey's question
about the Woollett product, so in the final scene his enjoyment of the melon
is blocked by the image of the old Berne clock: 'the charming melon, which
she liberally cut for him, and it was only after this that he met her question'
—the Woollett question. Strether in the image of the old Berne clock
identifies himself with time, and by implication the Woollett product is
also identified with time. For immediately following the identification of
Strether with the clock comes Strether's offer 'to name the great product
of Woollett. *It would be a great commentary on everything.*'

Time is the great commentary on everything; what we constantly talk
about is time; *The Ambassadors* talks about it on almost every other page.
Time is the mainspring theme of *The Ambassadors*—How to Live It.

The affair—I mean the affair of life—couldn't, no doubt, have been different for
me; for it's, at the best, a tin mould, either fluted or embossed, with ornamental
excrescences, or else smooth, and dreadfully plain, into which, a helpless jelly,
one's consciousness is poured—so that one 'takes' the form, as the great cook says,
and is more or less compactly held by it; one lives, in fine, as one can. [xi]

Life fixed in 'a tin mould'—*that* sums up Strether's life, at least up to his
encountering Miss Gostrey. At the same time that he meets her he encounters also an anonymous lady in a glass cage; she dispatches Waymarsh's telegram. Life fixed is what the lady in the glass cage symbolises.
She personifies Mrs Newsome. Woollett and its products—Mrs Newsome,
the Pococks, and Strether—they are all encased each in a hard shell. The
shell or tin mould is their obsession to fit things into a theory.

Woollett's fixed ideas preclude the possibility of the incalculable; consequently, the unpredictable upsets Woollett's previsions and theories and
untimely formulas; consequently, Strether—the Ambassador of the Woollett idea—finds himself again and again duped by the unpredictable, the
unexpected event. The unpredictable is 'the fate that waits for one, the
dark doom that rides. What I mean is that with such elements one can't
count' (ix). Paris contradicts Woollett because life—in Europe—isn't
fixed; it's fluid and unpredictable.

Woollett 'isn't sure it ought to enjoy', and neither is Strether; Miss
Gostrey shows him how: 'You see what I am', she tells him. 'I'm a
general guide—to "Europe," don't you know?' She is his Baedeker, the
guidebook to new vistas; she releases Strether from his fixed self. She is
the agent of 'illuminations', and that is why Strether admits 'I'm afraid

of you.' He's afraid of life, which is what Miss Gostrey represents. (The parallel is Mme de Vionnet with her capacity for 'revelations'.) In opposition is Mrs Newsome: 'I was booked by her vision', but her vision or programme is upset by what Strether experiences of Paris; it doesn't at all 'suit her book' (xxix).

Again and again Strether finds himself 'thrown back on a felt need to remodel somehow his plan' (ix). It is his sensibility, compounded of curiosity and indifference, that conditions him to adjust himself to the unpredictable, the surprise awaiting every plan. He was supposed to find Chad gone to the bad and his wicked woman 'horrible', but instead—contra Mrs Newsome's theory—he finds that Chad has improved; he is distinguished and the woman is delightful. Mrs Newsome had imagined stupidly; as Miss Gostrey puts it: 'she imagined meanly' (xxix).

Strether, unlike Mrs Newsome, possesses imagination enough to embrace contraries—to see things from opposite sides or contrary points of view. Making comparisons defines Strether's habit of mind. What distinguishes Strether is his double vision. While booked by Mrs Newsome's 'vision', Strether comes round to see that life is better lived unbooked— minus visions or theories about how to live it. Feeling duty-bound to Mrs Newsome's programme to retrieve her son from wicked Paris, his sense of commitment traps him. It provokes him into jumping to conclusions and, as it were, jumping the clock. In his first encounter with Chad it strikes Strether as imperative not to dawdle, 'not to lose another hour, nor a fraction of one'; his idea is to attack at once, 'to advance, to overwhelm, with a rush' (vii). His time sense is at odds with the Keatsian Moment Now; for either it lags behind the event, the flux of things, or on the contrary it anticipates the issue of the moment. His usual habit is to temporise. Whereas Chad had 'no delays', Strether procrastinates; lighting a cigarette 'gave him more time. But it was already sharp for him that there was no use in time' (xii). Although he is always on time, he is always too late. As Mamie Pocock is too late for Chad, so in another sense Strether is too late for Maria Gostrey.

Strether's way of keeping an eye on the hour is by keeping an eye on the clock, and that is why the fleeting hour escapes him. Reluctant to live except by the clock, he is always trying to regain time because he has lost so much of it in an 'empty present'. So is Lawyer Waymarsh. Hating Europe because everything there swims in uncertainties and delays or leisurely postponements, Waymarsh makes a 'sudden grim dash' into a jeweller's shop to make 'some extraordinary purchase'. Strether's theory is that Waymarsh is purchasing an article 'For nobody. For nothing. For freedom' (iii). Whereas Strether welcomes freedom from Woollett and questions Woollett's standards and fixed ideas, Waymarsh welcomes freedom from Europe and questions Europe's standards (or lack of them). Waymarsh at the start seeks freedom not from the clock but from the unclocked world of Paris. The tardy Waymarsh in that jeweller's shop presumably purchased a watch (I guess so) because he can't stand anything at loose ends and, being a lawyer, he's addicted to the law of the calculable, the Sacred Rage. However, Lawyer Waymarsh ends in the company of Miss Barrace and is freed thus from his Sacred Rage. To be freed from the Sacred Rage is 'to be whirled away'. Strether himself 'had never been whirled away . . .' (ix). Well, Waymarsh is finally whirled away, striking out for a different kind of 'freedom'—freedom from the Claim of the Ideal. Waymarsh thus cross-identifies with Chad, exemplar of the incalculable. Waymarsh—like Chad and Bilham and Strether—has his fling. Henry James fashions double versions of the Sacred Rage (commitment to the clock and freedom from it), and through Strether he renders his own double viewpoint and ironic insight.

In Strether's sensibility James projected his own: 'I have only to let myself *go!* So I have said to myself all my life—so I said to myself in the far-off days of my fermenting and passionate youth. Yet I have never fully done it' (*Notebooks*). Neither had William Dean Howells, and it was Howells' admonishment Not To Miss Life that provided his friend the germinal situation of *The Ambassadors*. Strether's mistake is James's mistake; it's Howells' mistake; it's everyone's mistake. It's the mistake of not letting yourself go; don't let life clock you. It's Woollett's mistake; it's Woollett's 'sacred rage' (as Strether calls it). It's Strether's terrifying obsession, and it's the obsession of us all; namely to look at the hour without enjoying its values, without experiencing time in its point-present nowness.

I rejoiced, says James in his Preface to *The Ambassadors*, in the prospect of creating a hero so mature, a man of imagination. Strether is brother to Isabel Archer of *The Portrait of a Lady* in that he shares her 'ridiculously active imagination'. But Strether has a questioning mind. No sooner is one question settled than another query and theory about it come to the front—'thanks to his constant habit of shaking the bottle in which life handed him the wine of experience, he presently found the taste of the lees rising, as usual, into his draught' (ix). Everything he wanted was comprised moreover in a single boon—'the common, unattainable art of taking things as they came' (v). Though time dominates Strether's imagination, the clocks of Woollett ticking off his fleeting hours, there are intervals when, as it were, the clock stops and the pendulum of his mind halts-for moments of enjoyment amidst its routine swing back to Woollett and Mrs Newsome. These intervals, moments of passive perception, constitute for Strether his happiest occasions.

One of these occasions occurs in the final scene of the book. *The Ambassadors*, ending where it began, circles back to England; once more Strether is with Miss Gostrey. There's been the breach with Mrs Newsome in the interim, and there's been a change in Strether: 'I *have* no ideas. I'm afraid of them. I've done with them' (xxxvi). His tin-moulded ideas have made a fool of him: namely Mrs Newsome's ideas, her programme. Mrs Newsome takes 'great views' of everything. but her grand views turn out to be too much for her. 'Everything's too much for her' (xxiii). Strether confirms Miss Gostrey's theory that Mrs Newsome 'was essentially all moral pressure' (xxvi). She's 'just a moral swell'. Europe contradicts fixed ideas and especially Mrs Newsome's because here in Paris the clock of freedom ticks rather than the mechanical and moral clock-time of Woollett. Strether's double consciousness swings back and forth from Paris to Woollett. When Chad declines to return home, Strether claims: 'I've stopped him' (xviii). Strether has stopped Chad's pendulum swinging homeward, and he 'took comfort, by the same stroke, in the swing of Chad's pendulum back from that other swing, the sharp jerk towards Woollett so stayed by his own hand. He had the entertainment of thinking that if he had for that moment *stopped the clock* . . .' (xix). That clocks are what Woollett produces seems evidenced in the double identification of Strether and Chad with clocks. Also, Mrs Pocock's salon in Paris is a gilded room all mirrors and 'clocks' (xxi).

That clocks are what Woollett produces seems furthermore evidenced by the fact that Woollett 'was essentially a society of women' (xix) and women are constantly imaged by the attribute of water—time itself. The time-theme in *The Ambassadors* is frequently evoked by water imagery— water frozen or fluid. That Mrs Newsome constitutes for Strether his 'whole moral and intellectual being or block', as Miss Gostrey sees it, brings out in Strether's imagination the apt image of Mrs Newsome as 'some particularly large iceberg in a cool blue northern sea' (xxix). Thus Mrs Newsome is identified with time. Sarah Pocock, Mrs Newsome's daughter, is a chip off the old iceberg: 'packed so tight she can't move. She's in splendid isolation', Miss Barrace remarks (xxvi). She is 'buried alive!' Miss Barrace concedes, 'She *is* free from her chin up'; but the rest of her is sunken ice; only 'She can breathe.' As that iceberg Mrs Newsome 'knows everything', so Mamie Pocock boasts: 'Oh, yes, I know everything.' As for Sarah Pocock, 'she's all *cold* thought' (xxix). It is Woollett's moral pressure that blocks Strether's sensibility. He's blocked by moral icebergs.

Mrs Newsome dominates the globe in the style of Queen Elizabeth (xvi), and 'Her tact had to reckon with the Atlantic Ocean, the General Post Office and the extravagant curve of the globe' (ix). Thus Mrs Newsome is identified with space. She is identified with time by the fact that her cabled ultimatum is weighted down by Strether's watch. When he receives her 'blue paper', Strether in order to keep it from blowing out the open window keeps it 'from blowing away by the superincumbent weight of his watch' (xvii). The fact that Strether weights down Mrs Newsome's cablegram *with his watch* rather than with any other object establishes the identity of Mrs Newsome with clock-time, establishes the identity of Woollett with clocked morality, establishes the identity of the very thing manufactured at the Newsome factory: watches or clocks.

Henry James manipulates his criticism of New England moral rigidity by the metaphor of submission to clock-time. At the literal level Strether disengages himself momentarily from his watch; at the symbolic level he disengages himself from the temporal and moral burdens impinging upon his double consciousness, from his habitual addiction to clocking the fleeting hour instead of enjoying it. Mrs Newsome's blue cablegram, commanding Strether from across the Atlantic, imposes Woollett's clocked morality— Live by the law of the clock—upon Strether's clocked consciousness. However, Strether succumbs to the Claim of the Ideal, which (to define it) is the impulse to renounce conformity and escape the conventions by which our lives are moulded. It's 'the impulse to let things be, to give them time to justify themselves or at least to pass' (XVI). The change in Strether is so pronounced that he becomes cross-identified with Bilham and with Chad. The Claim of the Ideal ('to be liable to strange outbreaks, belated, un-canny clutches at the unusual, the ideal', XXIII) opposes that side of the Sacred Rage which manifests Woollett's rage for commitments to the fixed order of things. *Now* in Strether's double consciousness the Sacred Rage embraces also the rage for freedom from fixed temporal and moral impingements. *The Ambassadors* builds on contraries, contradictions, con-fused or crossed identities, and ambivalences arising from poles of opposi-tion. Consequently, identities and scenes overlap, one image or scene impinging upon another. Thus we are in Paris even while we are simul-taneously in Woollett. It is Strether's capacity for seeing Woollett from the point of view of Europe that accounts for Strether's success (by the standards or code of Woollett) at being a failure.

Strether, the Ambassador of the Woollett theory, has found himself here in Europe duped again and again by the unexpected, the unpredictable upsetting the theory or fixed idea. ' "Call it then life"—he puzzled it out— "call it poor dear old life simply that springs the surprise." ' The surprise when it springs is for Strether 'paralysing, or at any rate engrossing . . .' (IX). The shock of recognition—at its most intense—occurs during Strether's excursion to the countryside (XXXI). His intent is simply to enjoy new vistas of countryside, which he had hitherto looked at 'only through the little oblong window of the picture-frame'. But Strether cannot refrain from fashioning his experience of French ruralism into a pretty picture, placing it into a picture-frame. And so he looks at the Parisian countryside as though it were a painting by Lambinet. Memories of Boston impinge thus upon his point-present sight of French countryside, and they frame it within a gilt frame. 'The oblong gilt frame disposed its enclosing lines; the poplars and willows, the reeds and river—a river of which he didn't know, and didn't want to know, the name—fell into a composition, full of felicity, within them. . . .' He began his journey with 'His theory of his excursion', and now seemingly 'not a single one of his observations but somehow fell into a place with it.' Everything seemingly fits into the frame of his theory. But Strether, obversely, gets framed. What frames him is his fixed idea.

When Strether first sights Chad and Mme de Vionnet coming round the bend of the river, he presumes that they are there simply for a day in the country, the same as he himself. But then his pretty picture gets upset on his recognition that the young man in the boat is 'in shirt sleeves' and that the young woman is 'easy and fair'. However, he is tardy in recognising the situation in that the lovers are first to spot Strether, and with *their* shock of recognition that the truth of their love affair risks being unmasked their boat drifts wide of its course and wavers. Even their boat, as it were, seems embarrassed.

Their excursion, in contrast to Strether's, involved no plan, no preconceptions, no fixed ideas. And that is what life is: it is what it unexpectedly *is*. Mme de Vionnet—like life itself—is unpredictable. The unnamed river —Strether did not want to know its name—is the river of life where the unexpected occurs. The lovers have come here by water, whereas Strether has come here by land, and so the lovers identify with the flux of time, water flowing (not frozen like that iceberg Mrs Newsome). This scene— the recognition scene—punctures Strether's notion that Chad's affair is an innocent relationship untainted by sensual or moral ambiguities, and here once more poor Strether ends disillusioned and simultaneously shocked. Madame de Vionnet, like Paris itself, 'had taken all his categories [his fixed ideas] by surprise' (xv).

II

As Strether cannot bring himself—except momentarily—to face into life or come to terms with the moment now, so neither can he bring himself (until the very end) to name the mysterious Woollett product.

'I'll tell you next time', Strether promised when Miss Gostrey first queried him about that Woollett product; but the next time doesn't occur until the very end of the novel. Strether at last offers to name the article. 'At this she stopped him off; she not only had no wish to know, but she wouldn't know for the world' (xxxvi). *The Ambassadors* begins and ends with this turnabout.

That James planned the Woollett article to be a 'distinctly vulgar article of domestic use (to be duly specified)' we know from the scenario preserved in *The Notebooks*. But James changed his mind and in the novel he intrudes to declare his intentions not to disclose its identity: 'It may even now frankly be mentioned that he [Strether], in the sequel, never was to tell her' (iv). Miss Gostrey at the start guesses it is 'Clothes-pins? Saleratus? Shoe-polish?' No, says Strether. 'No, you don't even "burn". I don't think, you know, you'll guess it' (iv). In *Aspects of the Novel* (1927), E. M. Forster put forth the whimsical notion that the article was button-hooks. If the article were something unimportant, such as button-hooks, there wouldn't be any purpose in having Strether so reluctant to name it. Why, then, does James make such a mystery about it? Forster ribs James for not telling us what the article is. 'For James to indicate how his characters made their pile—it would not do.' Like Miss Gostrey, E. M. Forster doesn't even 'burn'. It is not because the unnamed article is vulgar that James decided not to name it. In *The American* James is not squeamish about disclosing even such a vulgar object as bathtubs—on bathtubs and copper Christopher Newman made his fortune.

James's decision *not* to name the Woollett article was, of course, solely for artistic purposes. That he utilises the Woollett question as a riddle— *that* in itself hints at its importance. It seems logical to infer that what Woollett manufactures has to do with time even by the fact that the time-span of the narrative—exclusive of the Prologue (the first three chapters) —is spaced by the Woollett riddle. The Woollett question remains unanswered for the reason that to name it would constitute a spoiling of the riddle motif.

The Prologue (chapters I–III) concludes with lawyer Waymarsh rushing into a jeweller's shop to purchase an unnamed article. What is it? 'Then how will that jeweller help him?' Miss Gostrey speculates. 'Strether seemed to make it out, from their standpoint, *between the interstices of arrayed watches* [italics mine], of close-hung dangling gewgaws.' Waymarsh is late in keeping his appointment with Strether because—I presume —he lacks a timepiece. How will the jeweller help him out? Well, by providing him with a timepiece. Waymarsh is late at the appointed hour, whereas Strether is on time; and so *The Ambassadors*, which ends with a disengagement, begins with a misengagement.

At the start Strether meets not Waymarsh but Miss Gostrey, and she senses that something is amiss because Strether keeps looking at his watch. 'You're doing something that you think not right', says Miss Gostrey. He has misgivings about Europe's way of taking in the fleeting hour; it is to experience life unclocked. What's not right for Strether is to be strolling with his new acquaintance, and what's not right for Miss Gostrey is that Strether keeps looking at his watch. She educates Strether's time-troubled consciousness anew. Consequently, when once more Strether takes out his watch to check on the time, he does so mechanically, unconsciously, out of habit: 'He looked at the hour without seeing it'—as though indifferent now, for the moment, to clock-time, while simultaneously possessed nevertheless by what the watch-face says.

Although James never identifies the unnamed article, an alarm clock or watches—as I have elsewhere argued it[3]—is logically the answer to the never-answered riddle about the mysterious object Woollett produces. An alarm-clock fits all of Strether's specifications and not only is it literally the one possible article of appropriate fitness, but furthermore it elicits symbolically the time-theme of the whole book. In James's novels nothing is without interrelationship, all things in relationship being the keystone of the Jamesian canon. Ambiguity is the Jamesian aesthetic, and to resolve ambiguities is the critic's function.

III

When Strether calls on Chad to deliver Mrs Newsome's ultimatum that he return to Woollett, Chad is not at home, but Strether finds comfort in an image of a clock: 'He took comfort, by the same stroke, in the swing of Chad's pendulum back from that other swing, the sharp jerk towards Woollett, so stayed by his own hand. He had the entertainment of thinking that if he had for that moment stopped the clock, it was to promote, the next minute, this still livelier motion' (xix). Strether's own 'pendulum' swings likewise away from Woollett, halts as if time has had its stop, or swings towards Paris with livelier motion; what promotes its livelier motion is finally Mme de Vionnet, Chad's mistress. But then back it swings to Woollett again and again. In the interim come the pauses: 'Poor Strether had at this very moment to recognise the truth that, whenever one paused in Paris, the imagination, before one could stop it, reacted. This perpetual reaction put a price, if one would, on pauses. . . .' (v). In the pauses Strether lives in the Moment Now, and that much defines one pole of his double consciousness; the opposite pole, in conflict and in reaction, finds the pendulum of his imagination magnetised by Mrs Newsome and swinging back to Woollett. In the Luxembourg gardens, instead of enjoying them, Strether reads letters from Mrs Newsome: 'It filled for him, this tone of hers, all the air; yet it struck him at the same time as the hum of vain things' (v).

The hum of vain things, the roaring trade of the Newsome factory, signifies the Idea of Time, which (like the unnamable Woollett product) 'may well be on the way to become a monopoly' (iv). Chadwick Newsome declares, in his final interview with Strether, his intention of returning to Woollett and entering the advertising end of the roaring trade: 'The right man must take hold. With the right man to work it *c'est un monde.*' Chad knows the world, knows how to live time's fleeting hour; don't tamper with the clock, but rather advertise it. Having clocks in mind, Chad concludes: '*To wind up* where we began. My interest's purely platonic' (xxxv). In 'our roaring age,' Strether admits, 'Advertising is clearly, at this time of day, the secret of trade.'

Trading with time, that's what the novel is all about.

NOTES

[1] Text references are to *The Ambassadors*, ed. R. W. Stallman (1960); see for 'A note on the text of *The Ambassadors*', pp. 381–2. The 'Afterword' to this edition is a brief summary of the present essay which, recast anew and much shortened, derives also from the original version in *Modern Fiction Studies*, III (1957), pp. 41–56. This longer version was reprinted in my *Houses that James Built* (1961); see for 'A note on the text of *The Ambassadors*', pp. 51–3.

[2] Illustrations of the Berne clock are given in Alfred Chapuis's *Les Automates, figures artifielles d'hommes et d'animaux* (Neuchâtel, 1949).

[3] In 'Time and the unnamed article in *The Ambassadors*', *Modern Language Notes*, LXXII (1957), pp. 27–32.

José Ortega y Gasset†   MARCEL PROUST (1871–1922)

Time, distance and form
in Proust

This time death, in mowing down another's life, has in passing cut off our own pleasures. Many people, in every country, had based their budget of future delights upon new books by Proust. This phenomenon of a public which 'waits for' the next work of an author is extremely rare, and has been for some time. Certainly, there is no lack of worthy writers whom we receive in our libraries whenever they present themselves. But that we always receive their visit with politeness and respect does not mean that we desire it. For these gentlemen writing consists in forcing oneself to take positions or poses. With the most commendable constancy they subject us to their meagre repertory of stereotyped life-studies. After a few unavailing performances we feel no urgent need to undergo the spectacle again.

But there are writers of another sort: those who have the luck or genius to have stumbled upon a vein of 'things'. Their situation is very similar to that of discoverers in science. With simplicity and a stupefying effort-lessness they feel their foot gliding through a new region of aesthetic possibilities. If, using a vague and mystical term, we are accustomed to calling the first type of writers 'creators', we shall have to call the second type 'inventors' with everything that word signifies in its Latin root. For they have found the new and hidden fauna of an unknown countryside; at the very least, they have discovered a new way of seeing, a simple law of optics in which a certain unusual index of refraction is formulated. The status of such authors is much more assured: although their work is, in a sense, always the same, it promises us new things, fresh displays, and it is unlikely that eagerness to see them would be lacking in us. When Plato looks for a suitable classification by which to distinguish the philosophers, he fixes upon the class of *philotheamones*, or the friends of seeing, those who like to look. He thought perhaps that the most constant virtue in man is a certain visual enthusiasm.

† Reprinted from *Hudson Review*, XI (1958), pp. 504–13; translated by Irving Singer. This essay first appeared in January 1923, two months after the death of Proust. At that time *La Prisonnière*, *La Fugitive*, and *Le Temps retrouvé* had not yet appeared [I.S.].

Proust is one of these 'inventors'. In the midst of contemporary production, which is so capricious, so lacking in necessity, his work presents itself
with the stamp of something ordained. If it had never come into being,
there would have remained, in the literary evolution of the nineteenth
century, a specific gap with a clearly defined outline. One might even say,
in order to point up its inevitability, that it was created a little late, that
analysis would disclose a slight anachronism in its physiognomy.

The 'inventions' of Proust are of prime importance because they deal
with the most elemental ingredients of the literary object. Nothing less
is involved than a new way of treating time and situating oneself in
space.

But imagine that in order to give an idea of what Proust is we enumer-
ated his themes to someone who had never read him—summer life in a
familiar village; Swann's love; the emotional play of a boy and a girl with
the Champs-Elysées as background; a summer on the coast of Normandy,
in a deluxe hotel, before an unsettled sea on which glide the faces, with sea
nymph features, of *jeunes filles en fleurs*, etc. If we gave an account of this
sort, we would soon realise that we had said absolutely nothing and that
these themes, which novelists have elaborated any number of times, cannot
even suggest the character of what it is that Proust offers us. Some years
ago a poor hunchback used to visit regularly the library of San Isidro.
He was so short that he could scarcely reach the writing-desks. Invariably
he would repair to the librarian on duty and ask for a dictionary. 'Which
one?' the employee would ask in a friendly manner. 'Latin, French,
English?' And the little hunchback would reply: 'Look, it's all the same to
me: I just want it to sit on.'

We would make the same mistake as the librarian if we were to describe
Claude Monet by saying that he painted the Cathedral of Notre-Dame and
the Gare Saint-Lazare, or Degas by pointing out that he depicted laun-
dresses, ballerinas, and jockeys. For both painters these subjects, which
seem to be the theme of their paintings, are only the pretext: they painted
these things just as they might have painted others of an entirely different
sort. What they cared about, the real theme of their canvasses, is the aerial
perspective, the gauze of chromatic vibrations in which things, whatever
they may be, live sumptuously enclosed.

Something similar applies to Proust. The narrative themes that come
and go on the surface of his work have only a tangential and secondary
interest; they are like buoys adrift on the bottomless flood of his memories.
Before Proust, writers had commonly taken memory as the material with
which to reconstruct the past. Since the data of memory are incomplete and
retain of the prior reality only an arbitrary extract, the traditional novelist
fills them out with observations drawn from the present, together with
chance hypotheses and conventional ideas. In other words, he unites fraudu-
lent elements with the authentic materials of memory.

This method makes sense as long as the intention is, as it formerly was, to *restore* things of the past, i.e. to feign a new presence and actuality for them. The intention of Proust is the very opposite. He does not wish to use his memories as materials for reconstructing former realities; on the contrary, by using all conceivable methods—observations of the present, introspective analyses, psychological generalisations—he wants literally to reconstruct the very memories themselves. Thus, it is not things that are remembered, but the memory of things, which is the central theme of Proust. Here for the first time memory ceases to be treated as the means of describing other things and becomes itself the very thing described. For this reason Proust does not generally add to what is remembered those parts of reality which have eluded memory. Instead, he leaves memory intact, just as he finds it, objectively incomplete, occasionally mutilated and agitating in its spectral remoteness the truncated stumps that still remain to it. There is a very suggestive page in which Proust speaks of three trees on a ridge. He remembers that behind them there was something of great importance, something which has been effaced by time, abolished from memory. In vain the author struggles to recapture what has escaped him, to integrate it with that bit of decimated landscape—those three trees, sole survivors of the mental catastrophe which is forgetting.

The narrative themes in Proust are, then, mere pretexts and, as it were, *spiracula*, air-holes, tiny portals of the hive through which the winged and shuddering swarm of reminiscences succeed in liberating themselves. It is not for nothing that Proust gave his work the general title of *A la recherche du temps perdu*. Proust is an investigator of lost time as such. With utter scrupulousness he refuses to impose upon the past the anatomy of the present; he practises a rigorous non-intervention guided by an unshakeable will to avoid reconstruction of any sort. From the nocturnal depths of the soul a memory surges upwards, excitingly, like a constellation in the night which ascends above the horizon. Proust represses all interest in restoration and limits himself to describing what he sees as it arises out of his memory. Instead of reconstructing lost time, he contents himself with making an edifice of its ruins. You might say that in Proust the genre of Memoirs attains the distinction of a pure literary method.

So much for his treatment of time. But even more elemental and stupefying is the nature of his invention with regard to space.

Various people have counted the number of pages that Proust employs in telling us that his grandmother is taking her temperature. Indeed, one cannot talk about Proust without noting his prolixity and concern for minutiae. In his case, prolixity and minute analysis cease to be literary vices and become two sources of inspiration, two muses that might well be added to the other nine. It is necessary for Proust to be prolix and minute for the simple reason that he gets much closer to objects than people are accustomed to. He is the inventor of a new distance between us and things. This fundamental revolution has had such tremendous consequences—as I have said—that almost all previous literature appears to be grossly panoramic, written from a bird's-eye point of view, as compared with the work of this delectably myopic genius.

By virtue of our animal adaptation to the world, each thing imposes a determinate distance upon us. Seen from this distance, it seems to us to attain its best appearance. The man who really wants to see a stone gets close enough to discern the porousness of its surface. But if he really wants to see a cathedral, he will have to give up seeing the pores of the stones. Instead, he will step back in order to enlarge his visual field considerably. The standard for each distance is set by the organic utilitarianism which governs all the activities of life. But perhaps it was an error for the poets to have thought that this system of distances, excellent as it is for vital necessities, was equally good for art. Proust, no doubt tired of always seeing a hand drawn as if it were a momument, brings it close to his eyes and, hiding the horizon behind it, finds to his great surprise that in the foreground of his vision there appears an intriguing landscape in which the valleys of the pores undulate surmounted by the lilliputian forest of the hairs. This, of course, is only a manner of speaking: Proust was not interested in hands, nor things of the body in general, but in more intimate flora and fauna. He corrects the distance between ourselves and our human feelings, and he breaks completely with the traditional way of describing them as if they were monuments.

I think it would not be wholly amiss for us to investigate the matter a little further. We might then discover how this radical transformation of literary perspective originated in Proust.

When a primitive artist paints a vase or a tree, he starts with the assumption that everything really has an outline, that is to say, a definite contour or external form which, like a clearly marked frontier, separates it, isolates it, from all other things. To fix the outline of objects precisely, neatly, constitutes the major interest of the primitive. The impressionist, on the other hand, believes that this outline is illusory and not really given to us in actual vision. If we restrict ourselves to the tree as it is *seen*, in the most rigorous sense of the word, we shall discover that its outline is not clear-cut, that its silhouette is diffuse and imprecise, and that it is not distinguished from its surroundings by that non-existent contour but rather by the mass of chromatic tones interior to itself. For this reason, the impressionist does not *draw* the object; he attains it by accumulating tiny dabs of colour, each one of which is formless in itself, but all of which together, in combination, are able to engender before half-closed eyes the vibrant presence of the object. The impressionist paints a vase or a tree without there being anything in his canvas that has the figure of a tree or a vase. As a pictorial style, then, impressionism consists in negating the external form of real objects in order to reproduce their internal form, the inner chromatic mass.

This kind of art dominated European sensibility at the turn of the century. And it is interesting to note the parallelism with the philosophy and psychology of the period. The philosophers of 1890 maintained that reality is made up of our emotive and sensory states. Whereas the ordinary man, like the primitive in painting, interprets the world as something definite and fixed that is to be found outside of ourselves, something endowed with magnificent, immutable architecture, the impressionist philosopher considers the universe a mere projection of our feelings and sensations, a flux of odours, tastes, lights, pains, and desires, a never-ending procession of unstable inward reverberations. Likewise, primitive psychology supposes that our personality consists of an invariable core, a kind of spiritual statue that receives the changes in the environment with a steady, unaltering countenance. Such is the psychology of the man of Plutarch whom we find immersed in the sea of life, enduring his hardships as a rock endures the howling surge or a statue the inclemencies of the weather. By contrast, the impressionist psychologist denies the existence of what is ordinarily called character, i.e. the sculptural outline of the individual. Instead, he sees the self as in perpetual mutation, a succession of diffuse states, an ever changing sequence of emotions, ideas, sorrows, hopes.

These considerations help us to date the essential tendencies of Proust. The monograph on Swann's love is an example of psychological pointillism. For the medieval author of *Tristan and Iseult* love is a sentiment that possesses a clear and definite outline of its own: for him, a primitive psychological novelist, love is love and nothing other than love. As opposed to this, Proust describes Swann's love as something that has nothing like the form of love. All kinds of things can be found in it: touches of flaming sensuality, purple pigments of distrust, browns of habitual life, greys of vital fatigue. The only thing *not* to be found is love. It comes out just as the figure in a tapestry does, by the intersection of various threads, no one of which contains the form of the figure. Without Proust there would have remained unwritten a literature that must be read in the way that the paintings of Monet are looked at, with the eyes half-shut.

It is for this reason that when Proust is compared to Stendhal one must proceed with caution. In many respects they represent two opposite poles, and are antagonistic one toward the other. Above all else, Stendhal is a man of imagination: he imagines the plots, the situations, and the characters. He copies nothing: everything in him resolves to fantasy, into clear and concentrated fantasy. His characters are as much 'designed' as are the features of a madonna in the paintings of Raphael. Stendhal believes firmly in the reality of his characters and makes every effort to draw a sharp and unequivocal outline of them. The characters of Proust, on the other hand, have no silhouette; rather, they are changeable atmospheric condensations, spiritual cloud-formations that varying wind and light transform from one hour to the next. Certainly Proust belongs in the company of Stendhal, 'investigator of the human heart'. But while Stendhal takes the human heart as a solid with a definite though plastic shape, it is for Proust a diffuse and gaseous volume that varies from moment to moment with a kind of meteorological versatility. From the drawing of Stendhal to the painting of Proust there is the same distance as between Ingres and Renoir. Ingres has delineated beautiful women with whom one might fall in love. Not so Renoir; his method excludes the possibility. The living plasma of luminous points which is a Renoir woman may give us supremely the sensation of warm human flesh; but a woman, in order to be beautiful, must impose upon the mere expanse of her body the correct limit of a definite outline. Similarly, the literary and psychological method of Proust prevents him from modelling feminine figures of an attractive sort. Notwithstanding the predilection accorded her, the Duchesse de Guermantes seems to us ugly and self-assertive and nothing more. If, however, we were to enjoy anew the torrid years of our youth, it is certain that we would once again fall in love with la Sanseverina, that woman of so tranquil a face and so quivering a heart.

In short, Proust brings to literature what might be called a predominantly atmospheric purpose. The landscape and the characters, the inner world and the outer—everything has been volatilised into an aerial and diffuse palpitation. I would say that the world of Proust was made to be experienced like respiration, since everything in it flows like a current of air. In these volumes nobody does anything, nor does anything happen: there is just a passive succession of static situations. It could not have been otherwise: in order to do something, one must first be something definite. The action of an animal always develops like a line that originates by its own will and is reborn whenever impeded by obstacles, thereby revealing the existence of a self in opposition to intervening resistance. For this reason, the broken line that is the action of an animal—man or beast—is charged with a latent dynamism which lends dramatic impact to the very development of the action. The characters in Proust, however, belong to a vegetative order. For the plant, living is being, not doing. Submergerd in the atmosphere and incapable of opposing itself to it, the plant's passivity eliminates the dramatic. Likewise, the characters of Proust have a botanical being, inert within their atmospheric destinies, their lives with vegetative submission reduced to a chlorophyllian function, a chemical dialogue always the same, quasi-anonymous and one in which the plant tractably receives the imperatives of its environment.

In these books, the real agents of human change are not the characters so much as the winds and climates, both physical and moral, which successively envelop them. The biography of each character is dominated by certain spiritual trade winds that alternately sweep against it and polarise its sensibility. Everything depends upon the direction from which the squall happens to blow; and just as there are freezing blasts and balmy currents, winds from the north and from the south, so too the Proustian character varies according as the gust of existence blows from the Côté de Méséglise or the Côté de Guermantes. There is nothing surprising in the frequency with which this writer speaks of 'côtés' since, the world being a meteorological reality for him, he naturally thinks in terms of quadrants.

I suggest, then, that an inspired rejection of the external and conventional form of things forces Proust to define them by reference to their inner form, their internal structure. But this structure is of a microscopic sort, which explains why Proust had to get so abnormally close to things, and why he was led into poetic histology. More than anything else his work resembles those anatomical treatises that the Germans entitle, for example: *Über feineren Bau der Retina des Kanninchens*, 'On the microstructure of the retina of the rabbit'.

Microscopic interest signifies an interest in details. An interest in details requires prolixity. The atmospheric interpretation of human life, and the minute analysis used in describing it, inevitably impose upon the works of Proust an attribute that might well appear to be a defect. I am referring to the peculiar fatigue that the reading of these volumes produces in even their most devoted admirers. If it were merely a question of the usual fatigue that feather-brained books secrete, there would be nothing more to say about it. But the fatigue that comes with reading Proust has very special characteristics and has nothing to do with boredom. With Proust we never get bored. It is very rare that even a single page should be lacking in adequate, indeed ample, intensity. Nevertheless, we are always ready, at any moment, to leave off the reading of Proust. Moreover, throughout the work we feel ourselves constantly halted, as if we were not allowed to advance at will, as if the rhythm of the author were always slower than our own, imposing a perpetual *ritardando* upon our haste.

Therein consists both the drawback and the advantage of impressionism: in the volumes of Proust, as I have said, nothing happens, there is no dramatic action, there is no process. They are composed of a series of pictures extremely rich in content, but static. We mortals, however, by our very nature, are dynamic; we are interested in nothing but movement.

When Proust tells us that the little bell jangles in the gateway of the garden in Combray and that one can hear the voice of Swann who has just arrived, our attention lights upon this event and gathering up its forces prepares to leap to another event which doubtless is going to follow and for which the first one is preparatory. We do not inertly install ourselves in the first event; once we have summarily understood it, we feel ourselves dispatched towards another one still to come. In life, we believe, each event announces its successor and is the point of transition towards it, and so on until a trajectory has been traced, just as one mathematical point succeeds another until a line has been formed. Proust ruthlessly ignores our dynamic nature. He constantly forces it to remain in the first event, sometimes for a hundred pages and more. Nothing follows the arrival of Swann; no other point links up with this one. On the contrary, the arrival of Swann in the garden, that simple momentary event, that point of reality, expands without progressing, stretches without changing into another, increases in volume and for page after page we do not depart from it: we only see it grow elastically, swell up with new details and new significance, enlarge like a soap-bubble embroidering itself with rainbows and images.

We experience, thus, a kind of torture in reading Proust. His art works upon our hunger for action, movement, progression as a continual restraint that holds us back; we suffer like the quail that, taking flight within his cage, strikes against the wire vault in which his prison terminates. The muse of Proust could well be called 'Morosidad' (Sloth), his style consisting in the literary exploitation of that *delectatio morosa* which the Councils of the Church punished so severely.

We can now see with abundant clarity how the Proustian cycle of elemental 'inventions' is structured. We can now see how his modification of ordinary distance and form is the natural consequence of his fundamental attitude towards memory. When memory is taken as one material among others for the intellectual reconstruction of reality, we only avail ourselves of that bit of remembering which we can use. Instead of allowing it to grow according to its own inherent principle, we move beyond. In reasoning and in simple association of ideas, our soul effects a trajectory, passes from one thing to another, our attention progressing by means of successive displacements. But if, turning our backs on reality, we throw ourselves into the contemplation of memory, we see that it proceeds by mere expansion without our having moved, so to speak, from the initial point. To remember is not, like reasoning, to travel through mental space: it is the spontaneous growth of space itself.

I do not know what actual techniques Proust used in writing. But his paragraphs, which make such complex and sinuous patterns, seem to have undergone internal vicissitudes after their first writing. One can see that perhaps they were originally well-proportioned, but that the memory enclosed within them has subsequently put forth protuberances, even excrescences, which form strange and—to my taste—delicious grammatical nodules similar to the bony humps that form in the feet of Chinese ladies due to the tightness of their shoes.

Having started with these remarks about the most elemental and abstract dimensions of the work of Proust, it would seem that the moment has now arrived to talk about the work itself and about the author's temperament. We would then discover quite surprising correlations between the anti-dynamic attitude which controls his interpretation of time, distance, and form, on the one hand, and the rest of what is peculiar to him, on the other. It is intriguing to observe that a single organic principle, in its formulation very simple, suffices for explaining all the aspects of Proust's work—for example, the extraordinary perspicacity with which he describes the events of sanguinary circulation in the narrator, his sharp perception of hygrometric changes and of muscular sensations; finally, his transcendent, all-embracing snobbery. But this would require too many pages . . . and a kind of literary criticism which perhaps interests no one but myself.

Georges Poulet†   PAUL VALÉRY (1871–1945)

Studies in human time:
Valéry

I

In the beginning, according to Faust, was activity. Bolder, above all more rigorous than the Goethean hero, the Valerian hero—M. Teste without a doubt—goes back beyond all activity and puts the beginning of things at the place where the human being can hardly distinguish himself from the nothingness which he was and which he shall become again: 'In the beginning will be sleep.'[1] At this moment—if it is a moment?—the being is still only silence and absence, absence of himself from himself, silence of the consciousness, which shows him that death is situated not only at the end but at the beginning and in all the interstices of his existence:

> Man imagines he 'exists'. He thinks, therefore he is—and that naive idea of taking himself for a world subsisting of and by itself, is possible only through negligence.
> I neglect my hours of sleep, my absences, my deep, long, insensible variations.
> I forget that I possess in my own life a thousand models of death, of daily nothingnesses, an astonishing quantity of lacunae, of things suspended, of intervals unknowing and unknown.[2]

At the beginning, therefore, let us postulate non-being—that non-being in the purity of which the universe and myself are only a flaw—a non-being, however, which in a certain manner and by virtue of its very perfection, exists, and exists more than I. For I live only in what I was or what I shall be, that is to say, in a successive time in which the present escapes me: 'We see only the future or the past, but not the patches of the pure instant.'[3] Only that being who is entombed in his own sleep and in his own absence, really belongs to his present and possesses it: 'He is as if eternal, ignorant of himself.'[4] He alone is. That is why I cannot say: I am. But 'I was, you are, I shall be.'[5]

The time of sleep, therefore, primitive and primordial, is the earliest of

† Reprinted from *Studies in Human Time*, translated by Elliott Coleman (Baltimore, Md., 1956), pp. 280–90.

those 'durations independent of each other', whose 'contradictory coexistence' constitutes 'the strangest problem one can ever propose to oneself. . . .'[6] It is a time without past or future, above all without change, and it does not differ from that hypothetical thing called eternity except in this, that it is bounded on all sides by zones of consciousness. It is like an *island* of eternity, inserted into the intervals of pure time:

Sleep . . . mild and tranquil mass mysteriously isolated, sealed ark of life that transportest towards day my history, my hazards, thou ignorest me, thou preservest me, thou art my inexpressible permanence; thy treasure is my secret. . . . Thou hast made for thyself an island of time, thou art a time that is detached from the huge Time in which thy duration indefinitely subsists and is perpetuated like a ring of smoke.[7]

Meanwhile if this island of time shelters my sleeping presence, I know nothing of it; but I do know that when I emerge from it to pass into another time, I undoubtedly experience the feeling of leaving an environment of vacancy and beginning to be myself, but not the feeling of an absolute beginning suddenly taking form in a void. The mysterious non-being, from which the awakening snatches me, is composed of a certain spiritual matter. 'Atoms of silence',[8] 'atoms of time'[9]—these constitute it and preserve me in it. I leave it restored—restored in the full sense of the word—given back to the completeness of the possible. It is a state very close to that state of sleep Valéry alludes to in the following passage, in which he deplores the disappearance of *spare* time:

We lose that essential peace of the depths of being, that priceless absence during which the most delicate elements of life are refreshed and fortified, during which the being, in some way *is washed clean of the past and the future*, of present consciousness, of suspended obligations and ambushed expectations. . . . No interior pressure, but a kind of repose in absence, a beneficent vacancy which restores the mind to its proper freedom.[10]

'In the beginning will be sleep' is equivalent, therefore, to saying: In the beginning there will be a time that is free—time of pure virtuality in which the mind is withdrawn from all the obligations of wakefulness in order once more to become ready to seize new chances, for 'each atom of silence is the chance of a ripe piece of fruit'. And suddenly we understand why Valéry, in contrast to Goethe, did not speak of the *beginning* in the past tense, as of something which had taken place, but used the future tense, as of something that is still becoming, something that is going to happen: In the beginning *will be* and not *was*. The possible is not the being but prepares it, is simply its future. It is from the possible that one must seize hold of being, when it is becoming actual, in a sleep that becomes an awakening, in a future that is made present:

Chaos . . . primary disorder, in the ineffable contradictions of which space, time, light, possibilities, virtualities were still in latency.[11]

II

In the beginning will be sleep. But sleep is not consciousness. 'It is shifting, irresolute, still at the mercy of a moment, that the operations of the mind are going to be able to serve us.'[12] 'The idea, the principle, the lightning flash, the first moment of the first state, the leap, the bound out of sequence. . . . Throw the line there. That is the spot in the sea where you will get a catch.'[13] But what is there to catch except that lightning flash, that leap, that stunning presence of the moment of which consciousness is the prey? 'Disorder is therefore my first point. . . . It animates us.'[14] 'I consider the state close to stupor, the singular and initial point of knowledge.'[15] The first stage of Valerian consciousness is therefore consciousness of the moment at whose mercy it is; consciousness of a successive, discontinuous, disordered, anonymous plurality: 'Every instant the mind of the instant comes to us from the exterior.'[16]

The exterior is what extends about us; it is space. It is made up of an assemblage of patches incessantly changing. As far as the gaze may travel, as intense as may be the attention, they never meet anything outside except 'a chaos of lights and shadows . . . a group of luminous inequalities'.[17] 'So, in the enlargement of what is given, expires the intoxication with particular things.'[18] It expires in the consciousness of 'an inexpressible disorder of the dimensions of knowledge'.[19] There are nothing but 'ephemeral figures', 'enterprises interrupted . . . that are transformed one into the other'; the mind expresses them by means of 'an interior word, without person and without origin'.[20]

But if that exterior world made up of instantaneous spaces is only a chaos of particular things, what is the interior universe of him who reflects and observes it?

The observer is held in a sphere that is never broken. . . . The observer is first of all only the necessary condition of that finite space. . . . Neither memory nor any other force disturbs him as long as he identifies himself with what he beholds. And if ever I conceive him as lasting thus, I shall conceive that his impressions will never differ in the slightest from those he would receive in a dream.[21]

Thus, *identifying himself with what he beholds*, with an uninterrupted train of enterprises, thought is the prey of a Bergsonian universe whose reverse side is here found to be cruelly unveiled. To be abandoned to the transitory present and to the single flux of duration is to be abandoned to a double nothingness: a nothingness of the object which is never what it is; and a nothingness of thought which by its very spontaneity is made each instant into what it thinks, and thus becomes the object and sharer of its ephemeralness: 'Instantaneous nullity';[22] 'instantaneous and undivided state that smothers this chaos in nullity.'[23]

III

Nevertheless, from this chaos which is instantaneously annulled, it happens that a kind of unity emerges which will assemble things into more or less distinct structures:

That unity, which necessarily results from what I can see in an instant, that ensemble of reciprocal relations of figures or of patches . . . communicates to me the primary idea, the model and, as it were, the germ of the total universe that I believe to exist around my sensation, masked and revealed by it. I imagine invincibly that a vast hidden system supports, penetrates, nourishes, and reabsorbs each actual and sensuous element of my duration, impels it into being and into taking shape; and that therefore each moment is the knot of an infinity of roots that plunge to an unknown depth in an implicit extent—in the past—in the secret structure of our perceiving and calculating machine, which incessantly feeds back into the present.[24]

The first independent operation of the mind, the first gesture by which it frees itself from sensation, from space and from the immediate, is not a gesture of freedom but of bondage. King of the possible, the mind hastens to give itself masters and bounds. He who can invent all, invents the concept that he cannot produce his self-invention. Outside the present, he imagines the series of causes and determinations of that present:

In the mythical void of a time pure and bereft of whatever element may be similar to those that border us, the mind—assured only that there had been something, constrained by an essential necessity to suppose antecedents, 'causes,' supports of what is, of what it is—gives birth to epochs, states, events, beings, principles, images, or histories. . . .
That is why it came to me one day to write: In the beginning was the Fable![25]

It is a strange transformation that the mind now inflicts on the moment, and by which it radically changes the nature of it. In place of being, things now demand to have been. They wish themselves surrounded not simply by patches and places but by causes and time. They claim the right to be annexed to a duration. They entreat the mind to set them against the background of a past, one single past, of all possible pasts. Among all the floating images of what one could have been, one makes the choice of a certain image, one decides to think that this really represents what one has been:

The past is a thing entirely mental. It is only images and belief.[26]

The past is only a belief. A belief is an abnegation of the powers of our mind, which feels repugnance at forming for itself all the convenient hypotheses about absent things, and giving them all the same force of evidence.[27]

'Naïve and bizarre structure',[28] the belief in the past is none the less a structure. It supports its fables; it also supports the present. For that reason the present no longer appears to be suspended above non-being. Sustained from underneath and from behind, confirmed by analogies, strong in its ability to represent in the right place a continuing identity, the present makes every effort in order not to be different from what the past was and to present itself as a simple repetition. In this effort it finds itself aided, moreover, by the very coarseness of our perceptions:

> Not subtle enough, my senses, to undo that work so ingenious or so profound that the past is, not subtle enough for me to discern that this place or that wall are perhaps not identical to what they were the other day.[29]

Thanks to this simplification, the originality of each moment no longer risks confronting us without respite with a nature in which there is not one 'trace of past, repetition, similarity',[30] a nature, therefore, always instantaneous. By an audacious falsification of the latter, in giving ourselves a past, in giving ourselves similarities, we also give ourselves at the same time the chance and the means of finding constants, of constructing laws, of imagining the universal.

But if things are consolidated in this manner in a mental structure, they are found to lose, on the other hand, in variety and in authenticity. Instead of appearing as they are, they tend to resemble what they have been. They become impoverished; they become involved; they are reduced to signifying: 'Memory drives away the present.'[31] 'Each instant falls down instantly into the imaginary.'[32]

Thus the past is simply 'the place of forms without forces'.[33] It 'represents itself, but it has lost its energy'.[34]

Doubtless 'it is up to us to furnish it with life and necessity', and for that purpose 'to endow it with our passions and our values.'[35] Let us then recognise in this, not the presence of the 'past' but the presence of some 'present' in the past. Let us notice the *originality* of memory, 'that in which memory is not the past, but the act of the present':[36]

> There is thus engendered a state of mind that is curiously anti-historical, that is to say, a vivid perception of the completely actual substance of our images of the 'past' and of our inalienable liberty to alter them as easily as we are able to conceive them. . .[37]

But in this way the temporal point of view changes once more. Instead of imagining ourselves determined by what was, we now carry our indetermination backwards in order to make of the past a sort of anterior future. From this point of view 'to see again and to foresee, to recollect and to forebode, strongly resemble each other.'[38] But that is true only because it is possible to 'foresee' over all the points of time and in all the positions of the mind. What matters therefore above all is to understand what appears more and more as the 'time' of the mind, that is to say the future:

> The idea of the past takes on a meaning and constitutes a value only for the man who finds in himself *a passion for the future*.[39]

For Valéry, as for Vigny, the spiritual activity of man is fundamentally this passionate sense of the future.

IV

What is the future? To understand it, we must bring our attention to bear upon 'our most central sense, that intimate sense of the distance between desire and the possession of its object, which is nothing else than the sense of duration. . . .'[40] This sense of duration is the future. It is something that splits and yawns within ourselves, within the moment when we are ourselves; it is the feeling of a void, of a lack of a gap, with the need of filling it up. The future is first of all dissatisfaction and desire. If, on the one hand, we exist only in the present, if 'there is nothing of ourselves outside the instant',[41] on the other hand 'we consist precisely in the regret or in the refusal of *what is*, in a certain distance which separates us and distinguishes us from the instant.'[42] It is a distance that places before us a being that is still ourself and which seems to us to be more desirably ourself than ourself —the myth of Narcissus: 'Man is not all in one piece. One part of him precedes the other.'[43]

But this part of himself that precedes him does so only by an infinitesimal fraction of time: 'I feel the imminence.'[44] 'I love nothing so much as that which is going to happen.'[45] At this moment the being in whom the imminent future takes form—in whom 'the future is the most perceptible portion of the instant'[46]—discovers and experiences himself in his virtualities, in his expectations: 'Sweetness of being and not yet being.'[47] Man is less what he is than what he is on the point of becoming. He is in an exquisitely non-temporal position, where nevertheless all temporality takes refuge, 'between the void and the pure event'. Thought 'beats between the times and the instants'.[48] It desires, attracts, brings the possible to existence, and that almost by its ardour alone: 'I immolate myself interiorly to what I would wish to be.'[49]

Now in this function that is 'the simplest, the deepest, the most general of our being, which is to form the future',[50] in this state of presentiment in which the difference that separates and distinguishes us from the instant is a difference of pure thought—it is here precisely that thought becomes visible to itself; and, in creating time, the human being at the same stroke gives himself a consciousness of his consciousness. Thought of thought, the consciousness resides in that temporal distance that makes resound in the mind a depth that is always future. It is 'the distance between being and knowing',[51] the property of deviating from the instant, and even 'from one's own personality'.[52] It is the exact antitype of the figure of Narcissus. There issues from myself and from all the instants when I am myself, another self, a different self, a general and impersonal entity that contemplates me. He is, like M. Teste, 'the being absorbed in his variation'; he is the *measure of things*:

To say that man is the measure of things . . . is to oppose to the diversity of our moments, to the mobility of our impressions, and even to the particularity of our individualism, of our person . . . a Self that recapitulates it, dominates it, comprises it, as the law comprises particular cases, as the feeling of our strength contains all the acts that are possible for us. . . .

We feel ourselves to be this universal self. . . .[53]

Priceless creation, which is no longer that of an object engaged in time, but of that of a subject disengaged from time. Thanks to it,

. . . each life, particular as it is, possesses . . . deep within it a treasure, the fundamental permanence of a consciousness which nothing supports; and just as the ear discovers and loses again, throughout the vicissitudes of the symphony, a grave and continuous sound to be perceived—so the pure *self*, the unique and monotonous tone of the human being in the world, discovered by itself and then once again lost, eternally inhabits our senses; this profound *note* of existence dominates, as soon as one listens to it, the whole complication of the conditions and varieties of existence.[54]

But if the permanent consciousness is distinguished from the variable, if it is situated outside of duration, that is in order to give it its laws. The being who is lifted up to the thought of thought can then act on the thought as upon a mouldable material. 'Everything yields to this generality.'[55] 'The mind has the power to impart to an actual circumstance the resources of the past and the energies of the future.'[56] By pressures or relaxations it can slacken or accelerate the pace of the approach and the flight of whatever travels its roads. It knows what it must do for thought to come to light. It knows also, what is worth more, how it must be cultivated in order that it may not come too quickly to fruit. For the essential factor of this polity of the mind is less a presence among all the other presences than it is the power of keeping thought fixed for a long time, suspended before its attention, just as Joshua kept the sun motionless in the sky. From all sides ideas offer themselves to the mind, waiting on every hand for the chance of being plucked. But what matters is to profit to the maximum by this chance and, with that in view, to delay as long as possible the inevitable moment when all thought returns to chaos and the instantaneous state. It is a question of slipping into the break of continuity of the lode of time, between the moment of imminence and that of accomplishment, a whole new duration, in which the mind takes the time to foresee, to compose, to moderate, or to suppress,[57] that is to say, to make its work durable, as architect and engineer do. These thoughts that come to me, 'it is necessary that I stop them . . . that I interrupt the very birth of ideas. . . . It matters to me above all things that I obtain what will satisfy, with all the vigour of its novelty, the reasonable exigencies of what has been.'[58] It is necessary, therefore, 'to stay in a compulsory attitude'.[59]

In this duration which is compelled into being by the mind, and which the mind adorns by its operations, there is discovered a temporal frame of a nature precisely contrary to that of the duration-flux upon which the person leaving the state of sleep emerged in order to surrender himself to it —duration this time is no longer spontaneous but voluntary, no longer natural but artificial; 'delicate art of duration',[60] in which there is no longer an *infinity of interrupted enterprises*, but one sole enterprise having a beginning and an end, an orientation towards his proper achievement, the presence and the consciousness of a faith. That enterprise is the poem:

Even in the lightest pieces, it is necessary to think of duration—that is to say of *memory*, that is to say of form.[61]

One hundred divine instants do not construct a poem, which is a time of growth, and like a figure in time.[62]

A poem is a duration, during which, reader, I breathe a law that had been made ready.[63]

Duration is construction. . . .[64]

It follows then that, for Valéry, the true duration—contrary to the 'insupportable flight' and the 'happy surprise'—is a work of art, a creation of the mind which, in order to be formed, needs all its resources and all its vigilance; a creation, moreover, which has validity and reality only for the mind, and relatively to the object that it frames and forms—provisorily. There is no duration in itself, and there is no duration without work and object. And again, there is no infinite duration because there is no infinite work, because a work can begin to be only when anticipatively it has already been completed by the mind.

All true duration is, thus, like a grain of duration: something hard, closed in, a 'closed cycle',[65] that opposes its wall to emptiness, to chaos, to the formlessness of perceptible events, and within which there is found enclosed a life that is quivering and governed—governed toward its end.

Hardly has it ended when its creator abandons it—it and its duration. He rediscovers himself in his own duration, in a moment that is again an initial moment, *washed clean of the past and the future*, washed clean even of his own work, free, equal to the chances of the moment:

O moment, diamond of Time![66]
Here am I, the very present.[67]

NOTES

[1] *A.B.C., Commerce* (c. 1930).
[2] *Tel quel*, ii, p. 238.
[3] *Pièces sur l'art*, p. 146.
[4] *Variété III*, p. 104.
[5] *A.B.C., Commerce.*
[6] *Variété I*, p. 205.
[7] *A.B.C., Commerce.*
[8] *Poésies*, p. 201.
[9] *Variété II*, p. 195.
[10] *Variété III*, p. 284.
[11] *Mon Faust*, p. 55.
[12] *Variété I*, p. 220.
[13] *Monsieur Teste*, p. 125.
[14] *Variété III*, p. 205.
[15] *Tel quel*, ii, p. 243.
[16] *Ibid.*, i, p. 39.
[17] *Regards sur le monde actuel*, p. 22.
[18] *Variété I*, p. 234.
[19] *Ibid.*, p. 193.
[20] *Poésies*, p. 61.
[21] *Variété I*, p. 231.
[22] *Poésies*, p. 61.
[23] *Variété I*, p. 193.
[24] *Ibid.*, p. 134.
[25] *Variété II*, p. 254.
[26] *Variété IV*, p. 134.

27 *Mon Faust*, p. 25.
28 *Variété V*, p. 85.
29 *Monsieur Teste*, p. 129.
30 *Mélange*, p. 69.
31 *Pièces sur l'art*, p. 146.
32 *Variété II*, p. 253.
33 *Variété III*, p. 61.
34 *Mélange*, p. 84.
35 *Variété III*, p. 61.
36 *Mélange*, p. 69.
37 *Variété V*, p. 91.
38 *Variété IV*, p. 136.
39 *Regards sur le monde actuel*, p. 16.
40 *Variété III*, p. 283.
41 *Idée fixe*, p. 145.
42 *Mélange*, p. 80; *Tel quel*, I, p. 89.
43 *Idée fixe*, p. 100.
44 *Mélange*, p. 96.
45 *Eupalinos*, p. 35.
46 *Mélange*, p. 37.
47 *Poésies*, p. 128.

48 *Eupalinos*, p. 61.
49 *Monsieur Teste*, p. 124.
50 *Variété IV*, p. 191.
51 *Variété III*, p. 72.
52 *Ibid.*, p. 221.
53 *Ibid.*, p. 257.
54 *Variété I*, p. 204.
55 *Ibid.*, p. 192.
56 *Mélange*, p. 27.
57 *Variété I*, p. 176.
58 *Eupalinos*, pp. 114–15.
59 *Idée fixe*, p. 174.
60 *Monsieur Teste*, p. 28.
61 *Pièces sur l'art*, p. 91.
62 *Variété III*, p. 15.
63 *Poésies*, p. 62.
64 *Tel quel*, II, p. 334.
65 *Variété V*, p. 135.
66 *Mélange*, p. 96.
67 *Mon Faust*, p. 95.

Marcel Brion† JAMES JOYCE (1882–1941)

The idea of time in the work of James Joyce

Certain thinkers have at times wondered if the essential difference existing between man and God were not a difference of time. Space is not concerned here—God is everywhere—but, rather, this much more complex dimension which is generally inaccessible to human science. We measure time but we do not know what it is.

We often encounter in mystical literature the story of the monk or poet who has fallen asleep in the forest. When he awakes he no longer recognises either men or the countryside. His meditation or slumber, which to him has appeared very short, has in reality lasted hundreds of years. But during this moment in which he has been snatched from the tyranny of time he has caught a glimpse of the mysterious aspects of infinity, he has neared the laws of the Cosmos, the throne of God.

Theoretically, the difference in speed between two objects in motion is sufficient to make them imperceptible to each other; to destroy, practically, their existence.

The relations between human beings are those of time. All men are made similar by the nearly equal cadence of their heart-beats, but they are separated by the rhythms of their sensations or their thoughts. Only those walking at the same pace know each other.

The fourth dimension is actually the only one that matters. Space is nothing—it is reduced every day by mechanical means of communication—but consider two men seated side by side. They do not live in the same time. There is no possible communication between them. And it is often the tragedy of life to feel oneself only a few centimetres away from the beings among whom one lives, yet separated from them by all the infinity of time.

Time is not an abstract concept. On the contrary, it is perhaps the only reality in the world, the thing which is the most concrete. All the rest could only intervene in the form of its emanations.

† Reprinted from *Our Exagmination round his Factification for Incamination of Work in Progress*, by Samuel Beckett and others (1961), pp. 25–33; translated by Robert Sage (U.S. edition: *Finnegans Wake: A Symposium*). First published in Paris in 1929.

We may deduce from this that time is the essential factor in a work of art. (This appears quite evident when considered in one of its aspects—rhythm.) It is the law of architecture and of painting. The painters who have attained the greatest emotional power are precisely those whose work includes time—for example, Rembrandt. While we look at it, the picture seems always in the process of 'being made'. It seems to be constructing itself with the moments and it seems that if we were to return on the morrow we should find it changed. And, in fact, when we return on the morrow, it is changed. There are likewise masterpieces of sculpture which give the impression of a continual palpitation, of an uninterrupted succession of imperceptible movements. It is this that is ordinarily called life—but life is the consciousness of time.

A book's story may embrace several decades, several centuries without revealing time to us. Another imposes it in a brief moment. There are flat books and deep books (without metaphor and almost in a material sense); there are also books rich with time and books destitute of time. This is the reason that one of the greatest writers of our period, one of the most sensitive and most intuitive, made of time the essential dimension of his work—*temps perdu* and *temps retrouvé*.

Marcel's Proust's idea of time is extremely curious. In his books time is a character like the others—I might even say more than the others. Time is at the centre of his work like a sort of lighthouse with turning signals. The men who revolve around this luminous mass are suddenly illuminated by the beams of the projector in periodic flashes, and the moment the light abandons them they fall back into obscurity, nothingness.

It is in time that the characters of Proust become conscious of themselves. They seek themselves in it and are reflected in it. They complete their metamorphosis in it. But time remains exterior to them. They are not incorporated in it any more than they integrate it in themselves. They submit to it, as to gravity or the law of acceleration. But the author has conceived it so intensely that we feel this time to be materialised often like an object, applied like a thin and transparent pellicle on the face of men.

Perhaps because illness sheltered him from the customary rhythm of life, because it imposed upon him a different order of sensations, Proust understood time as a thing in itself, time which does not ordinarily separate us from our act and which we make simply a condition, an accessory of our existence.

With James Joyce it is another thing. I place James Joyce and Marcel Proust together intentionally because in my opinion they are the two greatest writers of our century, the only ones who have brought an original vision of the world to our epoch, who have renewed equally the universe of sensations and of ideas. The work of Proust and that of Joyce are the only ones between which a parallel may be drawn on an ideal plane of quality—and this for reasons which go far beyond questions of technique or talent—in the domain of literature and art. Perhaps it is because a sort of pure instinct of genius is likewise found here under a very elaborate art; but, above all, it is because with Joyce as with Proust time is a dominant factor.

On the absolute plane, the life of the ephemera and that of the animal endowed with the greatest length of life are equal. In the one case as in the other it is a life, and the fact that it stretches out for a few seconds or a few centuries has no importance. It is probable that both will be divided into a like number of units but that the unit will be long for the one and extremely short for the other. The idea of time being essentially that of the dissociation of moments, a hundredth of a second for an insect that lives for some minutes will be loaded with as many experiences as a year for the long-living animal. It is the same thing, all proportions retained, with men—some live at high speed, others at reduced speed; and they are separated, inexorably most often, by these different cadences.

We may thus account for the fact that eighteen hours of Bloom's life should give birth to *Ulysses*, and we can easily imagine that *Ulysses* might have been ten times as long, a hundred times as long, extended to infinity, that one of Bloom's minutes might have filled a library. This is the mystery of the relativity of time.

If time remains external to Proust, if he gives it an existence apart, isolated from his characters, for Joyce, on the contrary, it remains the inseparable factor, the primary element at the base of his work.

This is why he creates his own time, as he creates his vocabulary and his characters. He soon elaborates what he receives from reality by a mysterious chemistry into new elements bearing the marks of this personality. But even as he metamorphoses the countryside, the streets of Dublin, the beach, the monuments, he mixes all this into what appears to us at first sight as a chaos. This chaos is the condition necessary to all creation. The cards are shuffled to begin a new game and all the elements of a universe are mingled before a new world is made, in order that new forms may be given birth. A total refutation of man and his milieu, a rejection of combinations already used, a need of fine new instruments. Joyce dashes the scenes of the world down pell-mell to find an unhackneyed meaning and a law that is not outdated in the arrangement he is afterward to give them. To do this it is fitting that he should at the outset break through the too-narrow restraints of time and space; he must have an individual conception of these dimensions and adopt them to the necessities of his creation. In *Ulysses*, and still more in *Work in Progress*,[1] we seem to be present at the birth of a world. In this apparent chaos we are conscious of a creative purpose, constructive and architectural, which has razed every conventional dimension, concept and vocabulary, and selected from their scattered material the elements of a new structure. Joyce has created his language, either by writing words phonetically—and Heaven knows such a method is enough to discipline English—or by introducing foreign words and dialect forms, or finally by the wholesale manufacture of words which he requires and which are not to be had at second hand. And it is all done with an unprecedented creative power, with an almost unique fertility of imagination, inexhaustibly reinforced by the incredible extent of his culture. In the field of verbal richness Joyce has annexed the seemingly impregnable position of Rabelais; but whereas in Rabelais, form was under no direction other than that of an amused fantasy, in Joyce it is the handmaid of a philosophy. *Work in Progress* seems to be based on the historical theory of Vico[2]—an actual recreation of the world, its ideas and its forms.

Mr Elliot Paul well demonstrated recently[3] how Joyce in his composition of *Work in Progress* revealed an entirely individual conception of time and space.

This was already quite apparent in his first books. The stories in *Dubliners*, for example, seem entirely filled with the beating of a silent metronome. They unfold themselves in 'time'. Properly speaking, *Araby* is a drama of time, a drama of lost time; and we feel that each of the characters in *Dubliners* is rich or poor with his time, that the vibration of his life is hasty or slow.

In *Ulysses* the phenomenon is even more evident. To reduce the decades of the *Iliad*, the *Odyssey*, of Telemachus to eighteen hours in the life of a man—and of an ordinary man to whom nothing happens save the most ordinary events of existence—is one of the Einsteinian miracles of the relativity of time. And we understand it even better when we see the movement of the vibrations transformed in each chapter, changing rhythm and tempo, slowing up in the Nausicaa episode, blowing like the wind in that of Æolus, giving spacious and deep cadences to the gynaecological discussion. The chapter most powerfully demonstrating Joyce's mastery in expressing time is perhaps that in which Molly Bloom's reverie unrolls its rapid uninterrupted chain of ideas, memories and sensations, contrasting to her calm regular breathing.

Better than anyone else, Joyce has restored the sense of biological and intellectual rhythm. I imagine that he could write an unprecedented book composed of the simple interior physical existence, of a man, without anecdotes, without supernumeraries, with only the circulation of the blood and the lymph, the race of nervous excitations towards the centre, the twisting of emotion and thought through the cells. I imagine that Joyce could compose a book of pure time.

It sometimes seems that a page of Joyce is a strange vibration of cells, a swarming of the lowest Brownian movements under the lens of the microscope. In my opinion, if the recent books of Joyce are considered hermetic by the majority of readers it is because of the difficulty which the latter experience in falling into step, in adapting themselves to the rhythm of each page, in changing 'time' abruptly and as often as this is necessary.

But still more than to *Ulysses* these remarks apply to the book which *transition* is publishing and of which we as yet know only a part.[4] *Work in Progress* is essentially a time work. From a bird's eye view, time appears to be its principal subject. It begins in the middle of a moment and of a sentence, as if to place in infinity the initial disturbance of its waves. The concept of time here plays the principal role, not only by its concrete expressions but likewise by its abstract essence. It here takes on the significance of a creator-word and determines all the movements of the work.

The chronology of the story matters little to the author of *Work in Progress*. By his caprice, which in reality obeys a carefully studied and realized constructive will, characters most widely separated in time find themselves unexpectedly cast side by side; and, as for example Mr Elliot Paul recently wrote in *transition*, 'Noah, Premier Gladstone and "Papa" Browning are telescoped into one.' This image is perfectly accurate, and the optics of the work are so much the less accessible to the average reader as he does not always distinguish the moment in which the present episode is placed. When we are made to pass, without any transition other than an extremely subtle association of ideas, from Original Sin to the Wellington Monument and when we are transported from the Garden of Eden to the Waterloo battlefield we have the impression of crossing a quantity of intermediary planes at full speed. Sometimes it even seems that the planes exist simultaneously in the same place and are multiplied like so many 'over-impressions'. These planes, which are separated, become remote and are suddenly reunited and sometimes evoke a sort of accordion where they are fitted exactly, one into another like the parts of a telescope, to return to Mr Elliot Paul's metaphor.

This gift of ubiquity permits Joyce to unite persons and moments which appear to be the most widely separated. It gives a strange transparence to his scenes, since we perceive their principal element across four or five various evocations, all corresponding to the same idea but presenting varied faces in different lightings and movements.

It has often been said that a man going away from the earth at the speed of light would by this act relive in an extraordinarily short time all the events in the world's history. Supposing this speed were still greater and near to infinity—all these events would flash out simultaneously. This is what happens sometimes in Joyce. Without apparent transition, the Fall of the Angels is transparently drawn over the Battle of Waterloo. This appears to us as contrary neither to the laws of logic nor to those of nature, for these 'bridges' are joined with a marvellous sense of the association of ideas. New associations, created by him with amazing refinement, they co-operate in creating this universe, the Joycian world, which obeys its own laws and appears to be liberated from the customary physical restraints.

And we have, indeed, the impression of a very individual world, very different from our own, a world of reflections that are sometimes deformed, as in concave or convex mirrors, and imprinted with a reality true and whole in itself. I do not speak here only of the vocabulary which Joyce employs and which he transforms for his usage—which, one might, say, he creates—but especially of his manners of treating time and space. It is for this reason, much more than because of the work's linguistic difficulties, that the reader often loses his footing. This is related to the prodigious quantity of intentions and suggestions which the author accumulates in each sentence. The sentence only takes on its genuine sense at the moment that one has discovered its explanatory *rapprochements* or has situated it in time.

And if the books of Joyce are as difficult for many to read as those of Einstein it is perhaps because both of these men have discovered a new aspect of the world and one which cannot be comprehended without a veritable initiation.

NOTES

[1] i.e. the slowly-evolving *Finnegans Wake* which at the time of this essay's publication (1929) was appearing in *transition*. [Ed.].

[2] Giambattista Vico, author of *The New Science* (1725), often mentioned in connection with Joyce, e.g. in W. Y. Tindall's *James Joyce: His Way of Interpreting the Modern World* (1950), pp. 70 ff. [Ed.].

[3] In his essay 'Mr Joyce's treatment of plot', also included in *Our Exagmination* etc. (as above, p. 153, note). [Ed.].

[4] See above, note 1.

T. S. Eliot and the pattern of time

1

I do not suppose that *The Waste Land* was calculated to elicit agreement, much less to confound the ignorant and amaze the very faculties of eyes and ears. Rumour even has it that its quest ends where it began, the narrator lingering rather unnerved over his accumulated heap of broken images. The poem's detractors include Eliot himself, who sardonically called it 'just a piece of rhythmical grumbling'. Yeats also remarked of its verse cautiously, that 'there is much monotony of accent'; while a determined Amy Lowell concluded firmly: 'I think it is a piece of tripe.'[1]

So much for *The Waste Land*. But to clutch even by default at *Four Quartets* were equally hopeless, since no less a responsible critic than Donald Davie would place us, *ipso facto*, among 'the religiously inclined'—and who would not recoil from such a charge? Another responsible critic, Karl Shapiro, alerts us in addition to Eliot's 'insensitivity toward language'; and disclosing his own sensitivity remarks that 'the *Quartets* lie at the bottom of the literary heap'.[2]

That Eliot occasions judgements of this order I would be tempted to attribute, in part at least, to his espousal of the Christian faith in express opposition to the temper of his age. He went so far as to endorse Original Sin: in Agatha's formulation in *The Family Reunion*, 'A curse comes to being / As a child is born.'[3] We respond with scepticism for we believe in Progress; and we clothe our incredulity in parodies like *The Sweeniad* of 'Myra Buttle':

> Original Sin, Original Sin,
> Oh, what a terrible mess we are in!
> Adam condemned the whole of his kin,
> The young, the old, the fat, and the thin,
> Even a saint like Augustín,
> To suffer for ever the horrible in-
> Iquity of Original Sin!

To the extent that parodies partake of constructive humour, a committed Christian like Eliot would remind us that Christianity is not opposed to the joy of laughter. Kierkegaard indeed claimed that the Christian faith represents 'the most humorous point of view in the history of the world'.[4] We are not invited to think of the trite commonplace that a Christian's life is a comedy because it concludes with the ascent to felicity, for that were to reduce *The Cocktail Party* to a 'comedy' (as Eliot himself subtitles it) merely because Celia's crucifixion is said in that play to have been 'a happy death'. The Christian claim is infinitely more comprehensive. Best suggested by George Herbert when he wrote that 'All things are were bigge with jest', it avers the Christian's transcendence of the tragic through the abrogation of man's primal disobedience by the Christ. '*Jesus* is the true Seed of Mirth and Joy', declared Peter Sterry in the seventeenth century. 'When he cometh to us by his Spirit, . . . he bringeth back our Captivity from the Powers of Sin, Sorrow, and Death; he filleth our Mouths with laughter.'[5] Judged from such a vantage point, even the parody in *The Sweeniad* would be consciously misconstrued by a Christian to terminate beyond itself, beyond Original Sin, in the ecstatic sense of redemption shared among others by St Augustine. Such a mode of thought cannot but strike us as perverse in the extreme. But is it dramatically different from the claim often advanced on behalf of art? We think it perverse when Benjamin Britten asserts of Beethoven, that 'when he wrote some of his most tragic pieces, he was probably in a state of gaiety'. We think it equally perverse when Yeats assents to the presence of gaiety in *Hamlet* and *King Lear*: 'Gaiety transfiguring all that dread.' And we think it even more perverse when Joyce discourses on 'joy' in art, much as Christians posit 'Mirth and Joy' at the centre of their faith in full prospect of Calvary:

All art which excites in us the feeling of joy is so far comic and according as this feeling of joy is excited by whatever is substantial or accidental in human fortunes the art is to be judged more or less excellent: and even tragic art may be said to participate in the nature of the comic so far as the possession of a work of tragic art (a tragedy) excites in us the feeling of joy.[6]

The claim of Joyce on behalf of art was wedded by Eliot to the claims of the Christian faith. But these are never posited; they are felt along the pulses, in an impressive display of realism derived from 'the anguish of the marrow' he respected so much in Donne. *The Waste Land* may well be, as Eliot said of *Ulysses*, a 'continuous parallel between contemporaneity and antiquity';[7] but like *Ulysses* it is also a private vision, the harrowing experience of a mind existing in a dream transformed into a nightmare, impotent before the broken images, the meaningless fragments, the 'withered stumps of time' (l. 104). If the poem is a quest, it is quite unlike other tradition-bound pilgrimages whether in search of the Holy Grail, the New Jerusalem, the Vision Beatific. The quest in *The Waste Land* begins and ends in a world where connections are impossible because the past is obliterated by meagre memories, the present numbed by emaciated desire, and the future manacled by infirmity of purpose. From afar the thunder speaks, promising rain; but the narrator does not respond, immediately. The 'quest', there-fore, is not linear, since there is no unequivocal advance forward; nor is it circular, since the narrator steadily distances himself from the initial vision of the cruel spring. *The Waste Land* is indeed inconclusive, but in much the same way that *Waiting for Godot* is inconclusive. Apparent inaction and lack of resolution define the aim of Eliot's poem as of Beckett's play: confinement solely within the temporal order.

But even as both *The Waste Land* and *Waiting for Godot* are concerned with confinement solely within the temporal order, Eliot and Beckett alike alert us to another dimension which altogether escapes the consciousness of their protagonists. Eliot's method is well attested: it is a disposition toward multiplicity of reference, as in the explosion into secular time of the different pattern marked by the chimes of St Mary Wolonoth: 'a dead sound on the final stroke of nine' (*W.L.*, l. 68). The allusion to the Passion increases the distance between us and the narrator, ignorant as he remains throughout of the cumulative import of the fragments he records. Eliot makes the point even more forcefully, however, in that the style of *The Waste Land* is at odds with its nominal theme. The style violates time—clock time—at every turn; as we are often enough reminded, Eliot con-tributed to 'a breakdown of the tyranny of linear historic time'.[8] But the poem's theme, on the other hand, reasserts time on two levels: in the consciousness of the narrator, who exists solely within the temporal order; and in the consciousness of the reader, who is also made cognizant of the dimension of eternity. The final paradox is that 'linear historic time', so utterly repudiated through the poem's style, had to be admitted precisely because it is not a tyranny but a liberation. A Christian necessarily regards all time as sacrosanct, for the redemption of man was effected not outside history but within it, at the precisely datable 'historic time' that witnessed the life and death of the Christ. As Eliot was to say in *Four Quartets*, 'Only through time time is conquered' (*Burnt Norton*, l. 89).

II

We are all aware that the approach in *Four Quartets* is radically different from the 'continuous parallel between contemporaneity and antiquity' that is *The Waste Land*. Thematically, however, the Christian view of time and history was not only retained in *Four Quartets* but refined. Yet to insist on Eliot's Christian heritage is not to deny that he was exposed to other influences. The tendency to invoke Bradley and Bergson, for instance, is salutary, for not only were they crucial in the formative stages of Eliot's thought, but they assist us to define the nature of 'contemporaneity'. On the other hand it is difficult to credit that any single voice, however authoritative, had a decisive impact on Eliot, much less a lasting one.

Bradley's concept of 'immediate experience' apart,[9] his influence on Eliot remains diffuse, not precise—partly, no doubt, because Bradley is himself so elusive that, beyond the immediate context of his prose, he practically ceases to exist! Bergson's influence, which on the modern novel was extensive even if not exclusive,[10] was on Eliot entirely negative. Eliot had attended Bergson's lectures at the Sorbonne in 1911, but his initially enthusiastic response was short-lived. As he remarked on one occasion, 'My only conversion, by the deliberate influence of any individual, was a temporary conversion to Bergsonism.'[11] An appreciation for Bergson's revolutionary zeal did not finally coincide with an admiration for the ceaseless flux of his durational flow—*la durée réelle*—which denied the reality of clock time. For, 'Only through time time is conquered.'

The invocations of Bergson and Bradley are also salutary in that the endeavours of both form part of this century's obsession with time (see above, pp. 1 ff., 69 ff.). By the 1920s the pattern was obvious enough, its negative dimension best exemplified in Wyndham Lewis's zestful attack on 'time-mind'. Disconcerted because his immediate contemporaries had apparently conspired to create 'a sort of mystical time-cult', Lewis singled out Ezra Pound as the school's 'revolutionary simpleton'. Not to be outdone, Pound counter-attacked by describing Lewis as a mere 'chronological idiot'![12] The exchanges, for all the delight they generated, testify to the presence of sustained interests and even commitments. We expect Eliot to have been partial to Pound, and so he was; but it was not a partiality simply dependent on his respect for the recipient of *The Waste Land* ('il miglior fabbro'). It rested rather on Eliot's awareness of the manifold patterns which time had engraved on western thought, not exclusive of developments in science. Whether he actually understood Einstein is immaterial: suffice it that he appreciated the implications of the 'four-dimensional space-time continuum', witness in particular his translation for *The Criterion* in 1930 of Charles Mauron's essay 'On reading Einstein' (reprinted above, pp. 75 ff.). The essay constitutes evidence not of a novel but of a continuing interest on Eliot's part. Vital in this respect is Mauron's argument that Einstein's reorientation of space and time ('things-in-themselves') into one interdependent reality, has not annihilated mystical knowledge but fortified it in the sense that the elimination of every 'reality-in-itself' obliges us to revert to an awareness of our own lives in relation to something—or Someone—other than ourselves. We are far removed at this point from Bergsonism, which Mauron regarded with extreme scepticism; but we are well within sight of Eliot's conception of time which in *The Waste Land* is repudiated only to be reasserted in a new context, the context finally provided by *Four Quartets*.

Eliot's response to scientists like Einstein and philosophers like Bradley or Bergson, is commensurate to his response to his contemporaries in literature. In some instances, indeed, his attitude is plainly negative. D. H. Lawrence's commitment to 'the moment, the immediate present, the

Now', appeared almost calculated to demonstrate the polar opposite from Eliot's position which one might attain; and it may be that Eliot spontaneously sought to still Lawrence's turbulent metaphors ('The perfect rose is only a running flame, emerging and flowing off, and never in any sense at rest, static, finished' (above, p. 15)) when in *Four Quartets* the apocalyptic vision of 'the crowned knot of fire' terminates in the revelation that 'the fire and the rose are one'.

Yeats's attitude would have been judged equally restricted. *A Vision*, first published in 1925 and modified in 1937, is a reduction of select aspects of the Graeco-Roman conception of time and history to a personal and highly fanciful theory. Yeats's time-scheme is governed by the Platonic Great Year; and history, determined in some curious fashion by the Phases of the Moon, is conceived as a series of 'primary' conical spirals which on reaching their apex begin a downward movement intercepted by another rising ('antithetical') spiral. Each spiral, moreover, is divided into several wheels or 'gyres', and each gyre further subdivided into 'some two thousand odd years' that always end in chaos:

> Turning and turning in the widening gyre
> The falcon cannot hear the falconer;
> Things fall apart; the centre cannot hold:
> Mere anarchy is loosed upon the world . . . ['The Second Coming', ll. 1–4]

Eliot could not endorse a scheme which had so extensively mythologised reality. He would have been even less prepared to accept the sharp dichotomy once formulated by Yeats, that 'if it be true that God is a circle whose centre is everywhere, the saint goes to the centre, the poet and artist to the ring where everything comes round again'. We know Eliot's reply:

>   to apprehend
> The point of intersection of the timeless
> With time, is an occupation for the saint [*The Dry Salvages*, ll. 200–2]

—but a saint who is no less a part of the temporal order than are the poet and the artist. For, 'Only through time time is conquered.'

If the attitudes of Yeats and Lawrence were particular, what of Conrad's, Joyce's, Proust's? But in Conrad optimism is swallowed up by disillusionment since action terminates in the abyss of self-deception, and thence in death. The recurrent image of the river might have intimated the regular flow of history, but in Conrad it becomes a testimony to cosmic indifference. The river speaks 'always with the same voice as he runs from year to year, bringing fortune or disappointment, happiness or pain, upon the same varying but unchanged surface of glancing currents and swirling eddies'. Man is like the floating tree in *Almayer's Folly*:

The tree swung slowly round, amid the hiss and foam of the water, and soon getting free of the obstruction began to move down stream again, rolling slowly over, raising upwards a long, denuded branch, like a hand lifted in mute appeal to heaven.[14]

The burden of the argument here parallels the despondency that *The Waste Land* articulates in a fragment isolated from *Tristan und Isolde*: 'Oed' und leer das Meer.' But Eliot's pattern at that point is, like the fragment itself, incomplete. The quest continues.

Would Eliot have thought Joyce's attitude much more substantial? We know that he admired *Ulysses* as a technical achievement; but was he also favourably disposed toward its thematic patterns? Stephen Dedalus symbolises the restlessness that has come upon us, ever bent like the hawklike man whose name he bore to soar out of his Minoan captivity. 'History', Stephen remarks, 'is a nightmare from which I am trying to awake.'[15] Style and theme merge in Joyce to define a predicament that Eliot similarly suggested through the paralysed protagonist of *Gerontion*, lost in the labyrinthine mazes of the modern world:

> History has many cunning passages, contrived corridors
> And issues, deceives with whispering ambitions,
> Guides us by vanities . . . [ll. 35–7]

The sense of loss in *Gerontion* extends to the opening lines of *The Waste Land*, especially the phantasmagoric restlessness Eliot suggests through the brutal violation of the time sequence, the swiftly alternating tenses, the desperate accumulation of broken images. Memory is so impotent that it serves only to deepen the perturbation:

> [history] gives too late
> What's not believed in, or if still believed,
> In memory *only*, reconsidered passion . . . [*Gerontion*, ll. 40–2]

Proust's singular emphasis on intuitive memory is not Eliot's, unable as Eliot was to accept (as Beckett said in annotating Proust) that the past is 'agitated and multicoloured by the phenomena of its hours' while the future remains 'sluggish, pale and monochrome' (above, p. 2). To fragment time is to deny reality.

And so Eliot appears to linger among his contemporaries as amidst an alien people clutching their gods. Man seemed to him confined solely within the temporal order, like Prufrock measuring out life with coffee spoons, else seeking in the vulgar phrase of *The Waste Land* 'a good time' (l. 148). The immediate present is fruitless. As Celia remarks in *The Cocktail Party*,

> I abandoned the future before we began,
> And after that I lived in a present
> Where time was meaningless.[16]

To celebrate either past or future is equally as unproductive as to deny them. The past is not

> a mere sequence—
> Or even development: the latter a partial fallacy

> Encouraged by superficial notions of evolution,
> Which becomes, in the popular mind, a means of disowning the past.
>
> [*The Dry Salvages*, ll. 86–9]

Nor is the future promising; and therefore the large expectations of optimists ('April cold with dropping rain / Willows and lilacs brings again')[17] are firmly dismissed in the opening lines of *The Waste Land*:

> April is the cruellest month, breeding
> Lilacs out of the dead land . . .

In the end, Eliot's persuasion that

> the world moves
> In appetency, on its metalled ways
> Of time past and time future, [*Burnt Norton*, ll. 124–6]

obliged him first to abolish past, present and future as 'things-in-themselves', and then to reassert them within the pattern envisaged by St Augustine.

III

The Augustinian element in Eliot is in the first instance like the Bradleyan: diffuse, not precise. St Augustine exposed Eliot to a method of approach. Take Heraclitus. The ceaseless flux so fundamental to his vision of change and impermanence, is best symbolised by fire which is changeless in appearance yet ever-changing in fact. The thesis is adequately summarised in one of the two fragments quoted at the outset of *Four Quartets*:

> ὁδὸς ἄνω κάτω μία καὶ ὡυτή.
> The way up and down is one and the same.[18]

But Eliot did not simply christianise Heraclitus. True, the fragment just quoted is used to represent the continuity between the natural and the spiritual, between Nature and Grace; equally true, fire as an agent of the final purification is transmuted into a symbol of the ultimate permanence. But Eliot was primarily concerned to argue the contemporaneity of all thought by affixing it by way of typology to the concept of the Eternal Present (see above, pp. 5 ff.). Eliot's method is the general method of St Augustine—or, if we wish, of Hopkins in 'That nature is a Heraclitean fire and of the comfort of the resurrection.'

As with Heraclitus, so with Plato: Eliot's qualified response was dictated by a refusal to compromise his oft-quoted principle that 'Only through time time is conquered'. Plato's theory of the circularity of time was denied because it negated the irreversible and therefore unique nature of historical events. But to the extent that the theory is expounded within the mythological context of the *Timaeus*, Eliot responded all too favourably in that 'such experience' (as Harcourt Reilly observes in *The Cocktail Party*) 'can only be hinted at / In myths and images'.[19]

The circular movement ascribed to time by Plato as by Greek and Roman philosophers generally, Eliot found most categorically resisted by St Augustine who in *The City of God* had argued that history is progressively apocalyptic of the Divine Purpose in a linear pattern which, centred in the Christ, encloses all events from the creation to the Last Judgement (above, pp. 5 ff.). For Eliot, I believe, the Augustinian opposition to time's circularity alone provided him with a framework within which to deploy his consistently adverse images suggestive of recurrence and invariably linked to cyclical or spiral movement. To be 'whirled in a vortex' as the terrifying image in *East Coker* has it (1. 65), is to share Harry's despair in *The Family Reunion* that

> the things that are going to happen
> Have already happened

—which leads with mounting hysteria to:

> In and out, in an endless drift
> Of shrieking forms in a circular desert
> Weaving with contagion of putrescent embraces
> On dissolving bone . . .[20]

It is much the same with the first choric song in *The Rock*:

> O perpetual revolution of configured stars,
> O perpetual recurrence of determined seasons,
> O world of spring and autumn, birth and dying!
> The endless cycle of idea and action,
> Endless invention, endless experiment,
> Brings knowledge of motion, but not of stillness . . .

But the abrogation of meaning by the cyclical view of time is itself abrogated when the same chorus asserts at last:

> Then came, at a predetermined moment, a moment in time and of time,
> A moment not out of time, but in time, in what we call history; transecting, bisecting the world of time, a moment in time but not like a moment of time,
> A moment in time but time was made through that moment: for without the meaning there is no time, and that moment of time gave the meaning.
> [*The Rock* (1934), pp. 7, 50]

The advent of the Christ is unique and nonrecurrent like 'the one Annunciation' so pointedly stressed in *The Dry Salvages* (1. 84). It is also an event that occurred no less *sub specie aeternitatis* than *sub specie temporis*: 'in time and of time'. The Christian, Eliot argues, cannot deny the reality of time without denying the reality of the Christ. We live 'supra-historically in history', a modern theologian remarks. 'There is no way round the world, no way round history, but only a way through history.'[21] So, too, the four titles of the *Quartets* forcefully remind us that the Way is through particular places at particular times—'not out of time, but in time'.

Yet the acceptance in *Four Quartets* of the Christian view of time and history would appear to lack the conflict of the human situation in *The Waste Land*. As one reader observes, 'When the *Quartets* speak of the pattern of timeless moments, of the point of intersection, they speak *about* that pattern and that point; the true image of them is *The Waste Land*.[22] But as the approach is different, so is the nature of the conflict. In *Four Quartets* it is most in evidence where the rhythms echo the threnetic cadences of the Book of Ecclesiastes. This is not insignificant, for the author of Ecclesiastes alone among Biblical writers had departed from the commonly held views. His quest had led him to a vision not of any linear progress but of endlessly recurring cycles:

One generation passeth away, and another generation cometh: but the earth abideth for ever. The sun also riseth, and the sun goeth down, and hasteth to his place where he arose. The wind goeth toward the south, and turneth about unto the north; it whirleth about continually, and the wind returneth again according to his circuts. All the rivers run into the sea; yet the sea is not full: unto the place whence the rivers come, thither they return again. [Eccl. 1, 4–7]

The lines may remind us of *The Dry Salvages* where the river is gradually merged with the sea until both are said to move under the shadow of eternity:

> time not our time, rung by the unhurried
> Ground swell, a time
> Older than the time of chronometers . . . [ll. 36–8]

But I would argue that the appearance of the narrator's serene mood is repeatedly shattered by the reality of the counter-claims in Ecclesiastes. The nature of both appearance and reality is clearly defined at the very outset of *Four Quartets*, in the opening lines of *Burnt Norton*:

> Time present and time past
> Are both perhaps present in time future,
> And time future contained in time past.
> If all time is eternally present
> All time is unredeemable.

Here the assurance is nominal only, qualified by the hesitant word 'perhaps' as much as by the echoes of Ecclesiastes that later become explicit in the opening lines of *East Coker*:

> In my beginning is my end. In succession
> Houses rise and fall, crumble, are extended,
> Are removed, destroyed, restored, or in their place
> Is an open field, or a factory, or a by-pass . . .
> Houses live and die: there is a time for building
> And a time for living and for generation
> And a time for the wind to break the loosened pane
> And to shake the wainscot where the field-mouse trots
> And to shake the tattered arras woven with a silent motto.

We are intentionally put in mind of the celebrated cadences of parallelisms
in Ecclesiastes ('A time to be born, and a time to die; a time to plant, and
a time to pluck up that which is planted . . .'), for in Ecclesiastes as in
Four Quartets the cumulative repetitions suggest the despair of meaninglessness lest 'the things that are going to happen / Have already happened'.
But Eliot's narrator tries again:

> In my beginning is my end. Now the light falls
> Across the open field . . .

—yet no sooner does he think of festivities in celebration of wedlock than
the sombre cadences of Ecclesiastes again intervene to transform a musical
passage into a harshly arrested pattern of sounds:

> Keeping time,
> Keeping the rhythm in their dancing
> As in their living in the living seasons
> The time of the seasons and the constellations
> The time of milking and the time of harvest
> The time of the coupling of man and woman
> And that of beasts. Feet rising and falling.
> Eating and drinking. Dung and death. [*East Coker*, ll. 39–46]

The conclusion of *East Coker* does not hold much promise either, for the
apparently triumphant displacement of the first line ('In my beginning is
my end') by the last ('In my end is my beginning') tends not to negate
the cyclical reality but to reaffirm it. The quest will terminate only when
the sublunar cycles are subordinated to the apocalyptic vision of 'the still
point of the turning world'. Even here, one observes, Eliot refuses to deny
reality. Movement persists: 'The detail of the pattern is movement' (*Burnt
Norton*, l. 159). The 'still point' itself, is by no means still; for it is a dance
involving 'neither arrest nor movement', and a light which is 'a white
light still *and* moving' (*ibid.*, ll. 62–3, 72). In the end it emerges as the
metaphoric centre of a cosmic wheel which radiates outwardly into time
and history, and through the Christ connects the temporal world of becoming to the dimension beyond time and change.

The universe of *Four Quartets*, like the universe of *The City of God*, is
resolutely theo-centric. The Augustinian pattern transmuted into poetry
involves the progress of history toward the Last Judgement which the
fourth of the *Quartets* suggests through its perceptibly increased images of
fire. But the pattern involves other elements as well, notably St Augustine's influential approach to the problem of time in his *Confessions*.

St Augustine agreed with Plato and Plotinus that 'the world was made
with time, and not in time' (*The City of God* xi, 6, and *Conf.* xi, 30;
cf. *Timaeus* 38b, and *Enneads* iii, vii, 12). The agreement was intended
as an endorsement of the traditional differentiation between eternity which
abides, and the temporal world of flux where the heart of man—as Eliot
himself might have said—'flies about in the past and future motions of

created things, and is still unstable' (above, p. 5). But the agreement was also intended to lead to the conclusion that eternity is the state where 'the whole is simultaneously present' (*ibid.*). The Eternal Present in relation to the world of flux argues the omnipresence of the atemporal throughout the temporal order, the Dantesque 'ogni *ubi* ed ogni *quando*' (*Paradiso* xxix, 13), which Eliot asserts in terms of 'Eternity crossing the current of time':

> In every moment of time you live where two worlds cross,
> In every moment you live at a point of intersection.
>> [*The Rock* (1934), pp. 52–3]

In *Four Quartets* this ever-present 'point of intersection'—indirectly specified in the lines just quoted as the Cross—appears in Eliot's simultaneous insistence that history is 'now and England', yet 'not in time's covenant'. It appears also in the paradoxical statement that history is a pattern of 'timeless moments'. And it appears at last in the audacious lines:

> the intersection of the timeless moment
> Is England and nowhere. Never and always.
>> [*Little Gidding*, ll. 14, 52–3, 234–7]

A union utterly impossible on the human level is finally said to have been effected, as a gift:

> the gift half understood, is Incarnation.
> Here the impossible union
> Of spheres of existence is actual,
> Here the past and future
> Are conquered, and reconciled. [*The Dry Salvages*, ll. 115–19]

Reconciled *and* conquered: for, 'Only through time time is conquered.' Under the shadow of the Eternal present, moreover, the fragment from Heraclitus quoted earlier (p. 165) is recast to suggest the Augustinian pattern of the confluence of eternity and time: 'the way up is the way down, the way forward is the way back' (*The Dry Salvages*, l. 129). The Incarnation is datable in history; but it is also an experience in 'the immediate present'. To this extent it might be argued that the Incarnation is suggested even throughout *The Waste Land*.[23] It depends largely on the individual's response to the epiphany of 'the loop in time' which, Agatha remarks in *The Family Reunion*, 'does not come for everybody'.[24] But perhaps we should not underestimate Grace, especially the anticipatory Grace of 'the absolute paternal care / That will not leave us, but prevents us everywhere' (*East Coker*, ll. 160–1).

IV

St Augustine as we have seen formulated the definitive view of history's Christocentric movement in time to the end of time. But in arguing additionally that past, present and future are our dividual experiences reflecting their concurrent presence in God, he sought the meaning of reality in terms of God acting not only within history but in the mind of man (above, p. 5). The scheme accommodates neither Proust's singular emphasis on the past, nor Lawrence's on the immediate present; and it precludes also bright dreams of a resplendent time to come, as well as haunted nightmares of a future 'sluggish, pale and monochrome'. The developments Eliot discerned in the 'four-dimensional space-time continuum' not only of science but of modern art, and the tendencies he observed among his time-conscious contemporaries in literature, greatly contributed to the realisation of a style which in shattering traditional perspective asserted through a dislocation of time the modern predicament. But the syncopated style that testified to 'the anguish of the marrow' was in the end 'renewed, transfigured, in another pattern'—the pattern of timeless time articulated in *Four Quartets* under the shadow of St Augustine.

Yet our difficulties abide. Failing to comprehend how 'all things are bigge with jest', we do not respond to the 'comic' at the heart of Eliot's poetry. We respond even less to the opposition in *Ash Wednesday* between 'the unstilled world still whirled' and 'the silent Word' (v, 8–9), for the one argues reality, the other a mystery. We agree with Edward in *The Cocktail Party*: 'Nobody likes to be left with a mystery: / It's so . . . . unfinished.'[25] Yet that is precisely where Eliot like Augustine leaves us, with a mystery.

And unfinished at that.

NOTES

[1] *Seriatim*: Eliot, *The Waste Land: A Facsimile and Transcript of the Original Drafts*, ed. Valerie Eliot (1971), p. xxxiii; Yeats, in *T. S. Eliot: A Selected Critique*, ed. Leonard Unger (1948), p. 287; and Lowell, quoted by Jay Martin in his introduction to *The Waste Land: A Collection of Critical Essays* (Englewood Cliffs, N.J., 1968), p. 5.

[2] Davie, 'T. S. Eliot: the end of an era', in *T. S. Eliot: A Collection of Essays*, ed. Hugh Kenner (Englewood Cliffs, N.J., 1962), p. 192; and Shapiro, in *T. S. Eliot: 'Four Quartets'*, ed. Bernard Bergonzi (1969), p. 246.

[3] *Collected Plays* (1962), p. 108.

[4] *The Journals of Søren Kierkegaard*, ed. and trans. Alexander Dru (1938), p. 44.

[5] Herbert, 'The church porch', l. 239; and Sterry, *The Rise, Race, and Royalty of the Kingdom of God* (1683), p. 354.

[6] *Seriatim*: Britten, in an interview in *The Guardian*, 7 June 1971; Yeats, 'Lapis lazuli', ll. 9–20; and Joyce, *The Critical Writings*, ed. E. Mason and R. Ellmann (1959), p. 144.

[7] '*Ulysses*, order and myth' (1923), in *The Modern Tradition*, ed. R. Ellmann and C. Feidelson (1965), p. 681.

[8] Helen Williams, *T. S. Eliot: 'The Waste Land'* (1968), p. 20.

[9] See Hugh Kenner, 'Bradley', in his anthology (as above, note 2), pp. 36–57; also Eric Thompson, *T. S. Eliot: The Metaphysical Perspective* (Carbondale, Ill., 1963). Unduly inflated claims include Anne C. Bolgan's that 'it is Bradley's mind that lies behind the structuring principles of Eliot's poetry' (in *English Literature and British Philosophy*, ed. S. P. Rosenbaum (Chicago, Ill., 1971), chapter xii).

[10] The exclusiveness of this influence is forcefully argued by Kumar (see Bibl. 464). But I would not care to insist that any artist responds to one influence only.

[11] *A Sermon preached in Magdalene College Chapel* (Cambridge, 1948), p. 5.

12 On Lewis, see Bibl. *467*; and on his exchanges with Pound: *496*.

13 'Discoveries', in *Essays and Introductions* (1961), p. 287.

14 *Complete Works*, Kent ed. (1926), xi, 4; the first quotation is from xi, 162. Both passages are discussed within a comprehensive context by W. B. Stein (Bibl. *528*).

15 *Ulysses* (1960), p. 42.

16 *Collected Plays* (1962), p. 151.

17 Emerson, 'May-Day'; quoted by J. Clendenning (Bibl. *409*).

18 Fr. 60; in *Heraclitus: The Cosmic Fragments*, ed. and trans. G. S. Kirk (Cambridge, 1954), p. 105.

19 *Collected Plays* (1962), p. 210.

20 *Ibid.*, p. 89.

21 Karl Jaspers, *The Origin and Goal of History*, translated by Michael Bullock (1953), p. 275.

22 Frank Kermode, 'A Babylonish dialect', in *T. S. Eliot: The Man and his Work*, ed. Allen Tate (Penguin ed., 1971), p. 243.

23 As Florence Jones persuasively argues in 'T. S. Eliot among the prophets', *AL*, xxxviii (1966–7), especially pp. 285 ff.

24 *Collected Plays* (1962), p. 60.

25 *Ibid.*, p. 134.

Fritz Kaufmann†

Thomas Mann: the degradation and rehabilitation of time

THE DEGRADATION OF TIME

In Thomas Mann's early writings

It is the poet's peculiar gift to create symbols which show the universal in the individual and the essence of things in their impact on man. He brings to a focus the elemental and historical forces dispersed in time and space. Hence every view of things must be congenial to him which overcomes this disperson in some such way as mythical participation does. Kant and Schopenhauer had taught that spatio-temporal order is appearance and not ultimate reality, a doctrine confirmed by Thomas Mann's early feelings and impressions. Schopenhauer and Nietzsche, moreover, reverting to the Greek example, had changed the symbol of time from a straight line to a circle, the figure of eternity always turning back upon itself. In this way, the coincidence of the time-pattern as such with the pattern of its contents was brought to light. For it was precisely the recurrence of content that first suggested the circle as an adequate symbol of time.

The fading of our consciousness of time and space had been depicted by Thomas Mann in the traveller of his short story 'The wardrobe', as early as 1899. This Albrecht van der Qualen got off the same train (did he not?) that brought the young Thomas Mann to Italy. He lives, he really does not know where, for 'everything must be up in the air'. Actually he lives in the world of the story. 'He had no watch. . . . He did not like to be reminded of the hour, or the day of the week, and even with the calendar he had no truck. Some time ago he had dispensed with the habit of knowing the day of the month, the month of the year, or even the year itself.'

In *The Magic Mountain*

The last but one stage of life on the magic mountain is precisely a metaphysical dream-state in which Hans Castorp (not so young any more) goes

---

† Reprinted from *Thomas Mann: The World as Will and Representation* (Boston, Mass., 1957), pp. 125–32, which provides detailed references not shown here.

farther and farther astray from the order of normal life. 'He no longer carried a timepiece. His watch had fallen from his night-table; it did not run any more, and he had not bothered to have its measured rotation restored—for the same reason that he had long since given up using a calendar, whether to keep track of what day it was or to foresee a coming holiday—namely, for the sake of his "freedom," his strolling on the strand of eternity abiding now and forevermore. . . .'

Taken as a whole, *The Magic Mountain* is certainly a 'time novel', above all because time is its theme, the object of a thoroughgoing experiment and the victim of a shrewdly conceived plan of annihilation. Time's claim to being self-subsistent is challenged; its structure is shown to depend upon our way of dealing with it; thus, its contraction or distension is a complicated function of human concentration or distraction.

As a bourgeois, young Castorp had been familiar with time in the form of a straight line, a course no longer directed by divine providence but laid out in the plains of human life; the way of a finite being striving to obtain the means for certain temporal and always preliminary ends. To such a being time is precious, a commodity not to be wasted or squandered. Even up in the mountains time appears, at first, as weighty and fraught with interest. There are many new things to observe, the telling of which, it seems, will never end. But then, all at once, time begins to shrink and grow empty, thanks to a technique of living that makes boredom and diversion 'hard to distinguish sometimes'. For 'What we call boredom is rather a pathological waste of time resulting from monotony' and the tiresome routine of existence.

In the sanatorium pastime serves to kill time. The medical regime, on the other hand, articulates the day and cuts it up into small pieces, with the result that every day is of the same pattern as every other. Time thus loses more and more of its inner distinctions, as the individuality of its component elements is blurred. While the minutes are counted, for example, when taking one's temperature, time as such no longer counts. It is spent ever more carelessly. Life's progress is suspended in the circle of repetitions described by the hands of the clock. This repetition is no longer *experienced*. It lapses into mechanical, 'outer recurrence'. With all this monotony, such scrupulous care is taken to keep people occupied that this pseudo-life comes to absorb all their attention. At the beginning, the noisy machinery of life in the plains, protesting against the lofty existence of the 'mountaineers', had still sounded, though faintly and afar off; but now at last it has died away. To go back once more and serve in everyday life, one must 'desert to the colours'.

To all appearances, life has become carefree, because every other care is swallowed up by caring for one's bodily well-being, whose real precariousness is never admitted. All this would be of no more than pathological interest had not, in Hans Castorp's case, the whole process of loosening life's bonds led to a true 'ecstasy' and separation from the world. Obsessed by '*le goût de l'éternel*', under the growing spell of 'hermetic magic', he sees opening up to him a medium of life both new and deeply familiar, the medium of an aimless but strangely significant wandering:

> Du zählst nicht mehr, berechnest keine Zeit,
> Und jeder Schritt ist Unermesslichkeit.

> 'Time does not count, nor dost thou count the tide.
> Immeasurable is thine every stride.' [Goethe, 'Prooemion']

Lost in the sand dunes, the damp, salty taste from the sea of eternity on his lips, Hans Castorp is completely sidetracked from the normal march of time.

Time varies with and is determined by the time sense, while the time sense varies with and is determined by the sense of life. The very contents of time gain their temporal structure in accordance with one's basic outlook. In like manner the *nunc stans* of eternity is a mode of our experiencing things, of our taking them under a definite aspect, *sub specie aeternitatis*. In accordance with this mode, 'the eternal lights begin to sparkle' when 'the din of the day's battle has died away'. Out of enlightenment and intoxication grows Hans Castorp's sensual-supersensual vision of life; out of the polarities of existence grows his vision of man as a whole, man as both the source and the master of the oppositions between sense and the senses, liberty and piety, intellect and spirit.

But the growing breadth and depth of Hans Castorp's understanding remain within the sphere of his inner life and the 'pedagogical province' of *The Magic Mountain.* Neither his primarily receptive nature, nor his scientific studies and metaphysical experiences, nor the milieu of the sanatorium can supply that power of initiative which converts an idea into action and reality. Almost to the end, until the thunderbolt of the first world war awakens him from his mystic trance, the world remains for him a matter of 'intellectual representation' rather than a challenge to his responsible will. The Platonic man, having found his way out of the cave and into the light of day, goes back to share the discovery with his brethren. But the metaphysical dreamer is prone to continue his stroll along eternity's shore. Hans Castorp 'makes bold to take shadows for real things and things for shadows'.

This 'simple youth' turns into a kind of legendary figure, inhabiting a world of fairy-tale arithmetic and fairy-tale space-time. Like the Seven Sleepers (Hans Castorp spends seven years on the magic mountain), he forgets about actual time, and so do we, as we become immersed in the legend. For the world and time of the story as such are unreal—an imaginary world and an imaginary time, conjured up by the story-teller, 'the rounding conjurer of times gone by'. The story draws forth ideas and elements that were scattered in the past, and welds them into the immediate presence of a dream. 'De-realisation', removal from reality, and realisation of a new and wondrous order of life are here involved, both in the form of narration as such and as one of the themes in this particular story. It is a form of serious play, a revision of ancient wisdom in the light of modern analysis.

The metamorphosis of time, so characteristic a device of Thomas Mann's generation, occurs throughout his work. With him it is an evaluation of his philosophical heritage and leads into the myth. The magic of *Death in Venice*, for instance, lay precisely in its power of mythical transformation, a kind of holy distortion of the world in which we normally live. So it is in *The Magic Mountian*. What began apparently as a psychological time novel can pass almost imperceptibly into the realm of the mythical and mythological.

In *The Magic Mountain* the mythical time-form and the mythological categories of time transfigure first of all the recent past. In the medium of Hans Castorp's dream-life his outer world recedes from the daylight of practical consciousness into the twilight of the myth. In addition to this, the mythical epoch as such is already taken into consideration—experimentally, so to speak. The chapter 'By the ocean of time' contrasts two viewpoints, the microscopic view of the insect and the 'bird's-eye view that borders on the divine, where a thousand years are as a day':

It would not be hard to imagine creatures, perhaps living on one of the smaller plants, whose time-economy is a miniature of ours, creatures for whose abridged existence the brisk, tripping gait of our second-hand would have the sluggish, thrifty proceeding of the hand that marks the human hours. Contrariwise, one may conceive of people in a world so spacious that its time-system, too, has a vast, majestic stride, and the ordinary distinctions between 'just now' and 'in a little while', 'yesterday' and 'tomorrow', cover an immensely wider range.

In *Joseph and His Brothers*
One sees how a wide time span is associated in consciousness with a spatial span of comparable breadth. In latter-day 'little Europe', with its delicate organisation, its finely meshed patchwork of countries, and its peoples crowded together, nobody ever has any time, whereas the 'children of the East' can afford to be 'barbarically grandiose in their time-consumption'. In the early days of life, under the wide-arching eastern skies, mankind had a fabulous amount of time; hence the fable-telling poet loves those leisured ages and spacious lands. He is glad to linger with the men from the edge of the Arabian desert, to bear them company on the long, weary journey over the gleaming sands, the well-nigh endless journey through the almost endless space that lies between the patriarchs and their primeval world, and the highly sophisticated, 'contemporary' world of late Egyptian culture. To the impatient modern reader the account of this journey may seem a bit longwinded; but the author's very aim was to give his story the length and breadth of those periods that dispose of infinite time with a simple but imposing gesture. The poet's technique consists in the reproduction of that time-feeling that is akin to timelessness. And Thomas Mann's true reader is the one who freely yields to eternity's spell.

' "At the same time"—that is the very nature and essence of all things; they appear in disguise, one in the mask of the other; yet the beggar is nonetheless a beggar for the fact that a god may possibly be dissembled in him.' In those far-off times, 'there was no call to decide between being and significance.' When identity of being is determined by identity of meaning and function, the various distinctions introduced by time as *principium individuationis* fall by the wayside.

Thus, Abraham, ostensibly Joseph's great-grandfather, is in reality a multiple appearance who wanders through time, typifying the homeless man of God. Thus every generation is again given the same wise servant Eliezer; and thus Jacob's wicked brother, Esau the Red, is to be identified not only with Edom, his people (even as Jacob–Israel is one with his people); in addition he always reappears as foil or counterpart to the protagonists. He was Cain to Abel, Ishmael to Abraham and Isaac; he was Seth–Typhon, the murderer of Osiris; he will be Judas to Christ, and Hagen to Siegfried, he is Moloch, and in the starry heavens he is blood-red Mars, the fire-planet; he prefigures Goethe's Mephistopheles; he is the serpent, standing for Evil as such, not necessarily evil in his individual nature, but the Evil in the universal function assigned him as his role. Even today we use the singular exclusively in referring to such natural elements as the air and water that we breathe and drink, or the demon fire haunting this place and that. Just so, the old but ever new stories know nothing of individuals, but only of powers and potencies in different manifestations, powers that stage the same play over and over again.

In this way, the repetition of beings and events is clearly more than the mechnical swing of nature's pendulum. Certainly the idea of this recurrence is not wholly foreign to Thomas Mann. It is part and parcel of his creaturely feeling, as well as of his inheritance from Schopenhauer and Nietzsche, and appears as a structural element as early as in *Buddenbrooks*. This work begins (to mention but one or two instances) with the banquet in the spacious old house on the Mengstrasse, which the Buddenbrooks have just acquired from the Ratenkamps, that 'family, at one time so prosperous, who had built and lived in the house until a fatal decline and impoverishment forced them to move away'; and the book ends, again 'in order that fate may be fulfilled', after the sale of the house to the *nouveaux riches* Hagenströms, with Gerda Buddenbrook moving to Amsterdam: 'she took nothing with her and left as she had come.' In like manner, the precarious relationship between the brothers Johannes and Gotthold is to a certain degree recapitulated in that between Thomas and Christian: but the whole treatment is so objective and delicate that only very slowly does it dawn upon the reader that he is witnessing the same drama with a different cast of characters and in different settings of political economy. In both *Tonio Kröger* and *The Magic Mountain* an additional step is taken. Here the *hero* himself senses the typical in all appearances; he feels a more or less extensive identity between seemingly different persons of his world. But not until the Joseph saga are the figures themselves aware of their own identity with others who have played the same role, and of the obligations thus incurred.

THE REHABILITATION OF TIME

But is it not obvious that the character and the phenomena of life do not remain the same when coloured by however secret an awareness of recurrence and contrived by however dark a will of repetition? Following Schopenhauer, Thomas Mann insists, as late as 1943, in *Joseph the Provider*, on the present as being the very nature of life and on presentation as the very work and offering of art. The tenses of the past and the future are said to be only the mythical and even popular appearances of life. Now it is quite true that the living past and the living future are the depths which belong to the fullness of the present moment and into which this present (a *praesens de praeteritis*, a *praesens de praesentibus* and a *praesens de futuris*) extends. The development of the Joseph stories into the activism of *Joseph the Provider* belies, however, a ruling out of time in which time as passage and, therefore, the preciousness of the passing moment are definitely denied. What Thomas Buddenbrook once experienced is still true in the Joseph legends, in the dim realm of mythical consciousness—that in reality 'nothing began and nothing ceased'—but it is no longer the entire truth. If one wants to cling with Thomas Mann's Goethe to the circle as an image of time, the circle will be, at least, the symbol not of changeless eternity but of a permanence which is, at the same time, 'restless intensification, enhancement and perfection'. This would account for the time which yields results, though not fully for the stretch of time within which the works of the day have to be done.

In fact, *Joseph and His Brothers* shows just this duration in change and change in duration. While 'the well of the past' is certainly 'unfathomable' and some of the typical roles of the figures of the play, as for instance that of 'the Red One', go back to times immemorial, some others are newly created either within history or at what Japsers calls the 'axis time of world history', when the foundations of our modern world were laid. (Thomas Mann's Abraham is one of the founders of this new type of man.) And, as we mentioned before, they are all in a process of perpetual recreation and more or less free personal renewal. Joseph *is* what he represents ('I am it') and he is it *not* because he is it in his own individual way:

the general and the typical form vary when they find their fulfilment in the particular. . . . This is the very nature of civilised life that the taking and binding ground and the outlines of a classical pattern are filled out in the God-given freedom of the I; there is no human civilisation without both, the one as well as the other.

This is the insight Thomas Mann lends his hero; he has it 'since "he" came to years' and grew beyond the raptures and twitches of his youth;

secretly it is a personal confession of the author himself. The principle of individuation is still in disrepute. But the historical person represents more than the single individual is. We are our history in response to our personal and historical calling. History and the myth do not exclude one another. History passes into the myth, and the myth provides the ground patterns which history moulds in their personal appropriation. Understood in this sense, the myth bridges the gap between universal and individual life and between eternity and time. The plasticity of the mythical patterns does away with the rigidity of the *nunc stans*, the timeless now, and accounts for the enhancement of the moment in its both eternal and unique content and significance.

Margaret Church†   FRANZ KAFKA (1883–1924)

Time and reality
in Kafka's
The Trial and *The Castle*

I

Kafka once wrote in an aphorism that one of his most important wishes was 'to attain a view of life in which life, while still retaining its natural full-bodied rise and fall, would simultaneously be recognised no less clearly as a nothing, a dream, a dim hovering'.[1] This remark describes with some accuracy the style and mood of his two central works, *The Trial* and *The Castle*, and shows that it was Kafka's aim to employ in his fiction the idea that time and space are illusory.

The dreamlike quality of time values and the assumption of an interior time recognised alone by the officials and K. appear throughout *The Trial*. In a passage deleted from the first chapter, Kafka had written that the riskiest moment of the day is the moment when one awakes. 'For when asleep and dreaming you are, apparently at least, in an essentially different state from that of wakefulness.'[2] Because K. this morning has found his world different from the way it was the evening before, we understand that part of the dream world has intruded into his everyday world. The opening of 'The metamorphosis' may be compared with that of *The Trial* where Kafka writes that it takes great vigilance to see things in the same place that they were in the evening before.

On his first Sunday in court, K. hurries to arrive at nine o'clock 'although he had not even been required to appear at any specified time'.[3] Despite the fact that he is late he walks more slowly as he approaches the house of the examiners, as if now he had abundant time. If anything, Kafka is more adept at creating the dream than the 'full-bodied rise and fall' of life. When K. leaves the examining room, the magistrate mysteriously gets to the door before him as in a dream people appear at the beck and call of our fears and wishes.

In the unfinished chapter 'The house' we find the curious juxtaposition of dream upon dream. As K. lies down on the couch in his office, his thoughts

† Reprinted from *Twentieth Century Literature*, ii (1956), pp. 62–9.

hover between dream and reality, only here reality is that of K.'s waking life which is often like a dream to the reader. Thus Kafka makes us aware of various levels of reality—the dream within the dream. K.'s first dream represents his alienated situation as he views Frau Grubach's boarders, many unknown to him, for he had for some time not bothered himself about concerns of the house. Then as he turns from the group and hurries into the law court, corridors and rooms become 'familiar to him as if he had always lived there'.[4] As K. becomes more deeply implicated in the court, the details of living lose for him their significance, the dream becomes more like the inner dream.

In connection with the dream it should be noted that K. is often 'in the dark'. Heavy curtains hang over the windows in the advocates' bedroom; in the cathedral K. by mistake extinguishes his lamp and 'He stood still. It was quite dark and he had no idea which part of the church he was in.'[5] In this dream world one loses one's bearings; and since K. is lost inwardly, his physical relation to objects and places is an uncertain one, too.

When the student enters the examining room where K. stands alone with the woman who occupies the apartment outside, K. experiences his first meeting with a representative of the official group on human terms as a rival.[6] This meeting implies the recognition that the trial is on a different level from 'the full-bodied rise and fall'. Nevertheless it is interesting to note that the meeting takes place in the same examining room where K. had had his first hearing. Kafka thus creates a link between the two worlds (inner and outer), a link which gives artistic unity to the passage.

A scene in the lumber room[7] in the bank leads to further insights into the time experience in *The Trial*. When K. returns to the lumber room on the second evening, he finds everything exactly as he had left it the night before. The whipper is still standing in the same position in front of the warders. As K. opens the door, the warders at once cry out, 'Sir!' Time has not moved on this level of experience although K. has lived through a whole day of clock time. K. deals with this situation in the realm of action by asking the clerks to clear out the lumber room the next day although unconsciously he recognises that his experience is an inner one, for he would not ask them to do this if he thought that the whipper and the warders were there for the clerks to see. It is K.'s fault that the warders are being whipped, thus the scene represents hidden guilt. He asks the clerk to clear it out knowing that he cannot remove the imprint of the scene from his mind other than by the destruction of its outward symbols. Time has stood still in this back room of K.'s consciousness, a trick made possible by Kafka's concept of the idea of time as reality.

The appearance of K.'s uncle and the mention of his daughter, Erna, is one of the few insights we have into K.'s past. K.'s uncle understands, without being told, the facts of K.'s case. K. is aware that he has known all along that his uncle would turn up, for the uncle, like the rest of the characters in the book, has reality only in relation to K.'s inner life. As a moulder of K.'s past, the uncle, too, like the family of Amalia in *The Castle*, is implicated. The uncle is part of the everlasting present of K.'s mind time, neither past nor present having reality except as they are viewed by K. The Platonic character of Kafka's idea of time is clear when we observe that K. (as the initial suggests) is a symbol, not an individual, so we are dealing here not with a specific relationship of past and present but with a general one.

In the uncompleted chapter 'Journey to his mother' we find the same general relation between past and present. K.'s mother is almost blind, so unlike the uncle, she is ignorant of K.'s plight. Her refusal to be implicated in K.'s problem is further shown by her present indifference to K.'s visit, for earlier she had been anxious to see him. The mother, like the uncle, is part of K.'s mind, but the blind part, that which is suppressed: 'she believed him to be the Bank Manager and had done so for years.'[8] In another unfinished chapter 'Prosecuting Counsel' K. attributes to the early death of his father and the mistaken tenderness of his mother a childish quality he possesses.[9] Thus despite her 'blindness' the mother is implicated in K.'s fate. But the conscious recognition of his mother remains in the background; for several years he had intended to visit her, but he had never done so, and the fragment ends before the visit is made.

That the characters are projections of K.'s mind appears again in his interview with the advocate who at once knows all about K.'s case although, as K. reflects, this advocate is attached to the court at the Palace of Justice, not to the one with the skylight. As he ponders this incongruity the Chief Clerk of the Court (the one with the skylight) appears in a corner of the room where K. had not noticed him. The link between the two courts is thus inwardly established for K. The interview progresses and the advocate asks K. no questions; he either talks of his own affairs or strokes his beard. K. is, Kafka shows, his own advocate and as such the facts are known to him. K. learns that since the proceedings are not public, legal records are inaccessible to the accused and to his counsel—records of earlier acts which in life are often inaccessible because pressed into the unconscious. That this unconscious level is unreasonable and primitive is seen in Huld's remark that the officials are children.[10] The court and its officials exist in every life, in every time and place. And K. comments that 'so many people seem to be connected with the court'.[11] *The Trial* represents man's self-trial to determine his success or failure in the pursuit of an inner ideal. To claim, as critics have, that Kafka's books represent a specific theology, psychology, or philosophy seems to me to miss the point of Kafka's writing which was to embrace all quests without pointing to any one as *the way*. The search for and following of an inner ideal is an old theme in literature put into words by innumerable writers, but Kafka's distinction seems to lie not in his theme but in his technique which depends to a large extent on his abrogation of the time values of the outer world so that his odyssey is described in terms of the inner world where in the final analysis all our odysseys take place. 'You see, everything belongs to the Court',[12] the painter tells K.—even the girls on the stairs outside the painter's room. When Titorelli opens the door behind his bed, K. recognises the same Law-Court offices even though the painter lives in a different part of the city. 'There are Law-Court offices in almost every attic,' Titorelli explains. 'Why should this be an exception?'[13] And when Huld reflects that 'after a certain stage in one's practice nothing new ever happens', he is expressing in different terms the universal nature of the human quest.

The scene in the cathedral should, therefore, not be interpreted to mean that the end of the quest is to be found in orthodoxy of any kind. Rather the cathedral symbolises an inner spiritual goal which has no relation for Kafka to the cathedral as such.

As K. nears the end of his quest in the Cathedral Square, he is startled by the recollection that even when he was a child the curtains in this square had been pulled down. Inside the cathedral he watches the verger, whose limp reminds him of his childhood imitations of a man riding horseback. These two simple memories serve as touchstones of the world of objects and of the 'full-bodied rise and fall' of life. As links their existence in the passage is important, for through them Kafka reminds us that his purpose is to mirror life, but a life disguised so that it is in the semblance of all lives.

Clemens Heselhaus sums up the question of reality in *The Trial* by pointing out that the court itself is not real; only the reactions of K. to this unreality are real. Because the court is unreal all suppositions are possible, but only as suppositions, not as fixed truths. One cannot say that the court means this or that. One can only say that the physical realisation of the court is made concrete in the physical reactions and deeds which destroy a life.[14]

II

Kafka's extraordinary use of symbol, dream and parable[15] reaches its culmination in *The Castle*. Günther Anders writes that the strange element in K.'s experience is not that so much strange happens, but that nothing that happens, even the self-evident, is self-evident. There is no distinction between the ordinary and the extraordinary.[16] The reader fills in the emotional content, the philosophical content, directed by symbols like the beards of the assistants or the soft luxuriousness of the sleigh cushions. Normal space and time values are abrogated so that reality is that which exists within the mind, not independent of it.

When K. returns to the inn early in chapter 2 of *The Castle*, he is surprised to see that darkness has set in. 'Had he been gone for such a long time? Surely not for more than an hour or two, by his reckoning. And it had been morning when he left.'[17] As in Kafka's short story 'A common confusion', the length of the trip does not determine the time it takes. Kafka does not write: 'K'.s trip seemed to take a whole day.' Rather despite all of

K.'s outward reckonings, the inner time of the subconscious mind prevails, and it is *actually* dark when K. reaches the inn. *The Castle* is related in terms of the primitive, unreasoned drives and evaluations of our unconscious lives which for Kafka are more real than what appears on the surface as distorted reflections of these lives. Barnabas' speed in outstripping K. is so great that before K. can shout to him he has covered an impossible distance. Thus time again is observed through an unconscious estimate of it, and Barnabas is characterised in terms of a speed experience.

Another insight into Kafka's approach to time comes when dragged on by Barnabas, K. recreates a scene from childhood evoked by the difficulty of 'keeping up'. He finds himself by an old church in a market-place surrounded by a graveyard, in turn surrounded by a high wall. K. had failed to climb the wall until one morning in an empty market-place, flooded by sunlight, he had succeeded. The sense of triumph of that moment returns now to succour him. Evoked by a chance experience, the past becomes present.

One notices that throughout *The Castle* Kafka avoids measuring time.[18] For instance, when K. goes to see the superintendent in chapter 5 there is no mention of how many days or hours later this visit occurred after he left Frieda and the landlady. Thus the reader is shocked to learn from Pepi at the end of the book that only four days have elapsed since Frieda left her work at the bar. It is, of course, part of Kafka's technique to reveal this only at the end where it does not distort his time values which are not of calendar or clock. That the inner time of the mind prevails in the book is suddenly proved by Pepi's remark which is incredible except on the level of idea. Earlier in the book to learn the day would have only oriented us to conventional time values and spoiled the effect of the allegory. But now that hour and day have ceased to have meaning, to be reminded of them produces in the reader the surprise that Kafka wishes to induce so that they suddenly seem much more unreal than the flow of mind time in which the reader is immersed.

Telephone calls to the castle are of no avail, for the superintendent tells K. that all K.'s contacts with the castle have been illusory, 'but owing to your ignorance of the circumstances you take them to be real'.[19] All outside contact is illusion. K. mistakenly tries to use human logic and reason in dealing with the castle and its officials; therefore, he and the officials never talk on the same level, for their reasoning is incomprehensible to the human mind. That we interpret our dieties in human terms, however, is shown by Kafka when, for instance, Momus, the secretary, crumbles salt and carroway seeds on his paper.

Reality in the village is what the people make it. Thus Klamm's appearance fluctuates. He looks one way in the village and another way on leaving it. He looks different when he is awake from the way he looks when he is asleep. On one point only all the villagers agree—he wears always a black morning coat with long tails. The differences, Kafka explains, are the result of the mood of the observer—of his degree of excitement, hope, or despair. They are the varied impressions that the supplicant holds of the features of the oracle, the confessed of his confessor, or the patient of his psychoanalyst. The people's confusion of Momus and Klamm and Barnabas' doubts about the real Klamm are also explained by Kafka's concept of reality. Likewise, in a passage deleted by Kafka, K. feels as if Barnabas is two men whom only K., not outside judgement, can keep distinct. Barnabas, the messenger, and Barnabas, the brother and son, do not, therefore, ever really merge for the reader but remain, as for K., different, one of the castle, the other of the village. This points to the real nature of the Barnabas symbol—the man divided by having only partially attained his goal. Reality depends then on the observer, not on a set of unchanging values. Felix Weltsch sees in these double beings a comic element, 'a duality which is recognised as unity and a unity which is always falling apart into duality'.[20]

The castle dignitaries have the distinction of being freed from memory. Although K. challenges the landlady's remark about Klamm's memory as 'improbable and indemonstrable', we are told by Kafka that anyone whom Klamm 'stops summoning he has forgotten completely, not only as far as the past is concerned, but literally for the future as well'.[21] K. himself has practically no past; we hear hardly anything of earlier events in his life. The other characters as well are without childhood or ancestors. True, Frieda claims a childhood acquaintance with Jeremiah with whom she played on the slope of the castle hill, and K. accuses Frieda of having succumbed to the influence of memories, the past, in her 'actual present-day life',[22] but for all practical purposes there is no distant past in *The Castle*. With the exception of the story Olga tells K. or the hints of the landlady's affair with Klamm, there is little perspective in even the recent past. The larger racial past of the human species is, however, often implied in the allegory, for the subconscious level of the mind is, of course, much concerned with our primitive origins. Thus one sees, for example, in the connection of the villagers and K. to the castle the bafflement of man in relation to forces of nature and in relation to deity. Kafka, however, seems to imply that too much concern with the immediate and individual past clutters the mind, for the officials have no memory. It is well to note, nevertheless, that it is only the officials who lack memory of a dismissed case.[23] The villagers and K. do have memory, though it is little exercised because *The Castle* is written in the realm of dream where the past is disguised and integrated with the present.

Kafka probes deep by placing his entire story on the unconscious level. For Kafka past and present are bound together indistinguishably in the symbols and shadows of his dream world.[24] As in a dream, all that goes on is known at once by everyone in the village; for instance, the maids enter the room to move in with all their clothes hardly after K. has spoken the words accepting the post at the school. The landlady is aware of all that happens to K. as is everyone K. meets. This disconcerting state of affairs is further evidence of the dream atmosphere of the book, for in a dream our enemies and friends alike know with unerring certitude all the hidden embarrassments and decisions of our lives.[25]

As the culmination of Kakfa's work, *The Castle* depicts general themes: the alienation of man, the incomprehensibility of the divine, the quest of the hero for the fulfilment of an ideal. Behind Kafka's theme lies a concept of time based on the reality of idea which gives rise to his technique of parable couched in a dream world. Kafka's attitudes toward time and reality alone make possible his method of writing. Reality is of the mind; therefore the dream is real and our ideas are real.

Kafka's emphasis on the dream world and interest in an inner reality spring from a great many sources.[26] Primarily his whole attitude toward reality is deeply coloured by his personal problems of adjustment to life. His relationships to his family, to the women to whom he was engaged, to Milena were painful ones. As a Jew his relationship to the community was also an involved one. These unsolved relationships led to conflicts between the inner and the outer man, so that he eventually took cover in his writing behind the highly complicated screen of symbol and parable as a refuge from the impingements of the world of action and events. Thus the doctrine of ideas is for Kafka a successful defence. In *The Castle* we find Kafka's dream world and his idea of truth.

NOTES

[1] Franz Kafka, *The Great Wall of China, Stories and Reflections*, translated by Willa and Edwin Muir (1946), p. 267.

[2] Franz Kafka, *The Trial*, translated by Willa and Edwin Muir (1956), p. 318.

[3] *Ibid.*, p. 43. [4] *Ibid.*, p. 307. [5] *Ibid.*, p. 324.

[6] *Ibid.*, p. 69. [7] *Ibid.*, p. 103. [8] *Ibid.*, p. 294.

[9] *Ibid.*, p. 303. [10] *Ibid.*, p. 152. [11] *Ibid.*, p. 169.

[12] *Ibid.*, p. 188. [13] *Ibid.*, p. 205.

[14] Clemens Heselhaus, 'Kafkas Erzählformen', *Deutsche Vierteljahrsschrift für Literaturwissenschraft und Geistgeschichte*, XXVI (1952), pp. 353–76.

[15] Erich Kahler in his excellent discussion of Kafka's technique in 'Untergang und Übergang der epischen Kunstform', *Neue Rundschau*, LXIV (1953), pp. 1–44, points out that Kafka's stories move in a sphere which transcends the senses. His characters live daydreams in which vision and speculation are one. Since by the symbol thought is directed from the concrete to the abstract and by allegory thought is directed from the abstract to the concrete, Kahler rejects both these terms as descriptive of Kafka's works: he prefers to call them parables. Although Kafka's work does have many elements in common with the parable, I do not feel that this term is sufficient to describe the technical complexity of the novels which include symbols within the parables and in which the parable is often couched in terms of the dream. Furthermore, Kafka's parables lack the outspoken didactic purpose of most works in this form.

[16] Günther Anders, *Kafka Pro und Contra* (München, 1951), p. 25.

[17] Franz Kafka, *The Castle*, translated by Willa and Edwin Muir (1941), p. 23.

[18] For a discussion of Max Brod's arrangement of the material in *The Trial* (one which may cast doubt also on his arrangements in *The Castle*), see Hermann Uyttersprot, 'Zur Struktur von Kafkas "Der Prozess" ', *Revue des Langues Vivantes* (1953), pp. 332–76.

[19] *The Castle*, p. 95.

[20] Felix Weltsch, *Religiöser Humor Bei Franz Kafka* (Winterthur, 1948), p. 129 (quotation translated by Margaret Church).

²¹ *The Castle*, p. 109.

²² *Ibid.*, p. 325.

²³ Edwin Muir sees in Kafka's world the influence of 'Kierkegaard's doctrine of incommensurability of divine and human law' ('Franz Kafka', in *A Franz Kafka Miscellany: Pre-Fascist Exile* (1940), p. 62).

²⁴ The long winters in the village lend an atmosphere of darkness fitting to the dream. In fact, spring and summer seem to Pepi no longer than two days. ('Excerpts from final passages of *The Castle*', in *A Franz Kafka Miscellany*, p. 94).

²⁵ Kafka's dream is overburdened by anxieties, for almost all of the villagers seem to be hostile, indifferent, or fearful toward K.

²⁶ Emphasis on an inner reality may be seen in many of the authors read by Kafka. In both the biography by Brod and the diaries Kafka's interest in Plato is mentioned. In the movement of German 'poetic nihilism' Kafka read and admired Grillparzer and Stifter. He was especially interested in Grillparzer's 'Der arme Spielmann' with its theme of a transcendent reality for the artist, and in Stifter's *Der Nachsommer* in which the characters are scarcely more than incidental in the presenting of Stifter's idea of the permanent character of truth. Furthermore, he had read, Brod writes, Flaubert's *The Temptation of St Anthony* and *A Sentimental Education*. In both of these books Flaubert's devotion to idea is strong. In fact, in the former all movement is in the realm of the spirit. The theme of a transcendent reality may also be seen in Goethe, Kleist, and Hölderlin, all of whom Kafka read.

John Graham†　　　　　VIRGINIA WOOLF (1882–1941)

Time in the novels of Virginia Woolf

Virginia Woolf adhered to her own critical dictum, expressed most fully in *Mr Bennett and Mrs Brown,* that the business of the novelist is the exploration of character. The final goal of such exploration, however, is insight into the nature of human personality, and so into the meaning of life: for Mrs Brown, whose character is the proper subject of the novel, is really 'the spirit we live by, life itself'.[1] In struggling to reach that goal, Mrs Woolf is constantly preoccupied with problems relating to time, as most critics have noted. This preoccupation underlies her concern with the phenomena of memory, change, and death, and drives her to ask several ancient difficult questions. Her efforts to answer them result in two central views of life, neither of which is the result of mere neurotic nescience.

I

In the first three novels—*The Voyage Out, Night and Day,* and *Jacob's Room*—the questions are asked in a fairly obvious way, and focus on the apparent dichotomy between two kinds of time, two worlds: on the one hand, the world of linear time, of past, present, and future, in which we are subject to unremittent and uncontrollable flux; and on the other hand, the world of mind time, an inner world of thought and imagination, in which the chaotic flow of experience derived from our life in linear time is reduced to order and unity, and in which we are therefore liberated. In all of her novels, Mrs Woolf is troubled by the apparent dualistic conflict between these two worlds, in which the data of one frequently contradict the data of the other; but she is never willing to accept this dualism as absolute. In seeking to overcome it, two solutions are possible: either both worlds are parts of one larger reality, and are therefore integrally related to each other; or one of these worlds is unreal, and only in exploring the other shall we find a valid absolute.

In the first three novels neither of these solutions is worked out: the

† Reprinted from *University of Toronto Quarterly,* XVIII (1949), pp. 186–201.

problem is stated, not solved. These novels are dominated by what is called in one of them 'the profound and reasonless law' of linear time.[2] In this immense machine, human life is irrelevant; and against its tyranny the inner world of mind time opposes at best an escape into illusion. But in her next novel, Mrs Woolf does offer a solution, which depends to a great extent on the relationship between Clarissa Dalloway and Septimus Smith.

Although Clarissa exults in the 'triumph and the jingle' of 'life; London; this moment of June',[3] she is also afflicted by a sense of isolation and by the recurring terror of death. The nature of her predicament is expressed most powerfully when she stands in a little room, watching the old lady in the house across the way:

Big Ben struck the half hour.

How extraordinary it was, strange, yes, touching, to see the old lady (they had been neighbours ever so many years) move away from that window, as if she were attached to that sound, that string. Gigantic as it was, it had something to do with her. Down, down, into the midst of ordinary things the finger fell making the moment solemn. She was forced, so Clarissa imagined, by that sound, to move, to go—but where? . . . Why creeds and prayers and mackintoshes? when, thought Clarissa, that's the miracle, that's the mystery; that old lady, she meant, whom she could see going from chest of drawers to dressing-table. She could still see her. And the supreme mystery which Kilman might say she had solved, or Peter might say he had solved, but Clarissa didn't believe either of them had the ghost of an idea of solving, was simply this: here was one room, there another. Did religion solve that, or love? [pp. 192–3]

In many of Mrs Woolf's novels—as, for example, in *Jacob's Room*—the room symbolises the selfhood formed in time. From her small room, then, Clarissa looks out and sees the supreme mystery: that people exist in the same stream of time, each moving under the compulsion of the time-flow (as symbolised by Big Ben), visible to each other, yet unknown to each other and essentially alone. Not even the great unifying forces of love and religion can supply any pattern relating them to each other. Clarissa tentatively theorises about the unifying power of human personality, the 'unseen part of us, which spreads wide' (p. 232), and which persists after death, relating the individual to the dead and the absent, and to all that they have known. But this nebulous hope, held uncertainly, is clarified into a positive conviction only by the life and death of Septimus Smith.

In considering Septimus, we should remember what Mrs Woolf said in the preface: first, he was not included in the first draft of the novel; second, Clarissa was to die or commit suicide in the first draft; and third, Septimus is the 'double' of Clarissa.

'It might be possible, Septimus thought, looking at England from the train window, as they left Newhaven; it might be possible that the world itself is without meaning' (p. 133). With this thought begins his purgatorial voyage into insanity. Eventually he surrenders to his madness, seeking in no way to preserve any connection with humanity as it is directed in the ringing grooves of change by Doctors Holmes and Bradshaw. At this low point of his negative way into darkness Septimus receives his revelations. Sitting in Regent's Park, he looks at the trees:

But they beckoned; leaves were alive; trees were alive. And the leaves being connected by millions of fibres with his own body, there on the seat, fanned it up and down; when the branch stretched he, too, made that statement. The sparrows fluttering, rising, and falling in jagged fountains were part of the pattern; the white and blue, barred with black branches. Sounds made harmonies with premeditation; the spaces between them were as significant as the sounds. A child cried. Rightly far away a horn sounded. All taken together meant the birth of a new religion. [p. 32]

Septimus now sees, first, that there is a unifying reality hidden in the phenomena of time which gives them pattern and significance, and second, that the pattern is eternal because there is no death. These are revelations which, if he could communicate them to mankind, would save us, for he comes as a saviour, a redeemer: 'Look the unseen bade him, the voice which now communicated with him who was the greatest of mankind, Septimus, lately taken from life to death, the Lord who had come to renew society, who lay like a coverlet, a snow blanket smitten only by the sun, for ever unwasted, suffering for ever, the scapegoat, the eternal sufferer . . .' (p. 37). The ancient image of the sun, symbolising the power of divine revelation; the image of the Saviour; the emphasis upon the fact that he was lately taken from life (the life of self, the life in time) to death (the annihilation of self, the transcendence of time): all suggest that Septimus has had the vision of a cosmic unity which Clarissa, rooted as she is in the process of time, can receive only dimly and briefly. Her moment of vision, characteristically derived from her response to a human emotion, is described as 'a match burning in a crocus; an inner meaning almost expressed' (p. 47); but the illumination is for Septimus as agonisingly blinding as the blaze of a fiercely intense sun. He is a snow blanket smitten only by the sun; or again, he is a figure, lamenting the fate of man in the desert alone, who receives full on his face the light of the dawning sun, an 'astonishing revelation' (p. 106).

There is, however, a fatal and ironic flaw in the vision of Septimus: it is more than flesh can endure. His descent into the pit is naturally marked by agonised suffering; but even the saving revelations beat upon him with merciless intensity. The true terror of his vision is that it destroys him as a creature of the time-world. When he returns for a brief tranquil period to the world of actuality, it would seem that this is not so. He feels that Nature is redeemed for him by his vision; and certainly Rezia is redeemed through her love. The world is no longer without meaning, he feels: time is a garment of eternity now, a coverlet of flowers (p. 216).

But Septimus does not really return to sanity—the appalling sanity of Holmes and Bradshaw, our fallen selves. He has died to that sanity and cannot effect any Lazarean return. He is no longer human, and can only rise up from death to another level of being, after walking briefly in a world of flowers. Holmes and Bradshaw must crucify him: they seek dominion through the body (both are doctors), and their understanding is of the body. Therefore, when they finally return to kill him (as he believes), he cries, 'I'll give it you!' and with a dreadful casualness, which makes Holmes and the fallen reader turn pale, he leaps towards the central spiritual reality, leaving for them their due legacy—a piece of dead flesh, pierced and torn.

These points must be made in order to grasp the full significance of the relationship between Septimus and Clarissa. This relationship is progressively revealed in the long scene after she hears of his death and walks alone into the little room from which, earlier in the day, she had watched the old lady across the way. There she reflects on his death:

She had once thrown a shilling into the Serpentine, never anything more. But he had flung it away. They went on living (she would have to go back; the rooms were still crowded, people kept on coming). They (all day she had been thinking of Bourton, of Peter, of Sally), they would grow old. A thing there was that mattered; a thing, wreathed about with chatter, defaced, obscured in her own life, let drop every day in corruption, lies, chatter. This he had preserved. Death was defiance. Death was an attempt to communicate; people feeling the impossibility of reaching the centre which, mystically, evaded them; closeness drew apart; rapture faded, one was alone. There was an embrace in death. [pp. 280–1]

The meaning of Septimus' suicide is reinforced by the fact that he dies by leaping through a window. The window is used constantly as a symbol of the outlook of the self on the world around it. For the ordinary individual it is the aperture through which comes the only light he may receive; but it is also a barrier, hampering any movement of his being towards the source of that light. Clarissa makes no effort to break the pane of glass standing between herself and the sun; Septimus does: he casts off his temporal selfhood and leaps towards the centre. Clarissa realises that in his act is an integrity which she can never have entire, rooted as she is in time: for, as we have seen, in order to penetrate to the centre, like Septimus, one must either die, or go mad, or in some other way lose one's humanity in order to exist independently of time.

Clarissa, then, accepts actuality: she even invites Bradshaw to her party. And it is Clarissa who enters the room at the end of the book and triumphs over time. Septimus cannot triumph, but his is the complete vision which gives Clarissa the power to conquer time. This we see when Clarissa again watches the old lady across the way. The question asked in a similar situation earlier in the day is now answered:

The clock began striking. The young man had killed himself; but she did not pity him; with the clock striking the hour, one, two, three, she did not pity him, with all this going on. There! the old lady had put out her light! the whole house was dark now with this going on, she repeated, and the words came to her, Fear no more the heat of the sun. She must go back to them. But what an extraordinary night! She felt glad he had done it; thrown it away. . . . He made her feel the beauty; made her feel the fun. But she must go back. She must assemble. She must find Peter and Sally. And she came in from the little room. [pp. 283–4]

Clarissa intuitively grasps the meaning of Septimus' vision, which he could communicate only by death. She thereby absorbs into herself the significance which Septimus holds for the reader; and yet she retains her own special power to create in the imperfect fallen realm of human relations, a power which Septimus wholly lacked. In her handling of the conclusion to this novel Mrs Woolf attempts to make clear that we must retain the limiting protecting identity which is ours in time if we are to triumph over time.

A skilful subordinate device is used throughout the book to reinforce this relationship between Septimus and Clarissa. It is the recurring tag from Shakespeare, 'Fear no more the heat of the sun', in the passage quoted above. Early in the book Clarissa, looking into a shop window, sees a volume of Shakespeare opened at these lines, and as she reads them, she asks herself, 'What was she trying to recover? What image of white dawn in the open country?' (p. 12). Her failure to recover it suggests the inadequacy of her vague theories about the unifying power of personality, which I have already mentioned. Later on in the book Septimus, in his momentary tranquil return to the actuality of time, finds it transfigured. In the image which occurs to him, he had fallen through the sea of time into a flaming purgatorial world of terrifying and redeeming insight; now he returns to the surface to rejoice in the transfiguration of the world:

Every power poured its treasure on his head, and his hand lay there on the back of the sofa, as he had seen his hand lie when he was bathing, floating, on the top of the waves, while far away on shore he heard dogs barking and barking far away. Fear no more, says the heart in the body; fear no more. He was not afraid. At every moment Nature signified by some laughing hint . . . Shakespeare's words, her meaning. [pp. 211–12]

For Clarissa, standing in front of the shop window, this line meant an illumination lost and not yet recovered; for Septimus it means an illumination received in its fullest intensity. When Clarissa repeats this line near the end of the book, and after Septimus' death, she recalls for the reader the groping way in which she first read it; and this stresses the fact that she now repeats it as Septimus did, with a sense of peace and reassurance. So it is that the profounder vision of Septimus is given to Clarissa without its attendant agony.

When she leaves the little room, she returns to the larger room of human relations. Our knowledge of what she has just been thinking explains the sudden strange excitement of Peter Walsh as he sees her standing in the doorway of the drawing-room: for Clarissa, returning to the part, symbolises the transfiguration of time.

II

To the Lighthouse is the full and final expression of the synthesis first attempted in *Mrs Dalloway*. For that reason I shall limit myself to a few dogmatic remarks about the unifying and vitalising principle of the whole book, its symbolism. Sea images, colour images, the associations of memory, recurrent verbal patterns, all serve to spin around the actual events a subtle web of interrelated meanings. The structure also is significant: the first section is called 'The window', an image associated, as we have seen, with the individual's vision of life; the second, 'Time passes', portrays the assault of time on the integrity of that vision; and in the third, 'The lighthouse', the vision is triumphantly reaffirmed.

The lighthouse is the central symbol of the book, and what it means depends on who is looking at it: it has no single limited meaning, hence its power as a symbol. Its relation to Mrs Ramsay is of crucial importance, for Mrs Ramsay has the power to see what Clarissa finally saw, the transfiguration of time by eternity. But Mrs Ramsay's vision is more sophisticated. Although she sits in her room and undergoes the mystical experience of becoming the thing she looks at, the lighthouse, she nevertheless recognises that its meaning is paradoxical: it is 'so much her, yet so little her';[4] it stands firm and unchanging amid the seas of time, yet in a sense has no reality apart from the sea. Its beam revolves in a pulsing rhythm akin to that of the time process, and so, as she watches it, she calls it 'the pitiless, the remorseless' (p. 99); at the same time, however, it gives her a sense of stability, 'this peace, this rest, this eternity' (p. 96). It is not change, yet cannot be separated from change, and therefore represents a vital synthesis of time and eternity. This makes it an objective correlative for Mrs Ramsay's vision.

An objective correlative is lacking in *Mrs Dalloway*, where Clarissa, at the end, is herself the symbol of her own vision. That explains, I think, why Mrs Woolf could not follow her first intention of having Clarissa die: if she had, the meaning of Clarissa's vision would have vanished with Clarissa. It also explains why Mrs Woolf can and does let Mrs Ramsay die. It is important that she should die, for death is the most powerful assault which time can make on her vision. It sweeps her away, but it cannot destroy the lighthouse; and by the time she dies, the lighthouse has become the meaning of Mrs Ramsay.

The relationship of Mr and Mrs Ramsay repeats this idea on another level, for as husband and wife they are the lighthouse. Crudely put, Mrs Ramsay equals eternity, Mr Ramsay equals time; they are married. For Mrs Ramsay, though she triumphs in time, triumphs because she intuits eternity; and Mr Ramsay, though he loftily seeks a philosophical absolute which will solve the problem of 'subject and object and the nature of reality' (p. 38), cannot break his bondage to time without the aid of his wife. Together they fulfil each other, and are the creators of life.

In the last section the ancient mystic symbols of the Quest and the Love Union are related to the experiences of Mr Ramsay and Lily Briscoe. The completion of Lily's painting, her vision of life, involves a spiritual union with Mrs Ramsay; and Mr Ramsay's sea voyage is obviously a quest. But Lily's love union involves a quest, and Mr Ramsay's quest culminates in a love union; for as he voyages over an expanse of sea to the lighthouse, Lily voyages over an expanse of time, searching the past for the meaning of Mrs Ramsay. At the exact moment that Mrs Ramsay appears before Lily, enabling her to complete her painting, Mr Ramsay arrives at the lighthouse, and has rejoined his wife. The experiences of Lily and Mr Ramsay coincide therefore not only in time but in visionary significance as well. Both reaffirm the triumphant validity of Mrs Ramsay's vision of life.

Another set of symbols underlies this validity, performing the same function as the tag from Shakespeare in *Mrs Dalloway*. In the first section Mrs Ramsay goes to see if the children are asleep. They are not, for a boar's skull nailed on the wall has frightened Cam; and James, who loves its bare white severity, will not allow her to touch it. Mrs Ramsay drapes her green shawl over the skull, thereby creating for Cam a new world of heightened beauty and peace, at the same time assuring James that the skull is still there. In the middle section the little winds of corruption enter the abandoned house and separate shawl from skull, destroying the single visionary world created from both. In the last section Cam, staring at the rush of green waters beside her, slowly recalls that world of long ago, and feels suddenly at peace. James, gazing at the severe whiteness of the lighthouse, lonely on its bare rock, recalls how the lighthouse looked when he saw it years before with his mother, and concludes that both are true—the 'silvery, misty-looking tower' then, and the tower 'stark and straight' now (p. 276). Thus, in the children, the synthesis achieved by Mrs Ramsay is again vindicated.

If this analysis is right, then in these two novels Mrs Woolf has endeavoured to work out one of the two solutions to our original problem, for the world of mind time and the world of linear time are related to each other because both are related to a central and eternal reality.

III

The Waves is significant as a radical departure from *To the Lighthouse* in technique and thought. We are once again immersed in the sea of time, and its relentless pounding again opens the abyss between mind time and linear time which *To the Lighthouse* had triumphantly closed. The book is dominated by the rhythm of waves: in the poetic passages between sections, which I shall call lyrics, the sun slowly rises and falls in the sky, lending the shape of one enormous wave to the whole book; the thoughts of the characters eddy and swirl restlessly; the style surges and subsides with brilliant intensity. The dominant mood is one of anguished effort, suffering, and disillusionment: the total vision of the book is undoubtedly tragic.

While this is the major impact of *The Waves*, I wish to discuss certain other elements which redeem it from notoriety merely as a masterpiece of negation. It is my thesis that *The Waves* is an attempt to begin a new integration of the individual, not in terms of the cosmic unity found in *To the Lighthouse*, but in terms of a 'human communion'. What I mean by this phrase will, I hope, shortly appear.

When the six characters are all about twenty-five years old, they hold a farewell dinner for Percival, a seventh figure, who is leaving for India. We know Percival only through the eyes of the others, for his own consciousness is never explored; yet he is of central importance, because for the others he is a hero born to conquer the disintegrative powers of time. Like Mrs Ramsay and Clarissa, he is the unconscious doer, the artist in living, whose medium is action.

As the dinner progresses, the others feel their common love for Percival intensify until it becomes a creative act of communion with each other. Bernard, who is a novelist and the most important of the six characters, expresses clearly the significance of this communion as they leave the restaurant and he speaks of

... the swelling and splendid moment created by us from Percival; . . . . We have proved, sitting eating, sitting talking, that we can add to the treasury of moments. We are not slaves bound to suffer incessantly unrecorded petty blows on our bent back. We are not sheep either, following a master. We are creators. We too have made something that will join the innumerable congregations of past time. We too, as we put on our hats and push open the door, stride not into chaos, but into a world that our own force can subjugate and make part of the illumined and everlasting road.[5]

Now the 'moment' here is very different from what it was in *Mrs Dalloway* and *To the Lighthouse*: it no longer comes involuntarily or as a result of passive receptivity, it must be created; and it no longer consists in an intuition of a mystical cosmic reality unifying the worlds of mind and linear time. The world of the time-process 'out there' is now something which man must subjugate: he is a creator. This ardent sense of their own power is evoked by their communion, in which they lose the selfhoods imposed on them by time and become, in the image used, 'a whole flower, to which every eye brings its own contribution' (p. 137). Now the experience of creating this single identity lasts only a moment; but the communion created in that moment is eternal. It joins the 'innumerable congregations of past time' in a celebration of the larger body of Man, which exists in an everlasting Now.

It is significant that Percival is about to leave for India, a wasteland waiting to be rescued, where the luxuriant powers of nature reign triumphantly over impotent humanity. And it is significant that as he leaves he is bade by Bernard to look at the lights of the city blending in 'the yellow canopy of our tremendous energy' (p. 159), the material expression of the promethean powers of man. Percival, the protagonist of these powers, the god of action, is at the height of his glory; for in the lyric which follows his departure the sun is described as being at its zenith.

Once at its zenith, however, the sun must decline; Percival is thrown from his horse in India and is killed. Now time begins its subtle warfare of attrition against the meaning of Percival. The others forget him and go their separate ways: the 'tremendous energy' which he symbolised wanes within them, and in the world of action they feel that they have failed. Their lives have hardened into fixed patterns of behaviour, and they have lost any clear pattern of understanding. Then, when they are middle-aged, Bernard calls them together for a reunion dinner.

At this dinner the same communion is created, reaching its culmination as the six friends leave the restaurant and walk arm in arm into a park:

'The flower,' said Bernard, 'the red carnation that stood in the vase on the table of the restaurant when we dined together with Percival, has become a six-sided flower; made of six lives.'

'A mysterious illumination,' said Louis, 'visible against those yew trees.'

'Built up with much pain, many strokes,' said Jinny.

'Marriage, death, travel, friendship,' said Bernard; 'town and country; children and all that; a many-sided substance cut out of this dark; a many-faceted flower. Let us stop for a moment; let us behold what we have made. Let it blaze against those yew trees. One life. There. It is over. Gone out.' [p. 250.]

This needs little comment, in the light of the first dinner, but we might note that this single life is *cut from* the darkness of time, and burns *against* that darkness and *against* the yew trees of death.

At this dinner, however, Percival is absent; and the special promise which he symbolised has not been fulfilled. What was then the future has become the past, and as they lived it they did not conquer it. Now it is too late for action: as the sun declines it is time to seek understanding. That is why Bernard, the novelist, who seeks understanding before all things, dominates the dinner.

The new stature of Bernard is made clear in the last section of the book, where he sums up the meaning of his life. In it three notes are sounded. The first is that of constant warfare against the enemy, time. The second is the rejection of the idea that an eternal reality 'out there' underlies and unifies the time-process. As a youth, Bernard recalls, he believed that the willow tree pointed to something 'beyond and outside our own predicament; to that which is symbolic, and thus perhaps permanent'; now he dismisses the willow tree as one of those 'phantoms made of dust . . . mutable, vain' (pp. 272, 312). There is no trace of the vision of Septimus

Smith, for whom Nature's meaning was, 'Fear no more the heat of the sun'; and there is no lighthouse to guide us over the waves. Against his progressive disillusionment stands only the saving reality of the human communion enjoyed at the two dinner parties, and this is the third emphasis in his summing-up. He recalls the reunion dinner, and how 'we saw for a moment laid out among us the body of the complete human being whom we have failed to be, but at the same time, cannot forget' (p. 303). As the dinner progressed, however, their temporal identities fell away, this body rose, and they became it: 'I saw blaze bright Neville, Jinny, Rhoda, Louis, Susan, and myself, our life, our identity . . . we six, out of many million millions, for one moment out of what measureless abundance of past time and time to come, burnt there triumphant. The moment was all: the moment was enough' (p. 304). With a special awareness of their communion he leaves the restaurant in which he has been sitting and stands outside, watching the dawn rise, and facing the antagonist in the final passage of the book:

I am aware once more of a new desire, something rising beneath me like the proud horse whose rider first spurs and then pulls him back. What enemy do we now perceive advancing against us, you whom I ride now, as we stand pawing this stretch of pavement? It is death. Death is the enemy. It is death against whom I ride with my spear couched and my hair flying back like a young man's, like Percival's, when he galloped in India. I strike spurs into my horse. Against you I will fling myself, unvanquished and unyielding, O Death!
The waves broke on the shore. [p. 325]

Bernard dies. (That is the meaning of the final sentence, which is italicised, and so belongs with the lyrics, not with Bernard's final thoughts.) As an individual he breaks on the shore. But he is much more than an individual: throughout his summing-up, he slowly sheds his private selfhood and absorbs into himself the identities of the other six. This is indicated, for example, by the way in which he applies to himself images which, throughout the book, have been associated only with the others. They are types of humanity, and Bernard thus becomes the archetype of the race as it struggles with its creative powers against the tyranny of time. That is why Bernard, the archetype of vision, finally absorbs into himself the special significance of Percival, the archetype of action; for on both levels we must ride against the enemy. The visionary struggle is stressed, however, since Percival rises again only through the effort of Bernard. It is significant that the situation here is the reverse of what it was in *Mrs Dalloway*, where Clarissa, the artist in living, absorbed into herself the meaning of Septimus, the visionary.

Bernard thus assumes the proportions of the dragon-killer, the god who subdues chaos: he rides with 'spear couched' like some St George. He rides for the creation of the complete human being whom they have failed to be yet cannot forget, a spiritual being existing on a level untouchable by time. It is no accident that while the lyric at the beginning of the summing-up

described the disappearance of the sun in darkness, Bernard at the end sees
the dawn whitening the sky. While as an individual he sinks into death, he
also rises as the champion of the complete human being, and becomes that
being, like a dying-and-rising god.

I repeat that these remarks concern the effort which Mrs Woolf is
making to achieve a new integration of the individual. That effort is not
successful in *The Waves*, which is primarily the tragedy of the individual
wave breaking on the shore. Nevertheless, we find here the initial formulation of a synthesis which receives consummate expression in *Between the
Acts*.

IV

The main structural device of *Between the Acts* is simple and venerable,
that of a play within a play. There is the pageant, staged in the garden,
which I shall call the microcosmic pageant; and there is the macrocosmic
pageant which goes on before and after the small pageant, and between its
acts. My chief interest is in the relationship between these two pageants.
The proscenium arch for both of them is a carefully fostered sense of our
community in time, which frames and illuminates the present moment unfolding on the stage. All through the book a host of details sustains our
feeling that the past is present.

The macrocosmic pageant unfolds for the first hundred pages and centres
on the interplay of tensions between the main characters of the book. On a
sleepy afternoon these tensions increase until, by the time the pageant is
staged, each character is isolated from the rest. Bart Oliver, a disciple of
reason, is at odds with his sister, Lucy Swithin, a disciple of faith who is
addicted to an imaginative reconstruction of the past. Giles Oliver and his
wife Isa embody the impulses to creative action and creative vision which
we have noted in previous novels. Both are frustrated by the sequence of
events. Giles, a man of action, is oppressed by time past (his life as a
respectable stockbroker) and by time future (the impending war in
Europe, which he can oppose with no constructive or even destructive
action). His intense frustration leads him into childish infidelities to his
wife, who partly symbolises for him the oppressive sequence of events.
Ironically, he stands in part for the same thing in her eyes; but she longs
to escape from the burden of time into an inner world of absolute knowledge
and beauty, not into a world of liberating action, and so writes poetry about
her longing to 'lose what binds us here'.[6] They both love and hate each
other; the tension between them is not resolved until the last page of the
book, and even then only by implication.

The brooding and vague antagonism which settles over them all is partly
broken and partly aggravated by Mrs Manresa, a 'wild child of nature'
(p. 52), who surrenders eagerly to the flux of time, seeking no other
absolute.

The actors and audience now arrive and the microcosmic pageant begins. There are four acts in it, and I shall consider the first three in a general way only, in order to concentrate on the last. At the start of the pageant, and through most of the acts, a chorus of villagers winds its way among the trees which form the backdrop, chanting words half-audible to the audience; for the winds of time blow away the meaning of the anonymous millions of the past. The words we do hear emphasise the tenacious endurance of the race in time; whatever period of history is passing on the stage, the chorus is always the same.

The first act is an Elizabethan tragedy. Its details need not concern us, for along with all the other acts its value lies in the reactions of the audience to it. As Isa watches it she realises that the plot, a special pattern in past time, is unimportant, and that the meaning of the play is the timeless glory and terror of love, hate, and death. But immediately after the death of one of the characters in it, the actors come forward and join in a dance. As they whirl about, laughing and shouting, the audience too begins to laugh, to clap, to join in spirit this dance of irrepressible life. It is then, as the author of the pageant, Miss La Trobe, intended, that the members of the audience *become* the Elizabethans.

After the dance an interval occurs in which tea is served in the barn. Now the macrocosmic drama is resumed. The actors do not realize that they are acting it, and for this reason the connections between the two pageants must be implied. I have space to illustrate this technique only by one example. As Lucy Swithin sits in the barn, she watches the swallows flying among the rafters and thinks how they came to this spot 'before there was a channel, when the earth, upon which this Windsor chair was planted, was a riot of rhododendrons'. At this point Mrs Manresa bustles up.

'I was hoping you'd tell me,' said Mrs. Manresa. 'Was it an old play? Was it a new play?'
No one answered.
'Look!' Lucy exclaimed.
'The birds?' said Mrs. Manresa, looking up. [p. 130]

The birds are the answer to her question. The play is as old as the swallows coming ages ago; and as new as the swallows swooping above them in the present moment.

The next act is cut on the pattern of Restoration comedy, and ironically modifies the grand themes of hate and love and death presented in the previous act. Now we see vanity, deceit, lust, greed, the absurdity and pathos of old age, the power and cruelty of youth. One is struck by the vigour and humour of the satire; Mrs Woolf's hypersensitivity to the depravity of human beings is by this time less agonised and more robust. For Septimus Smith, the only terror he could not face was the depravity of men, and so he turned towards the centre. Now, however, *we* are the centre, and must face ourselves unflinchingly. Our depravity cannot be explained away; but in our attitude to it lies the measure of its power, which will be reduced to a minimum if met without illusion and with purifying laughter. No one in the book flees, as Septimus fled, to seek in death eternal life.

Yet we have eternal life, and its nature is suggested by Mrs Swithin between the acts, when she seeks out Miss La Trobe:

> She gazed at Miss La Trobe with a cloudless old-aged stare. Their eyes met in a common effort to bring a common meaning to birth. They failed; and Mrs. Swithin, laying hold of a fraction of her meaning, said: 'What a small part I've had to play! But you've made me feel I could have played . . . Cleopatra!' . . .
>
> 'I might have been—Cleopatra,' Miss La Trobe repeated. 'You've stirred in me my unacted part,' she meant. . . .
>
> 'You've twitched the invisible strings,' was what the old lady meant; and revealed—of all people—Cleopatra! Glory possessed her. Ah, but she was not merely a twitcher of individual strings; she was one who seethes wandering bodies and floating voices in a cauldron, and makes rise up from the amorphous mass a recreated world. [pp. 179–80]

Miss La Trobe, the artist in vision, arouses in Lucy the slumbering power of imaginative vision which enables Lucy to see that on a level of being of which she is rarely conscious she lives all lives at all times.

In the next act of the pageant the Victorians are exuberantly parodied. The two pageants are now merging in historical focus, since many among the audience have themselves lived during the age of faith in fossils and fossil faiths, now so ruthlessly mocked. But Mrs Swithin makes the truly significant comment when Isa asks incredulously if the Victorians were really like that:

> 'The Victorians,' Mrs. Swithin mused. 'I don't believe,' she said with her odd little smile, 'that there ever were such people. Only you and me and William dressed differently.'
>
> 'You don't believe in history,' said William. [p. 203]

Neither does Mrs Woolf. This pageant about history is a declaration that history itself is not leading us anywhere in particular. Yet we cannot escape it. Immediately after this act, a cloud-burst briefly showers the audience. Isa, for whom history is a burden to be lost in the attainment of some absolute, looks up: ' "O that our human pain could here have ending!" Isa murmured. Looking up she received two great blots of rain full on her face. They trickled down her cheeks as if they were her own tears. But they were all people's tears, weeping for all people' (p. 210). This is the darker aspect of our communion, the community in bondage which we share under the tyranny of time.

With sorrow still gripping them the audience watches a tableau; a man and a woman are shown rebuilding a ruined wall. This is the task which confronts them as the inheritors in time of a civilisation fallen into disrepair. As the audience watches in silence, all of the actors who have taken part in the pageant dance out onto the stage, each declaiming some lines from his particular role in history, and carrying a mirror or some shiny object which will reflect the members of the audience to themselves. At this effrontery only Mrs Manresa retains her poise: she seizes the opportunity to lean forward and powder her nose. This is only natural, since she has no other identity besides that reflected in the mirror of the present moment. The rest feel they have, however, and resent the necessity of facing their time-bound selves. Some of them are only restrained from leaving by a 'megaphontic, anonymous, loudspeaking' voice from the bushes.

The voice sums up both the satirical and constructive visions of Miss La Trobe, demanding first that the audience cast away illusions, and concluding with the bitter exhortation, '*Look at ourselves, ladies and gentlemen! Then at the wall; and ask how's this wall . . . civilization, to be built by* (here the mirrors flicked and flashed) *orts, scraps and fragments like ourselves'?* (p. 219).

Then the voice turns to another theme—our greatness, reflected, for example, in '*the resolute refusal of some pimpled dirty little scrub in sandals to sell his soul. There is such a thing—you can't deny it. What? You can't descry it? All you can see of yourself is scraps, orts and fragments? Well then listen to the gramophone affirming.*' Here is what the gramophone affirms:

Like quicksilver sliding, filings magnetised, the distracted united. The tune began; the first note meant a second; the second a third. Then down beneath a force was born in opposition; then another. On different levels they diverged. On different levels ourselves went forward; flower gathering some on the surface; others descending to wrestle with the meaning; but all comprehending; all enlisted. The whole population of the mind's immeasurable profundity came flocking; . . . from chaos and cacophony measure; but not the melody of surface sound alone controlled it; but also the warring battle-plumed warriors straining asunder: To part? No. Compelled from the ends of the horizon; recalled from the edge of appalling crevasses; they crashed; solved; united. . . .
Was that voice ourselves? Scraps, orts and fragments, are we, also, that? [pp. 220–1]

We all comprehend and enlist in the creation of a total harmony, the imaginative totality of our human life. This is the meaning of the music; and of Miss La Trobe's pageant; and of Mrs Woolf's book. This theme is emphasised later on when the Rev. G. W. Streatfield prosily remarks that 'To me at least it was indicated that we are members one of another. Each is part of the whole. . . . We act different parts; but are the same' (pp. 223–6). Or again, as the crowd leaves, amid idle chatter flash remarks luminous with the significance of all that has gone before. One of these is put in a question: 'The looking-glasses now—did they mean the reflection is the dream; and the music . . . is the truth?' (p. 234). The reflections in the glass are only our temporal identities; but the communion expressed in the music is our eternal identity.

After the crowd has gone we return to the macrocosmic pageant. But the characters in it are now enlarged: behind them is the meaning of the small pageant, and they are now acting in two plays. When Mrs Manresa walks towards her car 'like a goddess, buoyant, abundant, with flower-chained captives following in her wake', she is more than Mrs Manresa—she is one whole aspect of our life. So are the others as they linger in the dusk.

Miss La Trobe, the artist, prepares to leave, now that the audience is gone and she need not face them. Looking back on her pageant she sees that in the effort to communicate her vision lay glory; and that in her partial failure to do so lies the inevitable bondage of the artist to his own limitations, to his audience, and to his medium. Nevertheless, she cannot abandon the effort. As she leaves Pointz Hall another play begins to stir in her mind. ' "I should group them," she murmured, "here." It would be midnight; there would be two figures, half-concealed by a rock. The curtain would rise. What would the first words be? The words escaped her' (p. 246). Later, however, as she sits in a pub she again visualises the scene: 'There was the high ground at midnight; there the rock; and two scarcely perceptible figures. . . . She set down her glass. She heard the first words.'

Now as Miss La Trobe sits in the pub, the actors in the macrocosmic drama at Pointz Hall sit quietly in the drawing-room, reading the paper and looking over the day's mail. For the reader, however, they are playing two dramas at once. Slowly the light fades, the sky becomes cold and severe. Bart rises and stalks silently from the room. Lucy closes her outline of history and tiptoes out. Now Giles and Isa are alone: they rise.

Left alone for the first time that day they were silent. Alone, enmity was bared; also love. Before they slept, they must fight; after they had fought they would embrace. From that embrace another life might be born. But first they must fight, as the dog fox fights with the vixen, in the heart of darkness, in the fields of night.

Isa let her sewing drop. The great hooded chairs had become enormous. And Giles too. And Isa too against the window. The window was all sky without colour. The house had lost its shelter. It was night before roads were made, or houses. It was the night that dwellers in caves had watched from some high place among rocks.

Then the curtain rose. They spoke.

These are the concluding words of the book. They verify the vision of Miss La Trobe: for this is the scene which she visualises in the pub; and the words she hears are the words which are about to be spoken as the curtain rises and the book ends. In this way the pageant which she staged earlier in the day, like this new pageant she is creating, is revealed as not merely a reconstruction of the past but as a prophetic insight into the essential reality of human life at all times and places.

This effect is reinforced at the end because the reader suddenly is related to the novel in the same way that Giles and Isa are related to Miss La Trobe's pageant. Throughout the book Giles and Isa have watched a pageant which they consider an interpretation of their life; suddenly, at the end, the reader sees that it *is* their life. Then he feels in full the impact of the novel as an interpretation of *his* life, because the novel as a whole is a pageant occurring literally *between the acts* of the drama which the reader himself plays before and after reading it. The effect of this device is to intensify and to expand the vision of the book. Its scope suddenly widens from the limits of the events in it to include the events of the reader's own experience. When the curtain rises at the end, he closes the book and carries on the ensuing drama himself.

The vision of this book is clearly tragic. Shrouded in the sheltering and obscuring medium of time, Giles and Isa loom immense and representative, Man and Woman, gripped by the primeval necessities of love and hate and creation. From the complex interaction of these forces and others rises our eternal life; this is the deepest significance of Miss La Trobe's vision. The tragedy is that Giles and Isa, ourselves, do not retain their vision of this redeeming reality. They see in a glass darkly, a fact which leads Miss La Trobe to include comedy and tragedy in her pageant. But the comedy of our situation is so profound that it is almost indistinguishable from tragedy. Isa and Giles struggle and cleave together in 'the heart of darkness, the fields of night'. In the darkness their grandeur is titanic, but it is the grandeur of beings doomed to ceaseless struggle against the power on which they depend for existence.

There is no exultantly triumphant vision in *Between the Acts*; but it is a moot point whether a tragic vision of life can be called, as many have called this book, utterly negative. Although the luminous evasions of *To the Lighthouse* have been abandoned, the essential affirmation of the value of life remains, disenchanted but not destroyed.

Of the two solutions to our original problem it embodies the second— that is, the effort to find in one of the two worlds of mind and linear time an absolute. For *Between the Acts* means nothing less than this: if we would be saved, we must, through the exercise of our creative powers of mind and imagination, recreate time in our own image.

NOTES

[1] *Mr Bennett and Mrs Brown* (1924), p. 24.
[2] *The Voyage Out* (1929), p. 322.
[3] *Mrs Dalloway* (Modern Library ed., 1928), p. 5.
[4] *To the Lighthouse* (1937), p. 99.
[5] *The Waves* (1931), p. 158.
[6] *Between the Acts* (1941), p. 21.

Jean-Paul Sartre†   WILLIAM FAULKNER (1897–1962)

On *The Sound and the Fury*: time in the work of Faulkner

The first thing that strikes one in reading *The Sound and the Fury* is its technical oddity. Why has Faulkner broken up the time of his story and scrambled the pieces? Why is the first window that opens out on this fictional world the consciousness of an idiot? The reader is tempted to look for guidemarks and to re-establish the chronology for himself:

Jason and Caroline Compson have had three sons and a daughter. The daughter, Caddy, has given herself to Dalton Ames and become pregnant by him. Forced to get hold of a husband quickly . . .

Here the reader stops, for he realises he is telling another story. Faulkner did not first conceive this orderly plot so as to shuffle it afterwards like a pack of cards; he could not tell it in any other way. In the classical novel, action involves a central complication; for example, the murder of old Karamazov or the meeting of Edouard and Bernard in *The Coiners*. But we look in vain for such a complication in *The Sound and the Fury*. Is it the castration of Benjy or Caddy's wretched amorous adventure or Quentin's suicide or Jason's hatred of his niece? As soon as we begin to look at any episode, it opens up to reveal behind it other episodes, all the other episodes. Nothing happens; the story does not unfold; we discover it under each word, like an obscene and obstructing presence, more or less condensed, depending upon the particular case. It would be a mistake to regard these irregularities as gratuitous exercises in virtuosity. A fictional technique always relates back to the novelist's metaphysics. The critic's task is to define the latter before evaluating the former. Now, it is immediately obvious that Faulkner's metaphysics is a metaphysics of time.

Man's misfortune lies in his being time-bound.

. . . a man is the sum of his misfortunes. One day you'd think misfortune would get tired, but then time is your misfortune . . .

† Reprinted from *Literary and Philosophical Essays*, translated by Annette Michelson (1955), pp. 78–87. Written in July 1939.

Such is the real subject of the book. And if the technique Faulkner has adopted seems at first a negation of temporality, the reason is that we confuse temporality with chronology. It was man who invented dates and clocks.

Constant speculation regarding the position of mechanical hands on an arbitrary dial which is a symptom of mind-function. Excrement Father said like sweating.

In order to arrive at real time, we must abandon this invented measure which is not a measure of anything.

. . . . time is dead as long as it is being clicked off by little wheels; only when the clock stops does time come to life.

Thus Quentin's gesture of breaking his watch has a symbolic value; it gives us access to a time without clocks. The time of Benjy, the idiot, who does not know how to tell time, is also clockless.

What is thereupon revealed to us is the present, and not the ideal limit whose place is neatly marked out between past and future. Faulkner's present is essentially catastrophic. It is the event which creeps up on us like a thief, huge, unthinkable—which creeps up on us and then disappears. Beyond this present time there is nothing, since the future does not exist. The present rises up from sources unknown to us and drives away another present; it is forever beginning anew. 'And . . . and . . . and then.' Like Dos Passos, but much more discreetly, Faulkner makes an accretion of his narrative. The actions themselves, even when seen by those who perform them, burst and scatter on entering the present.

I went to the dresser and took up the watch with the face still down. I tapped the crystal on the dresser and caught the fragments of glass in my hand and put them into the ashtray and twisted the hands off and put them in the tray. The watch ticked on.

The other aspect of this present is what I shall call a sinking in. I use this expression, for want of a better one, to indicate a kind of motionless movement of this formless monster. In Faulkner's work, there is never any progression, never anything which comes from the future. The present has not first been a future possibility, as when my friend, after having been *he for whom I am waiting*, finally appears. No, to be present means to appear without any reason and to sink in. This sinking in is not an abstract view. It is within things themselves that Faulkner perceives it and tries to make it felt.

The train swung around the curve, the engine puffing with short, heavy blasts, and they passed smoothly from sight that way, with that quality of shabby and timeless patience, of static serenity . . .

And again,

Beneath the sag of the buggy the hooves neatly rapid like motions of a lady doing embroidery, *diminishing without progress*[1] like a figure on a treadmill being drawn rapidly off-stage.

It seems as though Faulkner has laid hold of a frozen speed at the very heart of things; he is grazed by congealed spurts that wane and dwindle without moving.

This fleeting and unimaginable immobility can, however, be arrested and pondered. Quentin can say, 'I broke my watch', but when he says it, his gesture is *past*. The past is named and related; it can, to a certain extent, be fixed by concepts or recognised by the heart. We pointed out earlier, in connection with *Sartoris*, that Faulkner always showed events when they were already over. In *The Sound and the Fury* everything has already happened. It is this that enables us to understand that strange remark by one of the heroes, '*Fui. Non sum.*' In this sense, too, Faulkner is able to make man a sum total without a future: 'The sum of his climactic experiences', 'The sum of his misfortunes', 'The sum of what have you.' At every moment, one draws a line, since the present is nothing but a chaotic din, a future that is past. Faulkner's vision of the world can be compared to that of a man sitting in an open car and looking backwards. At every moment, formless shadows, flickerings, faint tremblings and patches of light rise up on either side of him, and only afterwards, when he has a little perspective, do they become trees and men and cars.

The past takes on a sort of super-reality; its contours are hard and clear, unchangeable. The present, nameless and fleeting, is helpless before it. It is full of gaps, and, through these gaps, things of the past, fixed, motionless and silent as judges or glances, come to invade it. Faulkner's monologues remind one of aeroplane trips full of air-pockets. At each pocket, the hero's consciousness 'sinks back into the past' and rises only to sink back again. The present is not; it becomes. Everything *was*. In *Sartoris*, the past was called 'the stories' because it was a matter of family memories that had been constructed, because Faulkner had not yet found his technique.

In *The Sound and the Fury* he is more individual and more undecided. But it is so strong an obsession that he is sometimes apt to disguise the present, and the present moves along in the shadow, like an underground river, and reappears only when it itself is past. When Quentin insults Blaid,[1] he is not even aware of doing so; he is reliving his dispute with Dalton Ames. And when Blaid punches his nose, this brawl is covered over and hidden by Quentin's past brawl with Ames. Later on, Shreve relates how Blaid hit Quentin; he relates this scene because it has become a story, but while it was unfolding in the present, it was only a furtive movement, covered over by veils. Someone once told me about an old monitor who had grown senile. His memory had stopped like a broken watch; it had been arrested at his fortieth year. He was sixty, but didn't know it. His last memory was that of a schoolyard and his daily walk around it. Thus, he interpreted his present in terms of his past and walked about his table, convinced that he was watching students during recreation.

Faulkner's characters are like that, only worse, for their past, which is in order, does not assume chronological order. It is, in actual fact, a matter of emotional constellations. Around a few central themes (Caddy's pregnancy, Benjy's castration, Quentin's suicide) gravitate innumerable silent masses. Whence the absurdity of the chronology of 'the assertive and contradictory assurance' of the clock. The order of the past the order of the heart. It would be wrong to think that when the present is past it becomes our closest memory. Its metamorphosis can cause it to sink to the bottom of our memory, just as it can leave it floating on the surface. Only its own density and the dramatic meaning of our life can determine at what level it will remain.

Such is the nature of Faulkner's time. Isn't there something familiar about it? This unspeakable present, leaking at every seam, these sudden sudden invasions of the past, this emotional order, the opposite of the voluntary and intellectual order that is chronological but lacking in reality, these memories, these monstrous and discontinuous obsessions, these intermittences of the heart—are not these reminiscent of the lost and recaptured time of Marcel Proust? I am not unaware of the differences between the two; I know, for instance, that for Proust salvation lies in time itself, in the full reappearance of the past. For Faulkner, on the contrary, the past is never lost, unfortunately; it is always there, it is an obsession. One escapes from the temporal world only through mystic ecstasies. A mystic is always a man who wishes to forget something, his self or, more often, language or objective representations. For Faulkner, time must be forgotten.

'Quentin, I give you the mausoleum of all hope and desire; it's rather excruciatingly apt that you will use it to gain the reductio ad absurdum of all human experience which can fit your individual needs no better than it fitted his or his father's. I give it to you not that you may remember time, *but that you might forget it now and then for a moment* and not spend all your breath trying to conquer it. Because no battle is ever won he said. They are not even fought. The field only reveals to man his own folly and despair, and victory is an illusion of philosophers and fools.'

It is because he has forgotten time that the hunted negro in *Light in August* suddenly achieves his strange and horrible happiness.

It's not when you realize that nothing can help you—religion, pride, anything— it's when you realize that you don't need any aid.

But for Faulkner, as for Proust, time is, above all, *that which separates.* One recalls the astonishment of the Proustian heroes who can no longer enter into their past loves, of those lovers depicted in *Les Plaisirs et les jours,* clutching their passions, afraid they will pass and knowing they will. We find the same anguish in Faulkner.

. . . people cannot do anything very dreadful at all, they cannot even remember tomorrow what seemed dreadful today . . .

and

. . . a love or a sorrow is a bond purchased without design and which matures willynilly and is recalled without warning to be replaced by whatever issue the gods happen to be floating at the time . . .

To tell the truth, Proust's fictional technique *should have been* Faulkner's. It was the logical conclusion of his metaphysics. But Faulkner is a lost man, and it is because he feels lost that he takes risks and pursues his thought to its uttermost consequences. Proust is a Frenchman and a classicist. The French lose themselves only a little at a time and always manage to find themselves again. Eloquence, intellectuality and a liking for clear ideas were responsible for Proust's retaining at least the semblance of chronology.

The basic reason for this relationship is to be found in a very general literary phenomenon. Most of the great contemporary authors, Proust, Joyce, Dos Passos, Faulkner, Gide and Virginia Woolf have tried, each in his own way, to distort time. Some of them have deprived it of its past and future in order to reduce it to the pure intuition of the instant; others, like Dos Passos, have made of it a dead and closed memory. Proust and Faulkner have simply decapitated it. They have deprived it of its future, that is, its dimension of deeds and freedom. Proust's heroes never undertake anything. They do, of course, make plans, but their plans remain stuck to them and cannot be projected like a bridge beyond the present. They are day-dreams that are put to flight by reality. The Albertine who appears is not the one we were expecting, and the expectation was merely a slight, inconsequential hesitation, limited to the moment only. As to Faulkner's heroes, they never look ahead. They face backwards as the car carries them along. The coming suicide which casts its shadow over Quentin's last day is not a human possibility; not for a second does Quentin envisage the possibility of *not* killing himself. This suicide is an immobile wall, a *thing* which he approaches backwards, and which he neither wants to nor can conceive.

. . . you seem to regard it merely as an experience that will whiten your hair overnight so to speak without altering your appearance at all . . .

It is not an *undertaking*, but a fatality. In losing its element of possibility it ceases to exist in the future. It is already present, and Faulkner's entire art aims at suggesting to us that Quentin's monologues and his last walk *are already* his suicide. This, I think, explains the following curious paradox: Quentin thinks of his last day in the past, like someone who is remembering. But in that case, since the hero's last thoughts coincide approximately with the bursting of his memory and its annihilation, who is remembering? The inevitable reply is that the novelist's skill consists in the choice of the present moment from which he narrates the past. And Faulkner, like Salacrou in *L'Inconnu d'Arras*, has chosen the infinitesimal instant of death. Thus, when Quentin's memory begins to unravel its recollections ('Through the wall I heard Shreve's bed-springs and then his slippers on the floor hishing. I got up . . .') *he is already dead*. All this artistry and, to speak frankly, all this illusion are meant, then, merely as substitutions for the intuition of the future lacking in the author himself. This explains everything, particularly the irrationality of time; since the present is the unexpected, the formless can be determined only by an excess of memories. We now also understand why duration is 'man's characteristic misfortune'. If the future has reality, time withdraws us from the past and brings us nearer to the future; but if you do away with the future, time is no longer that which separates, that which cuts the present off from itself. 'You cannot bear to think that someday it will no longer hurt you like this.' Man spends his life struggling against time, and time, like an acid, eats away at man, eats him away from himself and prevents him from fulfilling his human character. Everything is absurd. 'Life is a tale told by an idiot, full of sound and fury, signifying nothing.'

But is man's time without a future? I can understand that the nail's time, or the clod's or the atom's is a perpetual present. But is man a thinking nail? If you begin by plunging him into universal time, the time of planets and nebulae, of tertiary flexures and animal species, as into a bath of sulphuric acid, then the question is settled. However, a consciousness buffeted so from one instant to another ought, *first of all*, to be a consciousness and then, *afterwards*, to be temporal; does anyone believe that time can come to it from the outside? Consciousness can 'exist within time' only on condition that it become time as a result of the very movement by which it becomes consciousness. It must become 'temporalised', as Heidegger says. We can no longer arrest man at each present and define him as 'the sum of what he has'. The nature of consciousness implies, on the contrary, that it project itself into the future. We can understand what it is only through what it will be. It is determined in its present being by its own possibilities. This is what Heidegger calls 'the silent force of the possible'. You will not recognise within yourself Faulkner's man, a creature bereft of possibilities and explicable only in terms of what he has been. Try to pin down your consciousness and probe it. You will see that it is hollow. In it you will find only the future.

I do not even speak of your plans and expectations. But the very gesture that you catch in passing has meaning for you only if you project its fulfilment out of it, out of yourself, into the not-yet. This very cup, with its bottom that you do not see—that you might see, that is, at the end of a movement you have not yet made—this white sheet of paper, whose underside is hidden (but you could turn over the sheet) and all the stable and bulky objects that surround us display their most immediate and densest qualities in the future. Man is not the sum of what he has, but the totality of what he does not yet have, of what he might have. And if we steep ourselves thus in the future, is not the formless brutality of the present thereby attenuated? The single event does not spring on us like a thief, since it is, by nature, a Having-been-future. And if a historian wishes to explain the past, must he not first seek out its future? I am afraid that the absurdity that Faulkner finds in a human life is one that he himself has put there. Not that life is not absurd, but there is another kind of absurdity.

Why have Faulkner and so many other writers chosen this particular absurdity which is so un-novelistic and so untrue? I think we should have to look for the reason in the social conditions of our present life. Faulkner's despair seems to me to precede his metaphysics. For him, as for all of us, the future is closed. Everything we see and experience impels us to say, 'This can't last.' And yet change is not even conceivable, except in the form of a cataclysm. We are living in a time of impossible revolutions, and Faulkner uses his extraordinary art to describe our suffocation and a world dying of old age. I like his art, but I do not believe in his metaphysics. A closed future is still a future. 'Even if human reality has nothing more "before" it, even if "its account is closed", its being is still determined by this "self-anticipation". The loss of all hope, for example, does not deprive human reality of its possibilities; it is simply a way of *being* toward these same possibilities.'[3]

NOTES

[1] The author's italics.
[2] Compare the dialogue with Blaid inserted into the middle of the dialogue with Ames: 'Did you ever have a sister?' etc., and the inextricable confusion of the two fights.
[3] Heidegger, *Sein und Zeit*.

W. P. Albrecht†   THOMAS WOLFE (1900–38)

Time as unity
in Thomas Wolfe

Attempting a form to correspond with his perception of reality, the modern novelist frequently has abandoned the framework of time and space familiar in the nineteenth-century novel. As the novelist tries to convey his own sense of the passing and duration of time, he is likely to link moments of time not by succession but by the continuity of personality, feeling, or development. And, especially if clock or calendar time seems hostile to a desired permanence or security, the novelist may look for a time pattern, or metaphor, more compatible with his desires.

It is largely through his effort to find permanence in flux that the novels of Thomas Wolfe may be considered 'modern' in their treatment of time. In Wolfe's novels time becomes a rushing, all-erosive river, which, nevertheless, may be arrested or turned back by the memory. Like Proust, Wolfe seeks to recapture the past through memory, including unconscious memory, and to show the sensations and moods that recollections of the past evoke in the present. Or again, like Joyce in *Finnegans Wake*, he opposes a linear concept of time with a cyclical one, wherein the eternal is repeated through apparent change.

In Wolfe's novels, however, the recollection of the past is clearly labelled as such and is not, as in Joyce or Virginia Woolf, fused almost indistinguishably with the present. Of course some scenes, like the cross-sectioning of Altamont through glimpses of simultaneous actions, show the influence of *Ulysses*, but unlike Joyce, Wolfe almost always orients the reader in time and space. In itself, the cross-sectioning technique does not make these scenes any more clearly part of the recaptured past than the more traditionally handled scenes; it does not provide a solution to Wolfe's time problem.

Wolfe's metaphor of time, therefore, is not a refocusing that sharpens the meaning or relevance of past and present actions while blurring the usual co-ordinates of time and space. The kind of unity that such refocusing gives the action of *Ulysses*, for instance, is lacking in Wolfe's novels.

† Reprinted from *New Mexico Quarterly Review*, xix (1949), pp. 320–9.

Nevertheless, the feeling of time—of flux and permanence—unifies each of the completed novels and the four novels considered as one. This is clearly not the temporal logic of *A Farewell to Arms* or even a unity of character development as found in *Of Human Bondage* or *Portrait of the Artist as a Young Man*, but a relating of action to time through image, myth, and symbol.

Wolfe was concerned, first, with the individual and, later, with society in relation to time. His first problem was that of the individual seeking stability in an ever-flowing river of time. This problem he temporarily solved through creativeness, which in a sense recaptured Eugene's past, made Eugene aware of growth, and gave his life direction. But creativeness was not enough. Eugene still faced the problem of his relations with other people. Neither as a man nor as an artist, he discovers in *Of Time and the River*, could he really escape behind the wall of creative solitude that he was building in *Look Homeward, Angel*. His relations with other people once more involved the time problem: time stood in the way of the relationships he desired; it did not let him gain from people all the love, pleasure, knowledge, significance that he wanted. But Eugene and, even more definitely, George Webber discovered, partly through the creative process, that certain human experiences are typical of all human experiences, that identification with the archetypical could bring a man something of the stability he desired. George could never achieve a sustained love for any individual, but sympathy with and understanding of many people enabled him to feel, at the close of *The Web and the Rock*, that he belonged to the great family of earth, no longer isolated in time and space.

Yet neither Eugene nor George, in recapturing the past through creativeness or identifying himself with the buried life of all men, could make time stand still. The forms of life—plant, animal, social—die when they cease to grow. Only growth with time permits life. At the end of *Look Homeward, Angel* Eugene realises that he can go home to the past only for the materials of his memory, not for his old ideas, his old loves, his old self. At the end of *You Can't Go Home Again* George knows that neither he nor the democracy of which he is a part can go home to its past, but can live only by flowing onward with the river of time.

In *Look Homeward, Angel*, to dramatise the individual lost and then found in relation to time, Wolfe uses the myth of pre-existence-and-return, and with it the usual Platonic contrasts of dark and light, many and one, isolation and union, imprisonment and freedom, shadow and reality. The images of loss or transience are dark, and in relation to lost or passing time man is only a 'ghost' or 'phantom', a 'stranger' isolated from others and even his own true self. 'Memory' links Eugene with a better time—a time of security and certitude suggested by the 'golden' abundance of Gant's Pennsylvania. (Similarly, in George Webber's memory the 'warmth and radiance' of his father's north has been woven with the 'darkness' of the Joyner's south.) But even while 'imprisoned in the dark womb' of his mother, Eugene began to lose the 'communications of eternity'. Those better times, like any past time, are in themselves irrecoverable, and in his loss Eugene becomes a 'stranger' in the 'insoluble prison of being', a 'phantom' destined to wander homeless and friendless in a darkened world.

Lostness in relation to past time cannot, of course, be completely separated from lostness in relation to what Wolfe, in *The Story of a Novel*, calls 'time immutable'; for it is the latter that sweeps away past, present, and future. In *Of Time and the River* and *The Web and the Rock* time immutable is symbolised by the river. The river of time suggests both transience and permanence. Time passes, and with it man's life. But time is also eternal in its flow, so that it becomes the immutable background for mutable life. In either sense it is usually dark, like the darkening or fading past, like the tragedy of man's strangeness and evanescence. It is a 'dark eternal river' in which man is a 'phantom flare of grieved desire'.

In *Look Homeward, Angel,* and less completely in *Of Time and the River*, permanence in flux and reintegration with the past are achieved through the act of literary creation. Wolfe's creative experience is best described by Wolfe himself in *The Story of a Novel*. In Paris in the summer of 1930 he felt the 'naked homelessness' that great cities always caused him, and in sheer effort to break the spell of time and distance that separated him from home, Wolfe's creative process began. Recalled by the image of some 'familiar, common thing' in the past, 'the million forms and substances' of Wolfe's life in America swarmed in 'blazing pageantry' across his mind, issuing even from the 'furthest adyt of his childhood before conscious memory had begun', yet transformed with the new wonder of discovery; and confronted by these blazing forms, Wolfe set himself the task of bringing them to life in a 'final coherent union'. In *The Story of a Novel* Wolfe explicitly names the 'door' of his search as the door to creative power.

This process is dramatised in *Look Homeward, Angel*. Again the images of dark and light suggest *lostness in time* and *being found*. Along with his sense of time as a fading light, Eugene finds within himself a living pattern of certain experiences originally separate in time but fused by imagination beyond the distinctions of the time-space world, the 'many' become 'one'. '. . . These images that burnt in him existed without beginning or ending, without the essential structure of time.' They have a 'white living brightness' compared with 'the ghostliness of all things else.' It is among these living images that Ben, although dead, takes his place, no longer a ghost, but bright and alive. 'And through the Square, unwoven from lost time, the fierce bright horde of Ben spun in and out its deathless loom.' A horde of Eugenes, too, 'which were not ghosts', troop past. 'And now the Square was thronging with their lost bright shapes, and all the minutes of lost time collected and stood still.'

Eugene has not only recovered Ben but found himself. The symbols of 'ghost' and 'angel' suggest, respectively, a spirit lost in death and a spirit secure in eternal life. In *Lycidas* the Angel is asked to look nearer home and to have pity on the drifting corpse of Lycidas. By analogy Ben's role in *Look Homeward, Angel* would seem to be the angel's, while Eugene is Lycidas; but throughout the novel Ben is also a ghost in that, like every person, he cannot be known even to his brother. In the last chapter he is restored to a 'life' that he did not have while alive; he is no longer a ghost because no longer a stranger. At the same time he is also an angel in the sense that he can now direct Eugene 'home'. '*You* are your world,' says Ben to Eugene, directing him to the bright world of fused experience. Ben is not explicitly named 'angel', but the identification is further implied by the stone angels' coming to life when Ben returns. In their marble deadness, the angels in the shop stand for Gant's frustration as an artist. But with Ben's return the angels come alive and with them Eugene's creative power. The title, therefore, is appropriately addressed to Eugene as well as Ben.

In 'God's lonely man' Wolfe again describes his creative process as breaking through the dark isolation of time to unite the creator with a bright permanence. But even in his creativeness the lonely man is still lonely. In *Look Homeward, Angel* Eugene has only exchanged one kind of loneliness for another: the inescapable, involuntary loneliness of all humanity for the voluntary loneliness of the creator. In *Of Time and the River* the 'spell of time', although it has recaptured the images of life in America, becomes an 'evil dream'. Remembered human relationships are not an adequate substitute for actual human relationships. Time immutable, as well as past time, denies Eugene and later George the complete and significant relationships they are seeking. The people they want to know, like the books they want to read and the women they want to love, are all too many and the time too short.

Wolfe's solution of this problem is necessarily symbolical; unable to know the plurality of experience, Eugene and George must choose the representative singular, the symbol. This solution is implied by the sub-title of *Look Homeward, Angel*: *A Story of the Buried Life*. The 'buried life' of Matthew Arnold's poem is the essential self wherein all men are one, brothers not strangers, a self which is realised in moments of love which, in turn, make man articulate and give his life direction. Again, as in the pre-existence-and-return myth, there are the familiar opposites of the many and the one, isolation and unity, false self and true self, weakness and power. The ending of *Look Homeward, Angel* stresses the articulateness and the direction rather than the unity and the love, although the latter are implicit in the final interview with Ben. In *Of Time and the River* Eugene clearly recognises, with pleasure, that the commonalty of man's experience resists the sorcery of time and space. Especially during those days in Tours when the bright images of America rush back into his consciousness, Eugene feels the similarity of all human experience everywhere and the consequent abridgement of the time and space that separate him from home.

Many a scene in the little French towns seems to Eugene 'intolerably near and familiar . . . and something that he had always known'. As he rediscovers 'the buried life, the fundamental structure of the great family of earth to which all men belong', he is filled 'with quiet certitude and joy'.

It is this archetypal quality that George feels deeply in the experiences of the little unknown men in *You Can't Go Home Again*. The face of his neighbour across the street, 'immutable, calm, impassive, . . . became for him the symbol of a kind of permanence in the rush and sweep of chaos in the city. . . . That man's face became for him the face of Darkness and of Time. It never spoke, and yet it had a voice—a voice that seemed to have the whole earth in it.' The man's face is a symbol of permanence in flux, but still dark in its suggestion of the pain and struggle that must precede a final peace. In the last two novels the frantic race with time gradually subsides as George realises that to know a part of the earth well and to understand the life of that part is to know the whole earth. Through a sympathetic identification George, in *The Web and the Rock*, has come to love 'life' and his 'fellow men' and to feel at one with 'the family of the earth'.

The permanent and qualitative, therefore, may be found in the temporary and quantitative, for transient multiplicity reflects timeless uniformity. The repetition of the archetype through numberless forms is a cyclical concept of time inherent in the pre-existence-and-return myth and in the metaphor of the buried life. With the cycle of time Wolfe frequently unites the earth symbol, which usually suggests permanence. Earth's cycle of growth preserves for Eugene something of past time, for after Ben's death Eugene knows that Ben will 'come again . . . in flower and leaf. . . .' Growth as the solution to the time problem is further suggested by the 'self' to which Ben directs Eugene's search, a self of accumulated experience which has not simply been but which is always becoming. Eugene's and George's delight in train rides and Wolfe's frequent image of the train rushing through the night in time with time imply even in the earlier novels that synchronism was the magic needed to break the spell of time. It is to this conception of permanence in change—of growth with time and the repetition of eternal forms through growth—that Wolfe turns for his final solution of the problem of the individual and society in relation to time.

Like the other novels *You Can't Go Home Again* ends with a kind of soliloquy, but compared with the endings of the other novels, George's closing letter to Foxhall Edwards is a straightforward piece of exposition with less than usual of narrative, scene, image, or symbol. But the symbols of time still appear. The river has its beneficent aspects; time as "Flow" has become definitely good. The earlier George, like Eugene, had tried to fix time or turn it back: to keep the past, to hoard up all experience within himself, to halt time until all space was his. Now George knows that 'the essence of faith is the knowledge that all flows and that everything must change'. Such immutability as man may attain is in growth.

This is the point at which, George discovers, he and Fox must part.

Fox is Ecclesiastes. He believes that mankind is condemned to irremediable evils. But George believes that 'man's life can be, and will be better' if 'Men-Alive', although only the creatures of Now and not forever, take up the battle of truth against 'fear, hatred, slavery, cruelty, poverty, and need. . . .' George finds his own life a symbol of the growth that synchronises man with the river of time, giving him the stability of time. Fox is 'the rock of life'; George is 'the web'. Fox is 'Time's granite'; George is 'Time's plant'. Fox is 'mankind'; George is 'Man-Alive'. George is not denying Fox's kind of stability, but he is claiming validity for his own kind. 'You and the Preacher may be right for all eternity, but we Men-Alive, dear Fox, are right for now.'

Wolfe never abandons the river symbol—in fact, it occurs in the closing sentence of *You Can't Go Home Again* and on the last page of Wolfe's last and uncompleted novel, *The Hills Beyond*—but the circular rather than the linear concept of time gains in emphasis. Like Fox, George reflects some of the wisdom of Ecclesiastes: '. . . Unto the place from whence the rivers come, thither they return again.' In his letter to Fox, George's life has become a 'circle', which George feels he has now rounded out. His 'whole experience' has swung round, 'as though through a predestined orbit. . . .' As the earth, the symbol of permanence, manifests its life in the cycle of growth, so has George made his life a symbol of permanence. He had to work himself out of the 'giant web' of the past, but with the 'plant' of this recollectiveness finally 'unearthed', the 'circle' of George's life has finally 'come full swing. . . .' This comparison of his cycle of creativeness with the earth's cycle of creation suggests the organic vitality of his accomplishment—'complete and whole, compacted of the very earth that had produced it, and of which it was itself the last and living part'—and the resulting sense of integration with the earth.

The ultimate resolution is not, however, in George's creativeness as a novelist. The themes of fraternity and movement with time are combined to resolve the problem of society. Like the individual, society must grow with time. To realise the brotherhood of man society must let its old forms die and create new ones. His visit to Hitler's Germany has shown George that the disease of Nazism is a virulent form of the disease that has struck America. America is lost, and only through growth may she find herself. '. . . The enemy is single selfishness and compulsive greed', and only through a change in the structure of society may the enemy be defeated.

I think that the life which we have fashioned in America . . . was self-destructive in its nature, and must be destroyed. I think these forms are dying and must die, just as I know that America and the people in it are deathless, undiscovered, and immortal, and must live.

The brotherhood of man, like the life-principle in nature, reveals its vitality in a succession of ever-changing forms.

Whether this pattern of man in relation to time is always adequately dramatised in character and action is, of course, an important question but one outside the scope of this article. The purpose here has been simply to analyse the time problem and to define the unity suggested by Wolfe's metaphors of time.

The unity of each novel, and of the four novels as one, is clarified by the opposition of the linear and the cyclical concepts of time. These two metaphors parallel each other throughout the four novels, the first representing the problem and the second its solution. The linear concept (that what passes is gone forever) is dominant in *Look Homeward, Angel*, *Of Time and the River*, and *The Web and the Rock*. In *Look Homeward, Angel*, however, it is combined with the pre-existence myth; pre-existence in bright, permanent unity and descent to a dark, transitory isolation. The return to brightness, permanence, and unity, which historically is implied in the pre-existence myth, is suggested only at the end of *Look Homeward, Angel*, by Ben's return in 'flower and leaf' and more definitely by Eugene's creative memory and, through it, his integration with the past. Likewise, in *Of Time and the River* and *The Web and the Rock*, the river symbol is opposed by the recurrent representations of the buried life, although, despite the ending of *The Web and the Rock*, the emphasis in both novels is on flux, and the dark images of transience and isolation are dominant. Finally, in *You Can't Go Home Again*, the cyclical concept solves the problem of both the individual and society in relation to time. To suggest the apparent evanescence and disunity of man's life, the images of darkness persist, but the now-beneficent river, the fecund earth, and the cycle of growth emphasise man's permanence and men's organic unity.

Richard Schechner†   SAMUEL BECKETT (1906–)

There's lots of time
in *Godot*

Two duets and a false solo, that's *Waiting for Godot*. Its structure is more musical than dramatic, more theatrical than literary. The mode is pure performance: song and dance, music-hall routine, games. And the form is a spinning away, a centrifugal wheel in which the centre—Time—can barely hold the parts, Gogo and Didi, Pozzo and Lucky, the Boy(s). The characters arrive and depart in pairs, and when they are alone they are afraid: half of them is gone. The boy isn't really by himself, though one actor plays the role(s). 'It wasn't you came yesterday', states Vladimir in Act Two. 'No Sir', the Boy says. 'This is your first time.' 'Yes sir.' Only Godot is alone, at the centre of the play and all outside it at once. 'What does he do, Mr. Godot? . . . He does nothing, Sir.' But even Godot is linked to Gogo/Didi. 'To Godot? Tied to Godot! What an idea! No question of it. (*Pause.*) For the moment.' Godot is also linked to the Boy(s), who tend his sheep and goats, who are his messengers. Nor can we forget that Godot cares enough for Gogo/Didi to send someone each night to tell them the appointment will not be kept. What exquisite politeness.

Pozzo (and we must assume, Lucky) has never heard of Godot, although the promised meeting is to take place on his land. Pozzo is insulted that *his* name means nothing to Gogo/Didi. 'We're not from these parts', Estragon says in apology, and Pozzo deigns, 'You are human beings none the less.' Pozzo/Lucky have no appointment to keep. Despite the cracking whip and Pozzo's air of big business on the make, their movements are random, to and fro across the land, burdens in hand, rope in place: there is always time to stop and proclaim. In Act One, after many adieus, Pozzo says, 'I don't seem to be able . . . (*long hesitation*) . . . to depart.' And when he does move, he confesses, 'I need a running start.' In Act Two, remembering nothing about 'yesterday', Pozzo replies to Vladimir's question, 'Where do you go from here' with a simple 'On.' It is Pozzo's last word.

The Pozzo/Lucky duet is made of improvised movements and set

† Reprinted from *Modern Drama*, IX (1966), pp. 268–76.

speeches (Lucky's has run down). The Gogo/Didi duet is made of set movements (they must be at this place each night at dusk to wait for Godot to come or night to fall) and improvised routines spun out of long-ago learned habits. Pozzo who starts in no place is worried only about Time; he ends without time but with a desparate need to move. Gogo/Didi are 'tied' to this place and want only for time to pass. Thus, part way through the first act the basic scenic rhythm of *Godot* is established by the strategic arrangement of characters: Gogo/Didi (and later the Boy) have definite appointments, a rendezvous they *must* keep. Pozzo/Lucky are free agents, aimless, not tied to anything but each other. For this reason, Pozzo's watch is very important to him. Having nowhere to go, his only relation to the world is in knowing 'the time'. The play is a confrontation between the rhythms of place and time. Ultimately they are co-ordinates of the same function.

Of course, Pozzo's freedom is illusory. He is tied to Lucky—and *vice versa*—as tightly as the others are tied to Godot and the land. In the scenic calculus of the play, rope=appointment. As one co-ordinate weakens, the other tightens. Thus, when Pozzo/Lucky lose their sense of time, there is a corresponding increase in their need to cover space. Lucky's speech is imperfect memory, an uncontrollable stream of un-consciousness, while Pozzo's talk is all *tirade*, a series of set speeches, learned long ago, and slowly deserting the master actor, just as the things which define his identity—watch, pipe, atomiser—desert him. I am reminded of Yeats' 'Circus animals' desertion' where images fail the old poet who is finally forced to 'lie down where all the ladders start / In the foul rag-and-bone shop of the heart.' Here, too, Pozzo will find himself (Lucky is already there). Thus we see these two in their respective penultimate phases, comforted only by broken bursts of eloquence, laments for that lost love, clock time.

The pairing of characters—those duets—links time and space, presents them as discontinuous co-ordinates. Gogo/Didi are not sure whether the place in Act Two is the same as that in Act One; Pozzo cannot remember yesterday; Gogo/Didi do not recall what they did yesterday, 'We should have thought of it [suicide] a million years ago, in the nineties.' Gogo either forgets at once, or he never forgets. This peculiar sense of time and place is not centred in the characters, but *between* them. Just as it takes two lines to fix a point in space, so it takes two characters to *unfix* our normal expectations of time, place, and being. This painting is not unique to *Waiting for Godot*; it is a favourite device of contemporary playwrights. The Pupil and the Professor in *The Lesson*, Claire and Solange in *The Maids*, Peter and Jerry in *The Zoo Story*: these are of the same species as *Godot*. What might these duets mean or be? Each of them suggests a precarious existence, of sense of self and self-in-the-world so dependent on 'the other' as to be inextricably bound up in the other's physical presence. In these plays 'experience' is not 'had' by a single character, but 'shared' between them. It is not a question of fulfilment—of why Romeo wants

Juliet—but of existence. By casting the characters homosexually, the author removes the 'romantic' element: these couples are not joined because of some biological urge but because of some metaphysical necessity. The drama that emerges from such pairing is intense and locked-in—a dream whose focus is internal without being 'psychological'. Internalisation without psychology is naked drama, theatre unmediated by character. That is why, in these plays, the generic structure of their elements—farce, melodrama, vaudeville—is so unmistakably clear. There is no way (or need) to hide structure: that's all there is. But still, in *Godot*, there are meaningful differences between Vladimir and Estragon, Pozzo and Lucky; but even these shadings of individuation are seen only through the couple: to know one character, you have to know both.

In Aristotelian terms drama is made of the linked chain: action > plot > character > thought. Connections run efficiently in either direction, although for the most part one seeks the heart of a play in its action (as Fergusson[1] uses that term). These same elements are in *Godot*, but the links are broken. The discontinuity of time is reflected on this more abstract level of structure. Thus what Gogo and Didi do is not what they are thinking; nor can we understand their characters by adding and relating events to thoughts. And the action of the play—waiting—is not what they are after but what they want most to avoid. What, after all, are their games for? They wish to 'fill time' in such a way that the vessel 'containing' their activities is unnoticed amid the activities themselves. Whenever there is nothing 'to do' they remember why they are here: To wait for Godot. That memory, that direct confrontation with Time, is painful. They play, invent, move, sing to avoid the sense of waiting. Their *activities* are therefore keeping them from a consciousness of the *action* of the play. Although there is a real change in Vladimir's understanding of his experience (he learns precisely what 'nothing to be done' means) and in Pozzo's life, these changes and insights do not emerge from the plot (as Lear's 'wheel of fire' does), but stand outside of what's happened. Vladimir has his epiphany while Estragon sleeps—in a real way his perception is a function of the sleeping Gogo. Pozzo's understanding, like the man himself, is blind. Structurally as well as thematically, *Godot* is an 'incompleted' play; and its openness is not at the end (as *The Lesson* is open-ended) but in many places throughout: it is a play of gaps and pauses, of broken-off dialogue, of speech and action turning into time-avoiding games and routines. Unlike Beckett's perfectly modulated *Molloy, Waiting for Godot* is designed off-balance. It is the very opposite of *Oedipus*. In *Godot* we do not have the meshed ironies of experience, but that special anxiety associated with question marks preceded and followed by nothing.

What then holds *Godot* together? Time, habit, memory, and games form the texture of the play and provide both its literary and theatrical interest. In *Proust*, Beckett speaks of habit and memory in a way that helps us understand *Godot*:

The laws of memory are subject to the more general laws of habit. Habit is a compromise effected between the individual and his environment, or between the individual and his own organic eccentricities, the guarantee of a dull inviolability, the lightning-conductor of his existence. Habit is the ballast that chains the dog to his vomit. . . . Life is a succession of habits, since the individual is a succession of individuals. . . . The creation of the world did not take place once and for all, but takes place every day. [Bibl. *398*]

The other side of 'dull inviolability' is 'knowing', and it is this that Gogo/ Didi must avoid if they are to continue. But knowledge is precisely what Didi has near the end of the play. It ruins everything for him:

Was I sleeping, while the others suffered? Am I sleeping now? To-morrow, when I wake, or think I do, what shall I say of to-day? That with Estragon my friend, at this place, until the fall of night, I waited for Godot? That Pozzo passed, with his carrier, and that he spoke to us? Probably. But in all that what truth will there be? [*Looking at Estragon*] He'll know nothing. He'll tell me about the blows he received and I'll give him a carrot.

Then, paraphrasing Pozzo, Didi continues:

Astride a grave and a difficult birth. Down in the hold, lingeringly, the grave-digger puts on the forceps. We have time to grow old. The air is full of our cries. (*He listens.*) But habit is a great deadener. (*He looks again at Estragon.*) At me too someone is looking, of me too someone is saying, He is sleeping, he knows nothing, let him sleep on. (*Pause.*) I can't go on! (*Pause.*) What have I said?

In realising that he knows nothing, in seeing that habit is the great deadener—in achieving an ironic point of view towards himself, Didi knows everything, and wishes he did not. For him Pozzo's single instant has become 'lingeringly'. For Pozzo 'the same day, the same second' is enough to enfold all human experience; Didi realises that there is 'time to grow old'. But habit will rescue him. Having shouted his anger, frustration, helplessness ('I can't go on!'), Didi is no longer certain of what he said. Dull inviolability has been violated, but only for an instant: one instant is enough for insight, and we have a lifetime to forget. The Boy enters. Unlike the first act, Didi asks him no questions. Instead Didi makes statements. 'He won't come this evening. . . . But he'll come to-morrow.' For the first time, Didi asks the Boy about Godot. 'What does he do, Mr. Godot? . . . Has he a beard, Mr. Godot?' The Boy answers: Godot does nothing, the beard is probably white. Didi says—after a silence—'Christ have mercy on us!' But both thieves will not be saved, and now that the game is up, Vladimir seeks to protect himself.

Tell him . . . (*he hesitates*) . . . tell him you saw me and that . . . (*he hesitates*) . . . that you saw me [. . .] (*With sudden violence.*) You're sure you saw me, you won't come and tell me to-morrow that you never saw me!

The 'us' of the first act is the 'me' of the second. Habits break, old friends are abandoned, Gogo—for the moment—is cast into the pit. When Gogo awakens, Didi is standing with his head bowed. Didi does not tell his friend of his conversation with the Boy nor of his insight or sadness. Gogo asks, 'What's wrong with you', and Didi answers, 'Nothing.' Didi tells Estragon that they must return the following evening to keep their appointment once again. But for him the routine is meaningless: Godot will not come. There is something more than irony in his reply to Gogo's question, 'And if we dropped him?' 'He'd punish us', Didi says. But the punishment is already apparent to Didi: the pointless execution of orders without hope of fulfilment. Never coming; for Didi, Godot has come . . . and gone.

But Didi alone sees behind his old habits and even he, in his ironic musing, senses someone else watching him sleep just as he watches Gogo: he learns that all awareness is relative. Pozzo is no relativist, but a strict naturalist. In the first act he describes the setting of the sun with meticulous hand gestures, twice consulting his watch so as to be precise. Pozzo knows his 'degrees' and the subtle shadings of time's passing. He also senses that when night comes it 'will burst upon us pop! like that! just when we least expect it'. And for Pozzo, once it is night there is no more time, for he measures that commodity by the sun. Going blind, Pozzo too has an epiphany—the exact opposite of Didi's:

Have you not done tormenting me with your accursed time! It's abominable! When! When! One day, is that not enough for you, one day he went dumb, one day I went blind, one day we'll go deaf, one day we were born, one day we shall die, the same day, the same second, is that not enough for you!

Of the light gleaming an instant astride the grave, Pozzo has only a dim memory. He has found a new habit to accommodate his new blindness; his epiphany is false. The experience of the play indeed shows us that there is plenty of time, too much: waiting means more time than things to fill it.

Pozzo/Lucky play a special role in this passing of time that is *Waiting for Godot*'s action. Things have changed for them by Act Two. Pozzo is blind and helpless, Lucky is dumb. Their 'career' is nearly over. Like more conventional theatrical characters, they have passed from bad times to worse. The rope, whip, and valise remain: all else is gone—Lear and the Fool on the heath, that is what this strange pair suggests to me. But if they are that in *themselves*, they are something different to Gogo/Didi. In the first act, Gogo/Didi suspect that Pozzo may be Godot. Discovering that he is not, they are curious about him and Lucky. They circle around their new acquaintances, listen to Pozzo's speeches, taunt Lucky, and so on. Partly afraid, somewhat uncertainly, they integrate Pozzo/Lucky into their world of waiting: they make out of the visitors a way of passing time. And they exploit the *persons* of Pozzo/Lucky, taking food and playing games. (In the Free Southern Theater production, Gogo and Didi pick-pocket Pozzo, stealing his watch, pipe, and atomiser—no doubt to hock them for neccessary food. This interpretation has advantages: it grounds the play in an acceptable reality; it establishes a first act relationship of double exploitation—Pozzo uses them as audience and they use him as income.) In the second act this exploitation process is even clearer. Pozzo no longer seeks an audience. Gogo/Didi no longer think that Pozzo may be Godot (Gogo, briefly, goes through this routine). Gogo/Didi try to detain Pozzo/Lucky as long as possible. They play rather cruel games with them, postponing assistance. It would be intolerable to Gogo/Didi for this 'diversion' to pass quickly, just as it is intolerable for an audience to watch it go on so long. What 'should' be a momentary encounter is converted into a prolonged affair. Vladimir sermonises on their responsibilities. 'It is not everyday that we are needed.' The talk continues without action. Then, trying to pull Pozzo up, Vladimir falls on top of him. Estragon does likewise. Obviously, they can pull Pozzo up (just as they can get up themselves). But instead they remain prone. 'Won't you play with us?' they seem to be asking. But Pozzo is in no playing mood. Despite his protests, Gogo/Didi continue their game. It is, as Gogo says, 'child's play'. They get up, help Pozzo and Lucky up, and the play proceeds. When they are gone, Estragon goes to sleep. Vladimir shakes him awake. 'I was lonely.' And speaking of Pozzo/Lucky, 'That passed the time.' For them, perhaps, but for the audience? It is an ironic scene—the entire cast sprawled on the floor, hard to see, not much action. It makes an audience aware that the time is not passing fast enough.

This game with Pozzo/Lucky is one of many. In fact, the gamesmanship of *Waiting for Godot* is extraordinary. Most of the play is taken up by a series of word games, play acting, body games, routines. Each of these units is distinct, usually cued in by memories of *why* Gogo/Didi are where they are. Unable simply to consider the ramifications of 'waiting', unfit, that is, for pure speculation (as Lucky was once fit), they fall back onto their games: how many thieves were saved, how many leaves on the tree, calling each other names, how can we hang ourselves, and so on. These games are not thematically meaningless, they feed into the rich image-texture of the play; but they are meaningless in terms of the play's action: they lead nowhere, they contribute to the non-plot. Even when Godot is discussed, the talk quickly becomes routinised. At one time Vladimir spoke to Godot. 'What exactly did we ask him for?' Estragon asks. Vladimir replies, 'Were you not there?' 'I can't have been listening.' But it is Gogo who supplies the information that Didi confirms: That their request was 'a kind of prayer . . . a vague supplication'. And it is both of them, in contrapuntal chorus, who confirm that Godot would have to 'think it over . . . in the quiet of his home . . . consult his family . . . his friends . . . his agents . . . his correspondents . . . his books . . . his bank account . . . before taking a decision'.

This kind of conversation populates *Godot*. A discussion or argument is transformed into routinised counterpoint. Much has been said about the beauty of Beckett's prose in this play. More needs to be said about its routine qualities. Clichés are converted into game/rituals by dividing the lines between Gogo and Didi, by arbitrarily assigning one phrase to each. Thus we have a sense of their 'pairdom', while we are entranced by the rhythm of their language. Beckett's genius in dialogue is his *scoring*, not his 'book'. This scoring pertains not only to language but to events as well. Whatever there is to do, is done in duets. By using these, Gogo/Didi are able to convert anxiety into habit. Gogo is more successful at this than Didi. For Gogo things are either forgotten at once or never forgotten. There is no 'time-span' for him, only a kaleidoscopic present in which everything that is there is forever in focus. It takes Didi to remind Gogo of Godot, and these reminders always bring Gogo pain, his exasperated 'Ah'. For Didi the problem is more complex. Gogo says 'no use wriggling' to which Didi replies, 'the essential doesn't change'. These are opposite contentions; that's why they harmonise so well.

A few words about Time. If waiting is the play's action, Time is its subject. Godot is not Time, but he is associated with it—the one who makes but does not keep appointments. (An impish thought occurs: Perhaps Godot passes time with Gogo/Didi just as they pass it with him. Within this scheme, Godot has nothing to do as the Boy tells Didi in Act Two and uses the *whole play* as a diversion in his day. Thus the 'big game' is a strict analogy of the many 'small games' that make the play.) The basic rhythm of the play is habit interrupted by memory, memory obliterated by games. Why do Gogo/Didi play? In order to deaden their sense of waiting. Waiting is a 'waiting *for*' and it is precisely this that they wish to forget. One may say that 'waiting' is the larger context within which 'passing time' by playing games is a sub-system, protecting them from the sense that they are waiting. They confront Time (i.e., are conscious of Godot) only when there is a break in the games and they 'know' and 'feel' that they are waiting.

In conventional drama all details converge on the centre of action. We may call this kind of structure centripetal. In *Godot* the action is centrifugal. Gogo/Didi do their best to shield themselves from a direct consciousness that they are at the appointed place at the prescribed time. If the centre of the play is Time, dozens of activities and capers fling Gogo Didi away from this centre. But events at the periphery force them back inwards: try as they will, they are not able to forget. We may illustrate the structure as in the diagram.

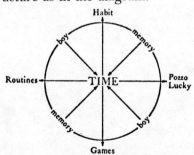

Caught on the hub of this wheel, driven by 're-minders' toward the centre, Gogo/Didi literally have nowhere to go outside of this tight scheme. The scenic counterpart is the time-bracket 'dusk–darkness'—that portion of the day when they must be at the appointed place. But even when night falls, and they are free to go, our last glimpse of them in each act is:

> ESTRAGON. Well, shall we go?
> VLADIMIR. Yes, let's go.
> *They do not move.*

As if to underline the duet-nature of this ending, Beckett reverses the line assignments in Act Two.

What emerges is a strange solitude, again foreshadowed by Beckett in his *Proust*. 'The artistic tendency is not expansive but a contraction. And art is the apotheosis of solitude.' In spinning out from the centre, Gogo/Didi do not go anywhere, 'they do not move'. Yet their best theatrical moments are all motion, a running helter-skelter, a panic. Only at the end of each act, when it is all over for the day, are they quiet. The unmoved mover is Time, that dead identicality of instant and eternity. Once each for Didi and Pozzo, everything is contracted to that sense of Time where consciousness is possible, but nothing else. To wait and not know *how* to wait is to experience Time. To be freed from waiting (as Gogo/Didi are at the end of each act) is to permit the moon to rise more rapidly than it can (as it does on *Godot*'s stage), almost as if nature were illegally celebrating its release from its own clock. Let loose from Time, night comes all of a sudden. After intermissions, there is the next day—and tomorrow, another performance.

There are two time rhythms in *Godot*, one of the play and one of the stage. Theatrically, the exit of the Boy and the sudden night are strong cues for the act (and the play) to end. We, the audience, are relieved—it's almost over for us. They, the actors, do not move—even when the Godot-game is over, the theatre-game keeps them in their place: tomorrow they must return to enact identical routines. Underlying the play (all of it, not just the final scene of each act) is the theatre, and this is exactly what the script insinuates—a nightly appointment performed for people the characters will never meet. *Waiting for Godot* powerfully injects the mechanics of the theatre into the mysteries of the play.

NOTES

[1] Francis Fergusson, in *The Idea of a Theatre* (1949) [Ed.].

The complex grammar
of Auden's time

Typological patterns, of which Auden speaks in *Homage to Clio*, constitute
an ineluctably mysterious framework for our human recognitions:

> Between those happenings that prefigure it
> And those that happen in its anamnesis
> Occurs the Event, but that no human wit
> Can recognize until all happening ceases.[1]

And elsewhere in 'The prophets' he tells of discovering the 'Good Place'—
the spatial equivalent of temporal 'Event'—only in anticipation and in
recollection. This is an apt paradox upon which to base an examination
of Auden's sense of time. That he always sought to recognise the 'Event'—
whether in Marx, Freud, Christ, Austrian kitchens, a cottage on Long
Island playing Buxtehude, or limestone caves of making—is imaged in his
dedication to the present moment, to the Now which may yield its discrete
mystery: 'sites made sacred by something read there, / a lunch, a good lay,
or sheer lightness of heart.'[2] But in his contrary surrender to the process
of time ('I, decent with the seasons') or of History ('Madonna of silences')
there lies both the realisation that while 'happening' is the syntax of time
'no human wit' will register the 'Event' and yet the confidence that in
prefiguration and anamnesis only will the 'Event' still occur.

I

New Year Letter, we may see with hindsight, was one 'Event' : 'O season
of repetition and return . . . the avenues of our longing.'[3] Between the
private recesses and richnesses of its notes (a commonplace book of
Auden's encounters—yet *Persicos odi, puer, apparatus*[4]) and the public,
neo-Augustan presence of its lithe octosyllabics echoes 'the Word which
was / From the beginning.'[5] It is a poem which canvasses all of Auden's
preoccupations with time and history, specifically the Christian injunction

that 'one must neither allow oneself to be ruled by the temporal moment nor attempt to transcend it but make oneself responsible for it, turning time into history'.[6]

In the poetry published before *New Year Letter* there was much celebration of the 'temporal moment'. 'Taller today' contrasts both the uneasy memory of glacier or 'Captain Ferguson' and the failure of nerve ('broke through, and faltered') with the almost timeless, yet precarious present:

> But happy now, though no nearer each other,
> We see the farms lighted all along the valley;
> Down at the mill-shed the hammering stops
> And men go home.
>
> Noises at dawn will bring
> Freedom for some, but not this peace
> No bird can contradict: passing, but is sufficient now
> For something fulfilled this hour, loved or endured.[7]

Much that is central to Auden's early poetry is there : the fragile certainties of a present achieved by love or endurance, the timeless spaces of the night, the distinct human advantage over bird or beast. In '1929' the poet regards the

>   colony of duck below
> Sit, preen, and doze on buttresses
> Or upright paddle on flickering stream,
> Casually fishing at a passing straw.
> Those find sun's luxury enough,
> Shadow know not of homesick foreigner
> Nor restlessness of intercepted growth. (p. 80)

The capacity for restlessness in face of the vicissitudes of growth (and '1929' is an Easter poem) is man's distinction, just as that of 'lovers and writers' is to find an 'altering speech for altering things' (p. 79). In his poetry of the 1930s Auden attends frequently to 'intercepted growth', to 'something immense in the past but now / Surviving only as the infectiousness of disease' (p. 83). It may be imaged in the decadence of country house life, the 'reserved tables' by 'plate-glass windows of the Sport Hotel' (p. 43), or sanatoriums along the 'savage coast' (p. 83). And he glorifies as their antinomies the strenuousness of present attention or of future resolution.

The imperative of 'now' is the more convincing syntax of Auden's thirties poetry. 'Lullaby' seeks to displace the recognitions of guilt and mortality (abstract consciousness of before and after) by the present vision that Venus sends: through a rapt attention to kiss and look the 'mortal world' is found to be blessedly sufficient (pp. 238–9). These sublunary lovers snatch from 'their ordinary swoon' the 'Certainty, fidelity' that eludes more grandiloquent asseverations of timeless passion:

> 'I'll love you, dear, I'll love you
>   Till China and Africa meet,
> And the river jumps over the mountain
>   And the salmon sing in the street.' [p. 227]

The exigencies of present attention are enacted in the verbal music of 'Look, stranger on this island now': though the visual items that the complex effects of sound rehearse for us *may* 'enter / And move in memory', it is their being 'now' that holds us (pp. 243–4). The fabulous world of 'A summer night 1933' may also constitute a resonant memory after the 'dykes of our content' are breached, but its power lies in the edenic longevity of that single summer evening upon the lawn:

> Fear gave his watch no look;
> The lion griefs loped from the shade
> And on our knees their muzzles laid,
>   And Death put down his book. [p. 110]

To surrender to such a passing moment is to secure at least an illusion of permanence. It commands more respect than the evasions man usually practises: 'And Time with us was always popular. / When have we not preferred some going round / To going straight to where we are?' (p. 130). This bias is not the lion's, whose leap never falters in face of the hourglass, nor the rose's, assured below the clock-tower: it is 'Our bias'.

Man's distinction is his awareness of time, his every action taken in its despite or favour:

> Not to lose time, not to get caught,
> Not to be left behind, not, please! to resemble
> The beasts who repeat themselves, or a thing like water
>   Or stone whose conduct can be predicted, these
> Are our Common Prayer . . .[8]

That intercession of 1948 restates the dilemmas of the previous decade within the fresh context of Auden's resolution of them. But the thirties saw less commonalty, less prayer. When the temporal moment was not honoured for its own sake, the need to transcend it authorised movement or the commitment to quest for *its* own sake. 'The watershed' (p. 183), despite a marvellous and lingering admiration for the industrial archaeology of a 'comatose' society and its attendant heroisms, urges the stranger to eschew a nostalgia for 'faces rather there than here'. The poem ends with a resumption of quest ('Beams from your car may cross a bedroom wall') and the rare, heroic stance of the hare ('taller than grass, / Ears poised before decision, scenting danger'). The lesson that is perhaps implied there from the animal kingdom is, however, fraught with problems: Auden rarely neglects to contrast animals' neglect of time with man's preoccupation:

> Fish in the unruffled lakes
> The swarming colours wear,
> Swans in the winter air

A white perfection have,
And the great lion walks
Through his innocent grove;
Lion, fish, and swan
Act, and are gone
Upon Time's toppling wave.

We, till shadowed days are done,
We must weep and sing
Duty's conscious wrong,
The Devil in the clock,
The Goodness carefully worn
For atonement or for luck;
We must lose our loves,
On each beast and bird that moves
Turn an envious look. [pp. 231–2]

Yet the practical implementation of envy is another example of 'our bias':
in 'Missing' (pp. 58–9) a melodramatic, fake heroism is borrowed from the
predatory kestrels floating above the moorland 'scar' : 'Leaving for Cape
Wrath tonight' is the gesture of 'skyline operations', of those heroes whom
the summer tourists seek as a vicarious adventure from city indolence.
What is needed is to fight for somebody's sake, to make oneself responsible for the present and not to die 'beyond the border'.[9]

But the poetry of the thirties also explored the larger contexts—evolution, historical process, memory, the urges of the spirit as well as of the
body—for what 'Kairos and Logos' called the 'rhetoric of time'. That poem
of 1941, like *New Year Letter*, presents a debate between the eternal and
timeless (the creating Word) and the Kairos, which is 'a vast self-love',
the selfish and specific moment of the individual. Like 'Merosis' (p. 96)
and 'Venus will now say a few words' (pp. 118–20), 'Kairos and Logos'
opposes the momentary bliss of mere bodily life, whether a good squash
stroke ('that flick of wrist') or sexual athletics ('the snare forgotten in the
little death'), to the exigencies and joys of larger time: 'another kind of
Death / To which the time-obsessed are all condemned' implies that those
who live for the small death of sexual orgasm[10] will indeed be condemned
to a Death, which—for 'The fair, the faithful and the uncondemned'—is less
punishment than solace. The determination of this larger order is enacted
by the marvellously certain, if aptly unimmediate, continuity of the sestina
form. And *logos*, poets' in recurrence, as in the beginning God's, is enabled
to break out

spontaneously all over time,
Setting against the random facts of death
A ground and possibility of order,
Against defeat the certainty of love. [p. 26]

Auden allows, then, for the moment to be understood in its infinite
context by the artist (God's vicar in the see of time) as well as by the

Creator. Such is the force of the well-known 'Musée des Beaux Arts': the significance of events, whether Icarus' fall or the Massacre of the Innocents, is never apparent when one is immersed in quotidian event. Though there are rival interpretations of Brueghel's vision,[11] Auden certainly provides a memorable version of the view that historical and mythological perspectives are lost in the artist's delighted absorption in the details of the human present. That there were other old masters in the same museum with a less oblique regard for larger time it does not seem Auden's business to mention. Yet, by implication, historical and mythopoeic vision *is* wittily allowed.

The inadequacies of a life, or town, devoted to the instant are imaged in 'Dover 1937'. 'Nothing is made in this town', because it is dedicated to transients: anyone who stops, like the soldiers in the pubs, is 'killing . . . time', while the majority of its visitors are 'departing migrants', pinning their hopes upon the unknown distance, or the 'returning', confident that in some English garden beneath the yew tree 'Everything will be explained.' At best 'Some are temporary heroes: / Some of these people are happy' (pp. 121–2). The alternative to such meagre and transitory content is explored more confidently in the sonnet sequence, 'In time of war' (pp. 271 ff.), originally published in 1939 in *Journey to a War*.

What emerges in 'Dover 1937' is only man's vague aspiration for future consolations: its inadequacy is reflected in the language of fortune-telling in some garish booth perhaps beside the 'historical cliffs':

> 'I see an important decision made on a lake,
> An illness, a beard, Arabia found in a bed,
>   Nanny defeated, Money.' [p. 121]

In contrast, the first sonnet from 'In time of war' makes the larger claim that only man is free to *shape* his future. Bee, fish or peach

> were successful at the first endeavour;
> The hour of birth their only time at college,
> They were content with their precocious knowledge,
> And knew their station and were good for ever. [p. 271]

Though the 'childish creature', man, is changed and shaken by the lightest wind, he can still 'choose his love'. From the lost Eden, whither poet or legislator would return us but cannot, there is only the way forward: the brass bands throbbing in the parks (are parks our present recollection of the once and future Eden?) 'foretell some future reign of happiness and peace'.

Various paradigms of human existence are offered, but the dominant emphasis is upon man's historical awareness—the glance behind to history ('Discovering the past of strangers in a library') and to myth ('an ancient South', for instance), together with the recognition of some evolutionary momentum:

> Certainly the growth of the fore-brain has been a success:
> He has not got lost in a backwater like the lampshell
> Or the limpet; he has not died out like the super-lizards. [p. 286]

Any society that thwarts man's place in this continuum and promises a refuge from time is anathema:

> *Leave Truth to the police and us; we know the Good;*
> *We build the Perfect City time shall never alter;*
> *Our Law shall guard you always like a cirque of mountains,*
>
> *Your ignorance keep off evil like a dangerous sea;*
> *You shall be consummated in the General Will,*
> *Your children innocent and charming as the beasts.* [p. 292]

The 'beats who repeat themselves' cannot be man's models. The authority to conquer time lies not in authoritarian regimes which beguile us with such panaceas but in man's own responsibility and in the possibilities of future grace:

> They hear their deaths lamented in our voice,
> But in our knowledge know we could restore them:
> They could return to freedom; they would rejoice. [p. 276]

The artist's celebration of life ('our buoyant song') is one distinct exercise of that responsibility, that voice. On the memorial stone to Auden recently unveiled in Westminster Abbey are inscribed his own lines on Yeats:

> In the prison of his days
> Teach the free man how to praise.

The paradox constitutes the most urgent and most moving theme in Auden's poetry. The artist will praise the present, the now—'life as it blossoms out in a jar or a face' (p. 277)—and his song will survive 'In the valley of its saying where executives / Would never want to tamper'. Yet, as Auden continues in that poem, poetry is still a part of time, surviving only to be 'modified in the guts of the living':

> it flows south
> From ranches of isolation and the busy griefs,
> Raw towns that we believe and die in; it survives,
> A way of *happening*, a mouth. [p. 65, my italics]

But in times of war the insistence of present horror radically challenges the artist's vision and confidence:

> History opposes its grief to our buoyant song:
> The Good Place has not been; our star has warmed to birth
> A race of promise that has never proved its worth. [p. 278]

'History' seems to serve Auden for the long-term perspectives, which he was to gloss as the 'silence' of Clio,[12] and, as here, the exigencies of *now*, the actualities of the Sino–Japanese war. The latter, through injury and mutilation, connects itself with the other in the image of hospitals, where the injured 'lie apart like epochs from each other'.

II

New Year Letter resumes and reconstitutes the poetry of the thirties. Its injunction—'We have no time until / We know what time we fill'[13] controls the whole meditation. And Time presents itself in various guises, various attitudes. It is the unalterable past, dreaded by the 'sleepless guests of Europe . . . Wishing the centuries away.' It is, in contrast to that historical burden, the attention to and context of current choices, imaged first as Time the surgeon:

> But up the staircase of events,
> Carrying his special instruments,
> To every bed-side all the same
> The dreadful figure swiftly came.

There is time which addresses itself to 'Collective Man' in the 'agora of work and news'. But there is time which 'can moderate his tone / When talking to a man alone': and it is this *ad hominem* address which Auden's *Letter*, itself a private form, records.

The imagery of Quest, never long absent in Auden,[14] is more than usually central to *New Year Letter*. Between the 'Prologue', where he recalls that nothing stops 'us taking our walks alone, / Scared of the unknown, unconditional dark', and the 'Epilogue' ('Will the inflamed ego attempt as before / To migrate again to her family place . .') there are various elements of archtypal Quest, which the poet has told us 'correspond to an aspect of our subjective experience'.[15] Since a new year enforces a special sense of happening, becoming, moving, it also authorises an imagery of quest—the wittily elaborated detective story metaphor of Part One, in Part Two the

> signpost on the barren heath
> Where the rough mountain track divides
> To silent valleys on all sides,

and the various attentions to maps and landscapes in the existential fabric of Part Three. The reader's own explorations among the footnotes, seeking their, sometimes dialectical, rapport with the verses, enacts the imagery of quest, which is finally established in the sequence of poems with that title.

Since Art, although a 'paradigm of the possible,'[16] is nevertheless a completed image of order outside time, typical and abstract, its aptness for personal voyages of discovery is limited:

> Though their particulars are those
> That each particular artist knows,

> Unique events that once took place
> Within a unique time and space,
> In the new field they occupy,
> The unique serves to typify.
> Becomes, though still particular,
> An algebraic formula,
> An abstract model of events
> Derived from dead experience,
> And each life must itself decide
> To what and how it be applied.

Only in brief and mystical moments, like 'A summer night 1933', does life assume the shape, order and abstract time of art. Only then, rarely, does our Becoming ('The seamless live continuum') have a glimpse of Being:

> Yet anytime, how casually,
> Out of his organized distress
> An accidental happiness,
> Catching man off his guard, will blow him
> Out of his life in time to show him
> The field of Being where he may
> Unconscious of Becoming, play
> With the Eternal Innocence
> In unimpeded utterance.

Such an 'Event', typical because typological, returns man briefly and (such is the implication of 'Out of his life in time') unrealistically to the Garden of Eden:

> But perfect Being has ordained
> It must be lost to be regained
> And in its orchards grows the tree
> And fruit of human destiny,
> And men must eat it and depart
> At once with gay and grateful heart.

Yet a rival emphasis is provided in 'The Quest' itself; for there the garden of XX—Alice's, Eliot's in *Burnt Norton*, that of the *Roman de la Rose*—is journey's end, 'The perfect circle time can draw on stone.'

Between those alternating visions of momentary and infinite the discourse of *New Year Letter* wavers. The endeavour has been described as Auden's 'last attempt to do without belief in God':[17] though it may be a critical anamnesis, retrospective identification of 'Event', I find that belief implied and available within the whole volume. It is characteristically Auden (and it is a topic to be returned to) that his apprehension of Christian Being is within the world of Becoming: 'Quando non fuerit, non est.' His conjunction of 'locality and peace' is thus central to the work's achievement: for it is the continuous sequence of localities of Nows and Heres, the

succeeding skylines of the mountaineering passage in Part Three, that
reveal larger patterns to the poet. Thus Auden shares with Blake, one of
his self-appointed judges in Part One, the instinct to hear 'inside each
mortal thing / Its holy emanation sing'. The distractions from locality and
present choices are the lures of various generalisations and our preference
for 'idées fixes to be / True of a fixed Reality'. We glance back at past
certainties and, like Lot's wife, 'shudder' our 'future into stone'. To ignore
the Adam in us (our 'crooked nose') and to resent the 'mere suggestion
that we die / Each moment' is to lose that joyous and innocent dedication
to the 'fact' that we 'live in eternity'. Our intuitions 'mock / The formal
logic of the clock' and even the Devil's energies paradoxically contrive to
thwart his aims and 'push us into grace':

> He has to make the here and now
> As marvellous as he knows how
> And so engrossing we forget
> To drop attention for regret.

Men's attention to 'themselves' and to asking 'What acts are proper to
their task' provides the context and the themes of Part Three and 'The
Quest' sequence. Human hell is to confuse Being and Becoming; heaven,
the Truth won 'out of Time' (ambiguously, 'outside' and 'from' Time), is
the collocation of 'Clock and Keeper of the years' with Unicorn, Dove, Fish,
Holy Spirit and Logos. The paradox is anchored and, indeed, enhanced
by its statement in America, where Auden has joined the 'patriots of the
Now'. The unique and solitary situation of each man, whether mountaineering or colonising the 'landscape of his will and need', is where 'Time
remembered' and 'time required'

> The positive and negative ways through time
> Embrace and encourage each other
> In a brief moment of intersection.

III

The echo in the 'Epilogue' there of Eliot provokes the question whether
Auden's commitment to Christian ideas, which necessarily he shares with
others like Eliot or Charles Williams, allows him any distinctive voice, any
personal contribution to their 'Common Prayer'. The writings that
succeeded New Year Letter, thirty years of poems, oratorio, libretti, all
refine and restate the epiphany of the Manhattan moment; and in so doing
they define the characteristic and best Auden: the combination of a way of
looking—Brueghel's vision, here, remains crucial—with a way of talking—
what might be called his limestone idiom.

There is, New Year Letter reminds us, little chance to sustain any 'unimpeded Utterance'. So the poet must achieve his intimations of redeemed
time, like Berkeley, 'more by grammar than by grace'. Auden increasingly

relies upon a syntax and form where his attention to both large and small, infinite and momentary, can be absorbed into a distinctly *arti*ficial whole. It is the tradition of Horace, englished by Pope, resumed by Byron, who drew it (dangerously) closer to conversational rhythms and habits of perception, and acquired by Auden, whose *Letters to Lord Byron* announced the convenience and scope of this mode:

> a form that's large enough to swim in,
> And talk on any subject that I choose,
> From natural scenery to men and women,
>   Myself, the arts, the European news . . .[18]

In *New Year Letter* this form became implicity linked to a certain Pennine landscape, and finally and explicitly in one of Auden's greatest poems—'In praise of limestone'—the geology lost its specific location and gained a larger territory.

Man's residence on earth has allowed him to 'make its flaws [his] own'. Auden's recognition of these imperfections, yet his sense of our potential heroism, is mirrored in one magnificent area of limestone landscape and in a style Horace termed 'sermoni propiora':[19]

> Whenever I begin to think
> About the human creature we
> Must nurse to sense and decency,
> An English area comes to mind,
> I see the native of my kind
> As a locality I love,
> The limestone moors that stretch from *Brough*
> To *Hexham* and the *Roman Wall*,
> There is my symbol of us all.
> There, where the *Eden* leisures through
> Its sandstone valley, is my view
> Of green and civil life that dwells
> Below the cliff of savage fells
> From which original address
> Man faulted into consciousness.

The movement of the verse, sometimes held awkwardly at line end, sometimes smoothly conducted through enjambment, is no 'unimpeded Utterance'; like the landscape and the human situation in time, it resists, gives way, and resists again. In two magnificent puns in the last couplet quoted the identity of habitat and mode of speech ('address') is plainly linked with a geological theology ('Man faulted into consciousness'). This insight, central to *New Year Letter*, but somehow lost among an American cityscape, occupies the centre of 'In praise of limestone'.

Such a landscape resists the grandiloquent and the lure of the infinite. It is composed of the here and now—'short distances and definite places'— and its inhabitants

> Adjusted to the local needs of valleys
> Where everything can be touched or reached by walking,
>   Their eyes have never looked into infinite space
> Through the lattice-work of a nomad's comb. [pp. 11-12]

Its particular *virtù* is the paradoxically consistent appeal that it makes to the 'unconstant ones'. It provides a necessary shape and order ('If it *form* . . .'); yet it produces 'a stone that responds' and its typical structures are 'a secret system of caves and conduits', 'private' pools. It seems, in fact, an appropriate country for Brueghel's attention to the 'human position': here the earnest poet is perturbed and the scientist, lost in his abstruse world, is rebuked by the street urchins' offers of (presumably) commercial sex. Yet this same landscape is the ineffable image of 'a faultless love' (that redolent pun again) and 'the life to come'.[20]

That is the country of most of Auden's subsequent verse. Its geological condition has not been prescriptive: man, for example, as we learnt last in *Thank You, Fog*,[21] can 'picture the Absent / and Non-Existent'. The world of Homeric heroism and 'most Feigning' poetry exists to mark a dedication to 'buoyant song'. But limestone conditions prevail sooner or later and we are reminded that it may be the 'right song / For the wrong time of year'.[22] The *Horae Canonicae* connect the absolute and eternal with the daily world of inattention and temporary evasion:

> [now] should come
>   The instant of recollection
> When the whole thing makes sense: it comes, but all
>   I recall are doors banging,
> Two housewives scolding, an old man gobbling,
>   A child's wild look of envy,
> Actions, words, that could fit any tale,
>   And I fail to see either plot
> Or meaning; I cannot remember
>   A thing between noon and three.

But in this 'green world temporal' of 'Lauds' or with the beauty of 'that eye-on-the-object look' of 'Sext' we are allowed the

>   time
> To misrepresent, excuse, deny,
>   Mithify, use this event
> While, under a hotel bed, in prison,
>   Down wrong turnings, its meaning
> Waits for our lives . . .[23]

Of course, this intricate weave of our lives with the Event has never tempted Auden, as it did Eliot in the *Four Quartets*, into any long solely poetic meditation after *New Year Letter*. It is perhaps the price paid to 'short distances and definite places'. His familiar stance has been *ad hoc*, opportunist in grammar as in grace:

> facing in four directions,
> outwards and inwards in Space,
> observing and reflecting,
> backwards and forwards through Time,
> recalling and forecasting.[24]

His homage to Clio, muse of history, is the tribute paid to the consciousness she enforces in him of being human in historical time: to her he prays forgiveness of our 'noises / And teach us our recollections'. But the *poet's* recollections are not those of 'serious historians [who] care for coins and weapons' but of those persons who 'are capable of deeds, of choosing to do this rather than that and accepting responsibility for the consequences whatever they may turn out to be'.[25] In this arena of responsibility Auden's art has chosen to flourish. If one of his final utterances seems a little austere, a little uncharacteristically abstracted from the fabric of the world, we can still recall the days of the willow-wren and the stare:

> But Time, the domain of Deeds,
> calls for a complex Grammar
> with many Moods and Tenses,
> and prime the Imperative.[26]

NOTES

[1] *Homage to Clio* (1960), motto to Part I.

[2] *City Without Walls* (1969), p. 123: from 'Prologue at sixty'.

[3] *New Year Letter* (1941), Prologue (originally entitled 'Spring in wartime'), lines 1 and 35.

[4] *Ibid.*, note to line 1316.

[5] *Ibid.*, Epilogue, final stanza.

[6] *Forewords and Afterwords* (1973), p. 25: from 'The Greeks and us'.

[7] *Collected Shorter Poems* (*1930–1944*) (1950), p. 123. All further references in the text are to this collection.

[8] *Nones* (1952), p. 13: from 'In praise of limestone'.

[9] This theme of commitment is augmented if we accept John Fuller's reading of 'Missing' as also a love poem—see *A Reader's Guide to W. H. Auden* (1970), p. 33.

[10] See Auden's remark in B.M. Notebook, fol. 11, that 'The course of every natural desire is that of the orgasm. Being satisfied they desire their own death'; quoted Fuller, *op. cit.*, p. 104.

[11] *Pictures from Brueghel*, for instance, which is discussed in my essay ' "Sight and song itself": painting and the poetry of William Carlos Williams', *Strivers' Row*, I (Baltimore, Md., spring 1974), especially pp. 77–88.

[12] Cf. 'From gallery-grave and the hunt of a wren-king / to Low Mass and trailer camp / hardly a tick by the carbon clock, but I / don't count that way nor do you' in *About the House* (1966), p. 13.

[13] From note to line 13.

[14] Still in his final, posthumous volume, *Thank You, Fog* (1974), Auden's title poem has him re-exploring Wiltshire countryside.

[15] See prefatory note to 'The quest', in *New Republic* (25 November 1940), in part quoted by Fuller, *op. cit.*, p. 143. 'The quest', of course, was incorporated into *New Year Letter* in 1941.

[16] An Audenesque phrase from Herbert Greenberg, *Quest for the Necessary. W. H. Auden and the Dilemma of Consciousness* (Cambridge, Mass., 1968), p. 102.

[17] *Ibid.*, p. 99.

[18] The text taken from *Longer Contemporary Poems* (Penguin Books, 1966), p. 19.

[19] *Satires* I, iv, 42. See Auden's invocation of Horace in his last volume, *Thank You, Fog*, p. 39.

[20] With a similar expression of the Eternal and Infinite in the local dialect of a schoolmaster's

cricket report Auden once hailed Martin Gardner's *The Ambidextrous Universe* as a work that had taught him that 'God was a weak left-hander'.

[21] *Thank You, Fog*, p. 25.

[22] From the motto to *The Shield of Achilles* (1955).

[23] *Ibid.*, pp. 78 and 72 respectively.

[24] *Thank You, Fog*, p. 15.

[25] Respectively, *Homage to Clio*, p. 22, and *Secondary Worlds* (1968), p. 120.

[26] *Thank You, Fog*, p. 16. Cf. the remark that 'All ethical statements are addressed to the will, usually a reluctant will, and must therefore appear in the imperative. "Thou shalt love thy neighbour as thyself" and "A straight line is the shortest distance between two points" belong to two totally different realms of discourse', in *Forewords and Afterwords*, p. 470.

Ambrose Gordon, Jr†   LAWRENCE DURRELL (1912–)

Time, space, and Eros:
the *Alexandria Quartet*
rehearsed

Near the intersection of Rue Nebi Daniel and Rue Fuad at the heart of Alexandria, somewhere deep beneath the flooring of the city's principal mosque, lies the Soma, the hidden tomb of the city's founder. According to a recently reissued guidebook, 'The cellars have never been explored, and there is a gossipy story that Alexander still lies in one of them, intact: a dragoman from the Russian Consulate, probably a liar, said in 1850 that he saw through a hole in a wooden door "a human body in a sort of glass cage with a diadem on its head and half bowed on a sort of elevation or throne. A quantity of books or papyrus were scattered about." '[1] The surrounding city, Alexandria itself, radiates from this point—this dead centre—'like the arms of a starfish', a spiral movement outward that we may think of as occupying both space and time. Alexandria as space (a small, hammer-headed seaport in the eastern Mediterranean), together with Alexandria as history, is generated from the dead body in the glass case with the books or papyrus scattered about. The body itself remains mysterious, equivocal, mystic, and dead.

This is the scene for Lawrence Durrell's long novel—or tetralogy—his *Alexandria Quartet*, a book that is haunted by both space and time—by space-time. Durrell tells us in a note introducing the first volume of the tetralogy, *Justine*: 'The characters in this novel, the first of a series, are all inventions together with the personality of the narrator, and bear no resemblance to living persons. Only the city is real.' We soon realise that the note is more than a perfunctory disclaimer. What reality the characters have is lunar, reflective; only by mirroring the city do they achieve their existence and become consubstantial with it. Thinking of Alexandria at a distance in space and time the narrator says:

I return link by link along the iron chains of memory to the city which we inhabited so briefly together: the city which used us as its flora—precipitated in us conflicts which were hers and which we mistook for our own: beloved Alexandria. . . . I see

† Reprinted from *Six Contemporary Novels*, ed. William O. S. Sutherland (Austin, Tex., 1962), pp. 6–21.

at last that none of us is properly to be judged for what happened in the past. It is the city which should be judged though we, its children, must pay the price.[2]

Alexandria is the key to the novel's content and to its form, since the pattern of the *Quartet* reflects the pattern of the city itself. It too spirals from the dead body of a young man—in space and time. The central act—the central fact—in this myriad book is the suicide (and the resulting dead body) of the brilliant, sardonic, and witty Irish novelist Ludwig Pursewarden. Pursewarden is presented as a remarkable, and remarkably buoyant and talented, person. The act seems out of character: it is a mystery. Thought clings to the fact. The narrator and the other characters seek to puzzle it out. Why did Pursewarden take his own life? What are we, his fellow mortals, to make of it? More generally, what motives can ever be imposed on brute dead fact itself? And living as we do in time, how can we ever breathe the breath of meaning back into the dead past or exhume its implications? These are among the central questions raised by the book. The answer is highly complex and is coterminous with the novel itself.

The proper stance to adopt before the dead fact of the past is perhaps suggested, on the memorable occasion of the body's discovery, by the doctor who has been hurriedly called to the scene. This is Balthazar, a man of great humanity and understanding, but not a great physician. He describes the finding of Pursewarden's body as follows:

The place was in the greatest disorder you can imagine. Drawers turned out, clothes and manuscripts and paintings everywhere; Pursewarden was lying on the bed in the corner with his nose pointing aloofly at the ceiling. I paused to unpack my big-intestine kit—method is everything in moments of stress—while Justine went unerringly across to the bottle of gin on the corner by the bed and took a long swig. I knew that this might contain the poison but said nothing—at such times there is little to say. The minute you get hysterical you have to take this kind of chance. I simply unpacked and unwound my aged stomach-pump which has saved more useless lives (lives impossible to live, shed like ill-fitting garments) than any such other instrument in Alexandria. Slowly, as befits a third-rate doctor, I unwound it, and with method, which is all a third-rate doctor has left to face the world with. . . .[3]

Here his account breaks off—and for us the novel begins, with the consideration of method. What is Lawrence Durrell's method, his way of handling the particular novelistic stomach-pump with which he extracts the poisons of the past? And how important is this method and how successful his handling of it? We have the author's own comment here as a pointer, given in an additional, and longer, note which prefaces the second volume of the series, *Balthazar*:

Modern literature offers us no Unities, so I have turned to science and am trying to complete a four-decker novel whose form is based on the relativity proposition.

Three sides of space and one of time constitute the soup-mix recipe of a continuum. The four novels follow this pattern.

The three first parts, however, are to be deployed spatially (hence the use of 'sibling' not 'sequel' [to describe them]) and are not linked in a serial form. They interlap, interweave in a purely spatial relation. Time is stayed. The fourth part alone will represent time and be a true sequel.

The subject-object relation is so important to relativity that I have tried to turn the novel through both subjective and objective modes. The third part, MOUNT-OLIVE is a straight naturalistic novel in which the narrator of JUSTINE and BALTHAZAR becomes an object, i.e. a character.

This is not Proustian or Joycean method—for they illustrate Bergsonian 'Duration' in my opinion, not 'Space-Time'. . . .

These considerations sound perhaps somewhat immodest or even pompous. But it would be worth trying an experiment to see if we cannot discover a mor-phological form one might appropriately call 'classical'—for our time. Even if the result proved to be a 'science-fiction' in the true sense.

Durrell's remarks—whether pompous, immodest, or not—may seem to be at best rather cryptic. One may well wonder what the relativity theory has to do with literary or artistic form. It is perhaps helpful to turn here for clarification to a critic not of literature but of architecture and the plastic arts, Siegfried Giedion. In attempting to elucidate the apparent aberrations of a painting by Picasso or the design of a building by Le Corbusier, Giedion explains that the key to their understanding is none other than our post-Einsteinian conception of Space-Time. He says:

The three-dimensional space of the Renaissance is the space of Euclidian geometry. But about 1830 a new sort of geometry was created, one which differed from that of Euclid in employing more than three dimensions. Such geometries have con-tinued to be developed, until now a stage has been reached where mathematicians deal with figures and dimensions that cannot be grasped by the imagination. . . .

The essence of space as it is conceived today is its many-sidedness, the infinite potentiality for relations within it. Exhaustive description of an area from one point of reference is, accordingly, impossible; its character changes with the point from which it is viewed. In order to grasp the true nature of space the observer must project himself through it. The stairways in the upper levels of the Eiffel Tower are among the earliest architectural expression of the continuous inter-penetration of outer and inner space.

Space in modern physics is conceived of as relative to a moving point of refer-ence, not as the absolute and static entity of the baroque system of Newton. And in modern art, for the first time since the Renaissance, a new conception of space leads to a self-conscious enlargement of our ways of perceiving space. It was in cubism that this was most fully achieved.

The cubists did not seek to reproduce the appearance of objects from one vantage point; they went round them, tried to lay hold of their internal constitution.[4]

In this respect, Lawrence Durrell is a cubist.

The 'spatial' form of the first three novels of Durrell's quartet is not, of course, something new to literature. Ford's *The Good Soldier* or Faulkner's *Absalom, Absalom* are at least as 'spatial' as *Justine*—not to mention the greater names of Proust and Joyce. The form depends upon the continual use of flashbacks as a structural principle, a brake upon the forward movement of the narrative. Among the 'Consequential Data' at the end of *Justine* is the following pertinent bit, entitled 'Pursewarden on the n-dimensional novel trilogy'. He explains:

The narrative momentum forward is counter-sprung by references backwards in time, giving the impression of a book which is not travelling from a to b but standing above time and turning slowly on its own axis to comprehend the whole pattern. Things do not all lead forward to other things: some lead backwards to things which have passed. A marriage of past and present with the flying multiplicity of the future racing towards one.[5]

Let us try to see what this might mean in practice by looking at the novels themselves: *Justine, Balthazar, Mountolive,* and *Clea.* Later we shall attempt to pass judgement upon Durrell's particular version of this method and evaluate the tetralogy as a whole.

In the opening pages of *Justine* we are introduced to Darley, the narrator, who has withdrawn some time earlier from Alexandria to a small island in the Aegean. This withdrawal is symbolic or ritualistic; Darley's island is to him what a cork-lined chamber had been to Proust: a refuge from the remorseless flow of time, a place for retrospection. *Justine* is a memory novel. In it, Darley's thoughts move back and forth over a past life that now apparently is complete and finished with. There seems little more about it to understand; writing is, he thinks, merely a matter of 'reworking reality'—a reality that seems to be fixed and fully comprehended. The story of Darley's past life in Alexandria, as it emerges, is commonplace enough, despite the exotic background: a sordid case of adultery, a love affair with the wife of a friend. The friend, named Nessim Hosnani, is a rich and refined Egyptian banker married to a passionate Jewess of relatively easy virtue, Justine. As might be expected, in time Nessim becomes aware of his wife's infidelities and shows signs of being insanely jealous, though at first his suspicion does not light on Darley. Later it is different. The denouement of the action in this first volume coincides with Nessim's great annual duck hunt on Lake Mareotis, to which everyone who is anyone in Alexandrian society is invited—Darley among others. Though warned by Justine against going, he feels he must accept the challenge. At duck hunts accidents sometimes happen; and sure enough there is a huntsman found dead at the end of the day, but it is not Darley, rather an apparently totally uninvolved third party. The event is decisive, for all that: at the end of the hunt it is learned that Justine—having found the tension intolerable?—has slipped away, left Alexandria, leaving Darley with his memories.

Nothing very surprising there—for the chief interest in the beginning volume of Durrell's quartet is found not so much in the story as in the atmosphere, the local colour, the gallery of assorted characters who throng Alexandria and the novel's pages. Among others, there is Darley's

tubercular mistress Melissa—a dancing girl from a cheap cabaret. There is
Balthazar, Jewish, a physician, a mystic, and an invert: a sort of Tiresias
who foresuffers all. There is his good friend Clea, a beautiful, blond
Egyptian girl, who is a talented but somewhat blocked painter—of good
family, intelligent, fastidious, and still a virgin. Far and away best of all,
there is Lieutenant Commander Scobie, O.B.E., a retired English merchant
seaman with (as he puts it) 'tendencies', now a Bimbashi of the Egyptian
police force. Scobie is irrepressible, often bawdy, mildly alcoholic, and
piously Roman Catholic; he brings to mind Djuna Barnes's Dr Matthew
O'Connor but is, nevertheless, an authentic comic triumph in his now right.
Later in the tetralogy he becomes, through a queer but typically Alexandrian mistake, a Coptic saint. And finally there is, of course, Pursewarden and his mysterious death.

It is only with *Balthazar* (the second volume) that Durrell's complex
method begins to show. For Darley comes to discover the past was not
quite as he imagined it. We are told that he sent to Balthazar the manuscript of *Justine* and has got it back together with considerable interlinear
comment. As Darley examines this 'great interlinear', the whole situation
slowly turns upon its axis. Darley begins to comprehend that, rather than
the perpetrator of a deception, he was himself the deceived. It was Pursewarden that Justine had loved—Balthazar says so flatly—and Darley had
been no more than a decoy to keep her jealous husband from guessing
the truth. In his own eyes he constantly cuts a rather sorry figure as the
past grows blurred beyond recognition. His first novel comes to seem to
him valueless. It is at this point that Balthazar offers him some sound
technical advice—advice that Darley is not yet ready to take. The advice
is the Method itself:

I suppose [writes Balthazar] that if you wished somehow to incorporate all I am
telling you into your own Justine manuscript now, you would find yourself with
a curious sort of book—the story would be told, so to speak, in layers. Unwittingly
I may have supplied you with a form, something out of the way! Not unlike Pursewarden's idea of a series of novels with 'sliding panels' as he called them. Or else,
perhaps, like some medieval palimpsest where different sorts of truth are thrown
down one upon the other, the one obliterating or perhaps supplementing another.
Industrious monks scraping away an elegy to make room for a verse of Holy Writ![6]

At the end of *Balthazar* the reader takes leave of Darley, still upon his
island, still brooding upon these dismaying revelations, too perturbed to
serve further as a narrator until he has come to terms a little more with
the past which, it appears, is not dead but is far more alive than he.

Volume three of the whole, *Mountolive*, is written in the third person.
In it a new 'sliding panel' is drawn back and a totally new vista opens
before us. This volume might be subtitled 'the politics of love'. It submits
an interesting proposition; namely, that our passional and erotic life is in
fact an extension of our political life. (The converse, it is suggested, is also
true, but less true.) There are further revelations. Justine, we learn, had

not in fact been true to Pursewarden any more than to Darley: her deepest passionate commitment had been to her husband, Nessim. More surprising still, we learn that she had made love to the two unsuspecting Anglo-Irish authors at her husband's express command—a kind of erotic espionage. But how had Nessim come to have such a hold upon her? The novel at this point takes us back to a period some years earlier when Nessim first sought Justine's hand in marriage, without success. We learn that he had at last offered her the supreme temptation, to which Justine succumbed: to become his partner in a vast political intrigue, anti-British and pro-Zionist, intended (in an obscure way not altogether clear to me) to further the aims of his own group, the Copts, who as ancient native Egyptian Christians were being increasingly dispossessed by the ruling Moslem elements. Durrell comments upon her acceptance as follows:

Oriental woman is not a sensualist in the European sense; there is nothing mawkish in her constitution. Her true obsessions are power, politics and possessions—however much she might deny it. The sex ticks on in her mind, but its motions are warmed by the kinetic brutalities of money. In this response to a common field of action, Justine was truer to herself than she had ever been, responding as a flower responds to light. And it was now, while they talked quietly and coldly, their heads bent toward each other like flowers, that she could at last say, magnificently, 'Ah, Nessim I never suspected that I should agree. How did you know that I only exist for those who believe in me?'[7]

Convincing? Perhaps. Before deciding, and before continuing with the synopsis, there are some matters which we must consider.

The *politics of love*—these considerations lead to the general question of love, the central topic of Durrell's book. As Pursewarden puts it (serving here, as elsewhere, as a kind of mouthpiece for the author):

Our topic . . . is the same, always and irremediably the same—I spell the word for you: l-o-v-e. Four letters, each letter a volume! . . . If I am wrong you have only to say so! But in my conception of the four-letter word—which I am surprised has not been blacklisted with the other three by the English printer—I am somewhat bold and sweeping. I mean the *whole bloody range*—from the little greenstick fractures of the human heart right up to its higher spiritual connivance with the . . . well, the absolute ways of nature, if you like. Surely . . . this is the improper study of man? The main drainage of the soul? We could make an atlas of our sighs![8]

And so it is, four letters and each letter a volume: the *Alexandria Quartet*. In Durrell's pages it is all there, the whole range: we are shown love hetero- and homo- and a-sexual; we note the peculiar activities of male, and female, and child prostitutes; observe the effects of Spanish Fly (upon Melissa); shake our heads over the brother–sister incest of the Purse-wardens and the Ptolemies; investigate the Oedipal feelings of Nessim and Mountolive; laugh at the caperings of transvestites like Scobie; query the behaviour of lesbians, succubi, and vampires; observe the hypocrisies of demonically driven maternal love (Justine); stand amazed before the desperate love of homunculi in bottles during an alchemical experiment; deplore the rape of minors; share in dazzling underwater erotic play; share the ambivalences of brotherly love (Narouz and Nessim); listen to Pursewarden on the relation of sexual release to laughter; and on and on and on. There is seemingly no end to it.

At the very beginning Darley tells us: '. . . Alexandria was the great wine-press of love; those who emerged from it were the sick men, the solitaires, the prophets—I mean all those who have been deeply wounded in their sex.'[9] Each of Durrell's four volumes is preceded by an epigraph from the Marquis de Sade, surely the prophet, if not patron saint, of all those who are deeply wounded in their sex—whence also apparently comes the title of the first volume and the name of its principal female character, Justine. The point to note is that love is seen as a sickness, and eventually in Durrell's Alexandria the sheer multifariousness of love leads to a kind of loathing—for the word at any rate, if not for the thing. Justine observes bitterly: 'Damn the word . . . I would like to spell it backwards as you say the Elizabethans did God. Call it *evol* and make it a part of "evolution" or "revolt". Never use the word to me.'[10] Hence Durrell's book is not a glorification of love as it is popularly conceived—far from it.

His diagnosis, however, though profound, is not original with him. How could it be? The classic objection to romantic love, to Eros, was formulated as long ago as the Greek myths: Love is blind! (Like Pursewarden's sister, Liza, and like the famous venereal beggar in *Madame Bovary*.) Love cannot see: Those who have been blinded by Cupid have lost the reality principle. They cannot look outward; they live in a mirror world. And it is such a world that Durrell presents to us. His characters, most of them, are inveterate mirror-watchers. They address remarks to themselves in mirrors, point pistols at their own reflected images in mirrors, preen and primp—whether at dress-makers, in barber shops, or in bathrooms. In particular, Pursewarden is given to writing on the surface of mirrors—usually with shaving-soap—notes to his friends, jeers directed at his mistresses, epitaphs to posterity, and, of course, most often merely little reminders to himself. The city itself has its mirror; Lake Mareotis that bounds Alexandria to the south is described as a vast mirror—while among the characters Justine is perhaps the most mirror-bound of them all. In this connection, the mirror world of love, Durrell remarks that the Alexandrians were particularly fond of caged songbirds, whose eyes they sometimes put out to make them sing the sweeter; while for the squeamish there was another method perhaps less cruel but equally effective. He tells us of 'the trilling of singing birds whose cages were full of mirrors to give them the illusion of company. The love songs of birds to companions they imagined—which were only reflections of themselves!' And he comments sagely: 'How heartbreakingly they sang, these illustrations of human love.'[11]

Such is the fascinating, horrible world (of Alexandria and of love) from which Darley must free himself; and in the last volume of the tetralogy, *Clea*, he does. His avenue of escape is a double one: Clea herself and laughter. Darley's last and deepest intimacy is with her, but it is quite different, we are asked to believe, from other intimacies that have gone before. For one thing, Darley and Clea never use the word 'love'. 'Between us', he explains, 'we had never used this dreadful word—this synonym for derangement or illness. . .'[12] And here there is no ache, no *angst*, nor any of the tedious, endless analysis of emotions that had clung to his earlier love affair with Justine. Physical love there is, and apparently of a very satisfying sort; but it is not compulsive. For days on end Darley and Clea can remain happily apart, while Clea devotes herself to her painting. Then they come together again freely and spontaneously. He tells us:

I might suddenly (sitting on a bench in a public garden, reading) feel cool hands pressed over my eyes and turn suddenly to embrace her and inhale once more the fragrance of her body through her crisp summer smock. At other times, and very often at moments when I was actually thinking of her, she would walk miraculously into the flat saying: 'I felt you calling me to come' or else 'It suddenly came over me to need you very much.'[13]

Though the new thing which is not love (l-o-v-e) is difficult to define or name, Pursewarden had a word for it. Two words. He said once: 'English has two great forgotten words, namely "helpmeet" which is so much greater than "lover" and "loving-kindness" which is so much greater than "love" or even "passion."' [14] It is 'loving-kindness' that Clea and Darley share. It is a revelation, and it comes not in tears but through laughter. Long before, Pursewarden had said to Darley: 'we know that the history of literature is the history of laughter and pain. The imperatives from which there is no escape are: *Laugh till it hurts and hurt till you laugh!* . . One day you are going to wake from your sleep shouting with laughter.'[15] Figuratively he does. For Clea is identified with her laughter summed up by it. When Darley meets Clea again (at the beginning of the final volume) upon his return to Alexandria from the island where he has been rusticating and writing, the following conversation takes place. It establishes the basis of their future relations.

She was more beautiful than I could remember her to have been slimmer, and with a subtle range of new gestures and expressions suggesting a new and troubling maturity.
    'You've grown a new laugh.'
    'Have I?'
    'Yes. It's deeper and more melodious. But I must not flatter you! A nightingale's laugh—if they do laugh.'
    'Don't make me self-conscious because I so much want to laugh with you.'[16]

We are asked, then, to believe that Clea has made some sort of breakthrough—but a breakthrough into what? Into what Pursewarden calls 'the heraldic universe'. This is the universe where there are no set identities, where there is no self. It alone is reality.

'We live,' writes Pursewarden somewhere, 'lives based upon selected fictions. Our view of reality is conditioned by our position in space and time—not by our personalities as we like to think. Thus every interpretation of reality is based upon a unique position. Two paces east or west and the whole picture is changed.'

And as for human characters, whether real or invented, there are no such animals. Each psyche is really an ant-hill of opposing predispositions. Personality as something with fixed attributes is an illusion—but a necessary illusion *if we are to love!*[17]

A necessary illusion if we are to love, an illusion we must banish if we are to live. The point is simply this: romantic lovers, at war with space and time, love out of weakness, out of a great hunger for self, a compulsion to define their identities. Like Justine, each feels himself to be hollow. 'Fill me,' they seem to say, 'fill me with *me.*' It has not occurred to these lovers that a certain hollowness perhaps is inherent in the human condition. They must, they feel, have a unique identity at all costs. But 'in the end', as Pursewarden says, 'everything will be found to be true of everybody.'[18] This, the romantic lover cannot believe. He lusts after his one-day-only: not the lady, though he may believe it is she—but himself.

I take it that something like this is the drift of Durrell's thought: an attack upon romantic love and an offering of something else in its place. This something else is the 'heraldic universe', a sort of beyond-the-looking-glass realm that we may hope some day to enter. It is, like the mystic's vision, ineffable. Behind the pages of the *Alexandria Quartet* the shades of Blake[19] and Lawrence[20] are caused to hover, perhaps to remind us (at the edge of happiness and on the brink of the absurd) that words, even at their best, are not much help in expressing ecstacy. Concerning his relations with Clea Darley remarks: 'Words, which were first invented against despair, are too crude to mirror the properties of something so profoundly at peace with itself, at one with itself. Words are the mirrors of our discontents merely . . .'[21] Words cannot reveal but (as the term implies) they can suggest the 'heraldic universe' through the creation of emblems or symbols, however inadequate. I think it well for us to remember this while considering the final events in *Clea*, where the action becomes very largely symbolic. Indeed one wonders whether Clea (like Sophia for the Gnostics) is not herself a kind of metaphor. Is she not perhaps more an 'archetypal pattern' than a girl?

The breakthrough that Darley and Clea achieve in their happiness together is rendered for us by Durrell through one of the most familiar of motifs from myth: the descent to the bottom of the sea, symbolic drowning, and rebirth. Viewed realistically, it is simply a matter of some scenes of underwater swimming; but the charged language of Durrell's treatment persuades us that much more is implied. For example this image of Clea seen from underwater and below:

A blinding parcel of light struck through the ceiling now and down flashed the eloquent body of Clea, her exploding coils of hair swerved up behind her by the water's concussion, her arms spread. I caught her and we rolled and sideslipped down in each other's arms . . .[22]

This is the 'objective correlative'—the kinetic image—for the moment of breakthrough into the heraldic world where we live not as little egos, Durrell believes, but as the reflections of archetypes. But he must not let this psychic breakthrough seem too easily achieved. It takes time and like all vital experience it hurts—laughter implying, as always, slow realisation and pain. So Durrell makes trouble here for his characters. At the realistic level of the narrative Clea's breakthrough is not yet complete; she is shown as suffering intermittently from some strange, and only partly under-stood, terror, a fear rooted in the past that breaks in upon their happiness. She suffers from visitations of ghosts of the past or—to alter the figure—from the psychic wounds that everyone must suffer until such time as the wounds are not so much cured as perhaps transcended—turned from a liability into an asset. From the point of view of realistic fiction the climactic scene is improbable in the extreme, but it comes symbolically apt. This is a scene of underwater crucifixion. There is an accident; someone aboard Clea's cutter drops a harpoon gun and it goes off. Clea, who at this moment is swimming deep beneath the surface, is pinned by the harpoon to an underwater wreck (the dead past itself?). It is then that Darley goes into action. Like Theseus pursuing Ariadne's thread through the labyrinth, he follows the harpoon line down. He releases Clea in the only way possible, by hacking at her hand with an old bayonet (kept apparently for such purposes). Not a moment too soon they surface, Clea unconscious. Darley drags the body to land.

With a heave now I straightened her out and fell with a thump upon her, crashing down as if from a very great height upon her back. I felt the soggy heavy lungs bounce under this crude blow. Again and again, slowly but with great violence I began to squeeze them in this pitiful simulacrum of the sexual act—life saving, life giving.[23]

Slowly Clea revives. She is rushed to the hospital and eventually recovers. But her hand is lost—along with the unnamed horror. And with the loss of the hand and the horror goes the last of her work-blocks; she is at last free to create. For the new, artificial hand—since it *is* art, though anchored firm in the flesh—can create art. Clea explains all this in a letter to Darley:

Of course I was frightened and disgusted by it at first, as you can imagine. But I have come to respect it very much, this delicate and beautiful steel contrivance which lies beside me so quietly on the table in its green velvet glove! Nothing falls out as one imagines it. I could not have believed myself accepting it so completely —steel and rubber seem such strange allies for human flesh. But the hand has proved itself almost more competent even than an ordinary flesh-and-blood member! In fact its powers are so comprehensive that I am a little frightened of it.

It can undertake the most delicate tasks, even turning the pages of a book, as well as the coarser ones. But most important of all—ah! Darley I tremble as I write the words—IT can *paint!*[24]

The long novel ends with an exodus from Alexandria, permanent this time. For by accepting the limitations of the continuum, the city whose creatures they have been, and by paying with hand and heart for the experience, Clea and Darley are released from its bonds. Together they leave the selected fictions of a dead past—and leave behind romantic love—to enter into the creative laugh of the 'heraldic universe', and the amazing and unpredictable future.

Such is the pattern of Durrell's *Alexandria Quartet*, at least to the extent that I understand it. But what are we to think of all of this? From his reviewers Durrell has received high praise. He has been compared to Proust; he has been called (by Clifton Fadiman, to be sure) 'the finest English novelist of his generation.' Now no one could deny that Durrell's novel is entertaining in the extreme—and doubtless edifying. But is it more? Is it one of the great novels of our time? Reluctantly, I think not.

Durrell's narrator Darley probably puts his finger on the book's weakness—one of its weaknesses—when he says:

I was suddenly afflicted by a great melancholy and despair at recognising the completely limited nature of my own powers, hedged about as they were by the limitations of an intelligence too powerful for itself, and lacking in sheer-word-magic . . .[25]

This seems to me quite just. The *Alexandria Quartet* does reveal a powerful intelligence; but novels are composed of words, and the words of this novel —whether we think of them as Darley's or Durrell's own—in the life they lead leave much to be desired. The phrase 'sheer-word-magic' neatly illustrates the point: it belongs to the language of publishers' blurbs. It is a cliché, and Durrell's style is, alas, riddled by clichés, clogged with stock phrases, and characteristically vitiated by overwriting. It tends to be somewhat fruity—as Pursewarden himself says, a style 'touched with plum pudding'.[26] (Incidentally, I can see no difference here between *Mount-olive*, which is written in the third person, and the other three volumes supposedly penned by Darley.) Plum pudding is all right now and then but not as a steady diet.

What is harder to decide is whether these stylistic deficiencies are the consequence or the cause of others more serious. Style is largely a matter of tone and, I believe, necessarily depends upon an ingrained belief in the stability of one's own identity and that of one's listener—both social norms, fictive though each may be. Now when the individual psyche is broken down prismatically (and that is exactly what Durrell attempts)— when every quality that may be predicated of anyone becomes true of everyone—the soft iridescence that results, though endlessly suggestive, finally becomes extremely monotonous. There is in Durrell's pages much bright sensation, much bright thought, but little in the way of solid humanity to give it substance[27]—perhaps the inescapable limitation of the 'heraldic universe', the world beyond the looking-glass. This is perhaps simply another way of saying Alexandria is *too* exotic a city; in the *Quartet* there is too much excitement of every sort. If Pursewarden could step out of its pages to judge the book, I believe he would say: 'Too much plum pudding and fixings, not near enough meat and potatoes.'

But these reservations sound carping—and probably are. When all is said, we should be thankful for what the *Alexandria Quartet* gives us: a rather dazzling display of intelligence, method, and humour—more humour than I have suggested. And it is impossible to feel anything other than admiration for the tenacity of purpose that permits Lawrence Durrell to hold on to a single situation for over a thousand pages—and to hold our attention too—so that on the last page, the wheel slowly turning, he may set down the definitive words of his fairy tale, 'Once upon a time . . .'

NOTES

[1] E. M. Forster, *Alexandria. A History and a Guide* (1961), pp. 112–13.

[2] Lawrence Durrell, *Justine* (1957), p. 13.

[3] Lawrence Durrell, *Balthazar* (1958), p. 149.

[4] Siegfried Giedion, *Space, Time and Architecture* (Cambridge, Mass., 1954), pp. 431–2. The passage is given in a fuller context on p. 85, above.

[5] *Justine*, p. 248.

[6] *Balthazar*, p. 183.

[7] Lawrence Durrell, *Mountolive* (1959), p. 201.

[8] Lawrence Durrell, *Clea* (1960), pp. 131–2.

[9] *Justine*, p. 14.

[10] *Ibid.*, pp. 74–5.

[11] *Mountolive*, p. 285.

[12] *Clea*, p. 256.

[13] *Ibid.*, pp. 160–1.

[14] *Balthazar*, p. 128.

[15] *Clea*, p. 138.

[16] *Ibid.*, p. 77.

[17] *Balthazar*, pp. 14–15.

[18] *Ibid.*, p. 15.

[19] See *Mountolive*, pp. 64 ff.

[20] For example, *Clea*, p. 127: 'Dear D.H.L. so wrong, so right, so great, may his ghost breathe on us all!'

[21] *Clea*, p. 222.

[22] *Ibid.*, p. 227.

[23] *Ibid.*, p. 251.

[24] *Ibid.*, p. 278.

[25] *Ibid.*, p. 177.

[26] *Balthazar*, p. 245.

[27] Save for Scobie. Unfortunately considerations of space do not permit me to pay due respects to his shade.

Bibliography

The present list is suggestive not exhaustive. Its introductory nature dictated that only works available in English be cited; for had an effort been made to include for example French studies of Proust or German studies of Thomas Mann, I would have needed space several times the length of this volume. The reader should hold in mind that, where foreign authors are concerned, studies in languages other than English are necessarily of primary importance, witness among countless examples:

Bianquis, Geneviève: 'Le Temps dans l'œuvre de Thomas Mann', *Journal de Psychologie*, XLIV (1951), pp. 355–70.
Camón Aznar, José: *La idea del tiempo en Bergson y el impresionismo* (Madrid, 1956).
Joukovsky, F.: *Montaigne et le problème du temps* (Paris, 1972).
Pablos, Basilio de: *El tiempo en la poesta de Juan Ramón Jiménez* (Madrid, 1965).
Vial, Ferdinand: 'Le Symbolisme bergsonien du temps dans l'œuvre de Proust', *PMLA*, LV (1940), pp. 1191–1212.
Zimmerningkat, Martin: 'Das Tempus bei Sartre', *Die Neueren Sprachen*, XVII (1968), pp. 27–35.

The standard bibliographies on each given author should therefore be consulted even where some of the specialised publications cited here provide references further afield. It should nevertheless be noted that the headnotes to each section occasionally provide a few general studies in languages other than English.

The present list is further restricted by the exclusion of studies which I was unable to locate.

CONTENTS

ABBREVIATIONS

AL   *American Literature*
AS   *Arts in Society* (University of Wisconsin)
CE   *College English*
CL   *Comparative Literature*
CLit   *Contemporary Literature*
DHI   *Dictionary of the History of Ideas*, ed. Philip P. Wiener (1973)
EC   *Essays in Criticism*
ELH   *Journal of English Literary History*
ELR   *English Literary Renaissance*
F & F   *Films and Filming*

GLL	German Life and Letters
GR	Georgia Review
HR	Hudson Review
HT	History and Theory, v (1966), Beiheft 6
HTR	Harvard Theological Review
JAAC	Journal of Aesthetics and Art Criticism
JEGP	Journal of English and Germanic Philology
JJQ	James Joyce Quarterly
JP	Journal of Philosophy
JWCI	Journal of the Warburg and Courtauld Institutes
MD	Modern Drama
MFS	Modern Fiction Studies
MLN	Modern Language Notes
MLQ	Modern Language Quarterly
MQR	Michigan Quarterly Review
MT	Man and Time, Papers from the Eranos Yearbooks, ed. Joseph Campbell, III (1957)
NCF	Nineteenth Century Fiction
PBA	Proceedings of the British Academy
PMLA	Publications of the Modern Language Association
PQ	Philological Quarterly
PR	Philosophical Review
PRev	Partisan Review
RES	Review of English Studies
SAQ	South Atlantic Quarterly
SDR	South Dakota Review
SP	Studies in Philology
SQ	Shakespeare Quarterly
SR	Sewanee Review
SRev	Southern Review
SS	Shakespeare Survey
TCL	Twentieth Century Literature
TM	Time and its Mysteries, Series I, by Robert A. Millikan et al. (1936)
TSLL	Texas Studies in Literature and Language
TSP	Time in Science and Philosophy, ed. Jiří Zeman (Amsterdam, 1971)
UCPP	University of California Publications in Philosophy
UTQ	University of Toronto Quarterly
VT	The Voices of Time: A Cooperative Survey of Man's Views of Time as expressed by the Sciences and by the Humanities, ed. J. T. Fraser (1966)
WSCL	Wisconsin Studies in Contemporary Literature
YFS	Yale French Studies

The place of publication is given only if other than London or New York.

I GENERAL STUDIES

In addition to the studies listed below, consult available collections such as *Il Tempo*, ed. Enrico Castelli (Padua, 1958), which contains twelve wide-ranging essays. J. T. Fraser's *Of Time, Passion and Knowledge* (1975) is authoritatively comprehensive. See also *TSP* and *VT*, and below, *113, 120, 161, 174, 191*, etc.

1   Bercovitch, Sacvan: *Horologicals to Chronometricals: The Rhetoric of the Jeremiad* ('Literary Monographs' III, Madison, Wis., 1970).
2   Boas, George: *The Acceptance of Time*, in *UCPP*, XVI (1950), pp. 240–70.
3   Breasted, James H.: 'The beginnings of time-measurement and the origins of our calendar', in *TM*, pp. 59–94.
4   Cipolla, Carlo M.: *Clocks and Culture 1300–1700* (1967).
5   Clarke, Arthur C.: 'About time', in his *Profiles of the Future: An Inquiry into the Limits of the Possible*, 2nd ed. (1973), ch. XI. Speculations by a science-fiction writer.
6   Fraser, J. T.: 'The study of time', in *VT*, pp. 582–92.
7   Glasser, Richard: *Time in French Life and Thought*, trans. C. G. Pearson (Manchester, 1972).

8 Guye, Samuel, and Henri Michel: *Time and Space: Measuring Instruments from the 15th to the 19th Century*, trans. Diana Dolan (1971).

9 Hall, Edward T.: 'The voices of time', in his *The Silent Language* (1959), ch. I.

10 Le Lionnais, François: *The Orion Book of Time*, trans. William D. O'Gorman (1960).

11 Loyd, H. Alan: 'Timekeepers: an historical sketch', in *VT*, pp. 388–400.

12 Merriam, John C.: 'Time and change in history', in *TM*, pp. 23–36.

13 Smith, John E.: 'Time, times, and the "Right Time" ', *The Monist*, LIII (1969), i, pp. 1–13.

14 Thompson, E. P.: 'Time, work-discipline and industrial capitalism', *Past and Present*, XXXVIII (1967), pp. 56–97.

15 Toulmin, Stephen, and June Goodfield: *The Discovery of Time* (1965). Cf. *43*.

16 Whitrow, G. J.: *What is Time?* (1972). Introductory. Cf. *190*.

II HISTORICAL DEVELOPMENTS: CLASSICAL, BIBLICAL, EARLY CHRISTIAN, ORIENTAL

17 Armstrong, A. H., and R. A. Markus: 'Time, history, eternity', in their *Christian Faith and Greek Philosophy* (1960), ch. IX.

18 Barr, James: *Biblical Words for Time* ('Studies in Biblical Theology' XXXIII, 1962), Cf. *72*.

19 Benjamin, A. Cornelius: 'Ideas of time in the history of philosophy', in *VT*, pp. 3–30.

20 Berdyaev, Nicolas: *The Meaning of History*, trans. George Reavy (1936), esp. ch. IV, 'Of celestial history: time and eternity'.

21 Bober, Harry: 'In principio: creation before time', in *De Artibus Opuscula XL*, ed. Millard Meiss (1961), I, pp. 13–28, and II, pp. 5–8.

22 Boman, Thorlief: 'Time and space', in his *Hebrew Thought Compared with Greek*, trans. Jules L. Moreau (1960), pt. III.

23 Brabant, Frank H.: *Time and Eternity in Christian Thought* (1937).

24 Brandon, S. G. F.: *Time and Mankind: An Historical and Philosophical Study of Mankind's Attitude to the Phenomena of Change* (1951).

25 Brandon, S. G. F.: *History, Time and Deity: A Historical and Comparative Study of the Conception of Time in Religious Thought and Practice* (Manchester, 1965). See also his 'Time as God and Devil', *Bulletin of the John Rylands Library*, XLVII (1964), pp. 12–31.

26 Brandon, S. G. F.: 'The significance of time in some ancient initiatory rituals', in *Initiation*, ed. C. J. Blecker (Leiden, 1965), pp. 40–8.

27 Bultmann, Rudolf: 'Man in his relation to time', in his *Primitive Christianity in its Contemporary Setting*, trans. R. H. Fuller (1956, repr. 1960), pp. 214–23; also his *History and Eschatology* (Edinburgh, 1957).

28 Callahan, John F.: *Four Views of Time in Ancient Philosophy* (Cambridge, Mass., 1948). On Plato, Aristotle, Plotinus, and Augustine.

29 Christian, William A.: 'Augustine on the creation of the world', *HTR*, XLVI (1953), pp. 1–25; repr. in *A Companion to the Study of St Augustine*, ed. Roy W. Battenhouse (1955), pp. 315–42.

30 Chroust, Anton-Hermann: 'The metaphysics of time and history in early Christian thought', *New Scholasticism*, XIX (1945), pp. 322–52.

31 Conzelmann, Hans: *The Theology of St Luke*, trans. Geoffrey Buswell (1960). The more appropriate title of the original: *Die Mitte der Zeit*.

32 Coomaraswamy, Ananda K.: *Time and Eternity* (Ascona, 1947). On Hindu, Buddhist, Greek, Mohammedan and Christian views.

33 Corbin, Henry: 'Cyclical time in Mazdaism and Ismailism', in *MT*, pp. 115–72.

34 Cullmann, Oscar: *Christ and Time: The Primitive Christian Conception of Time and History*, trans. F. V. Filson (Philadelphia, 1950; 1951). Cf. *48*.

35 Cushman, Robert E.: 'Greek and Christian views of time', *Journal of Religion*, XXXIII (1953), pp. 254–63. On Plato, Aristotle and Augustine.

36 Daniélou, Jean: *The Lord of History*, trans. Nigel Abercrombie (1958).

37 Eliade, Mircea: *Cosmos and History: The Myth of the Eternal Return*, trans. W. R. Trask (1954).

38 Eliade, Mircea: 'Time and eternity in Indian thought', in *MT*, pp. 173–200; also 'Indian symbolisms of time and eternity', in his *Images and Symbols*, trans. Philip Mairet (1961), ch. II.

39 Frank, Erich: 'Creation and time', in his *Philosophical Understanding and Religious Truth* (1945), ch. III.

40   Gilson, Etienne: 'Creation and time', in his *The Christian Philosophy of Saint Augustine*, trans. L. E. M. Lynch (1961), pt. III, ch. I.

41   Gunn, J. Alexander: *The Problem of Time: An Historical and Critical Study* (1929).

42   Gunnell, John G.: *Political Philosophy and Time* (Middletown, Conn., 1968).

43   Haber, Francis C.: *The Age of the World: Moses to Darwin* (Baltimore, Md., 1959). Cf. *15*.

44   Heschel, Abraham J.: 'Space, time, and reality: the centrality of time in the Biblical worldview', *Judaism*, I (1952), pp. 262–9; see also pp. 44–51, 277–8.

45   Hooke, S. H.: 'The time element in the coming of the spirit', in his *Alpha and Omega* (1961), pp. 225–9.

46   Löwith, Karl: *Meaning in History* (Chicago, Ill., 1949).

47   Mánek, Jindřick: 'The Biblical concept of time and Our Gospels', *New Testament Studies*, VI (1959), pp. 45–51.

48   Marsh, John: *The Fulness of Time* (1952). Cf. *34*.

49   Massignon, Louis: 'Time in Islamic thought', in *MT*, pp. 108–14.

50   Milburn, R. L. P.: *Early Christian Interpretations of History* (1954).

51   Momigliano, A. D.: 'Time in ancient historiography', in *HT*, pp. 1–23.

52   Mommsen, Theodor E.: 'St Augustine and the Christian idea of progress: the background of *The City of God*', in his *Medieval and Renaissance Studies* (Ithaca, N.Y., 1959), ch. XII.

53   Muilenburg, James: 'The Biblical view of time', *HTR*, LIV (1961), pp. 225–52.

54   Nakamura, Hajime: 'Time in Indian and Japanese thought', in *VT*, pp. 77–91.

55   Needham, Joseph: *Time and Eastern Man* (1965). A lecture on Chinese attitudes to time. See also his 'Time and knowledge in China and the west', in *VT*, pp. 92–135.

56   O'Brien, D. (ed.): *Empedocles' Cosmic Cycle* (Cambridge, 1969), ch. IV, 'Time'.

57   Onians, Richard B.: *The Origins of European Thought* (Cambridge, 1951), ch. IX, 'Time—Ἦμαρ'.

58   Pike, Nelson: *God and Timelessness* (1970).

59   Puech, Henri-Charles: 'Gnosis and Time', in *MT*, pp. 38–84.

60   Quispel, Gilles: 'Time and history in Patristic Christianity', in *MT*, pp. 85–107.

61   Russell, J. L.: 'Time in Christian thought', in *VT*, pp. 59–76.

62   Rust, Eric C.: 'Time and eternity in Biblical thought', *Theology Today*, X (1953), pp. 327–56.

63   Salmon, Wesley C. (ed.): *Zeno's Paradoxes* (Indianapolis, 1970). Twelve essays; with full bibliography.

64   Sambursky, S., and S. Pines (eds.): *The Concept of Time in Late Neoplatonism* (Jerusalem, 1971). The text, with translation, of the theories of Iamblichus, Proclus, Simplicius, *et al.*

65   Shinn, Roger L.: 'Augustinian and cyclical views of history', *Anglican Theological Review*, XXXI (1949), pp. 133–41.

66   Shotwell, James T.: 'The discovery of time', *JP*, XII (1915), pp. 197–206, 253–69, and 309–17; also his 'Time and historical perspective', in *Time and its Mysteries*, series III, by H. N. Russell *et al.* (1949), pp. 63–91.

67   Snaith, Norman H.: 'Time in the Old Testament', in *Promise and Fulfilment*, ed. F. F. Bruce (Edinburgh, 1963), pp. 175–86.

68   Suter, Ronald: 'Augustine on time, with some criticisms from Wittgenstein', *Revue Internationale de Philosophie*, XIV (1962), pp. 378–94.

69   Torrance, Thomas F.: *Space, Time and Incarnation* (1969).

70   Wayman, Alex: 'No time, great time, and profane time in Buddhism', in *Myths and Symbols*, ed. J. M. Kitagawa and C. H. Long (Chicago, Ill., 1969), pp. 47–62.

71   Wicksteed, Philip H.: *The Religion of Time and the Religion of Eternity* (1899; repr. 1932).

72   Wilch, John R.: *Time and Event: An Exegetical Study of the Use of ʿēth in the Old Testament . . .* (Leiden, 1969). Cf. *18*.

73   Wilhelm, Hellmut: 'The concept of time in the Book of Changes', in *MT*, pp. 212–32. On Chinese thought.

74   Yaker, Henri: 'Time in the Biblical and Greek worlds', as below (*191*), ch. I.

III   PHILOSOPHY, SCIENCE, AND THE PHILOSOPHY OF SCIENCE

In addition to the studies listed below, see the wide-ranging essays—and the bibliographies appended to each—in 'Interdisciplinary perspectives of time', *Annals of the New York Academy of Sciences*, CXXXVIII (1967), pp. 367–915, and in *TSP*. Other philosophical discussions of time are listed in the extensive bibliographies of the collections edited by Gale (*113*), pp. 495–507, and Smart (*174*), pp. 427–36. On biology see especially the lengthy bibliography in Luce (*144*). Extremely technical expositions—i.e. on space-time—are excluded.

75   Alexander, Samuel: *Space, Time and Deity* (1920), 2 vols.
76   Alexander, Samuel: *Spinoza and Time* (1921).
77   Baker, John Tull: *An Historical and Critical Examination of English Space and Time Theories from Henry Moore to Bishop Berkeley* (Bronxville, N.Y., 1930).
78   Bergson, Henri: *Creative Evolution*, trans. Arthur Mitchell (1911); *Duration and Simultaneity, with Reference to Einstein's Theory*, trans. Leon Jacobson (Indianapolis, 1965); *Matter and Memory*, trans. N. M. Paul and W. S. Palmer (1911); *Time and Free Will: An Essay on the Immediate Data of Consciousness*, trans. F. L. Pogson (1910).
79   Bernstein, Jeremy: *Einstein* ('Modern Masters', ed. Frank Kermode, 1973). With bibliography.
80   Bochner, Salomon: 'Space', *DHI*, IV, 294–306. With bibliography.
81   Bonnor, William: *The Mystery of the Expanding Universe* (1964).
82   Borel, Emile: *Space and Time*, trans. A. S. Rappoport and J. Dougall (1926, repr. 1960).
83   Bosanquet, B.: *The Value and Destiny of the Individual* (1913), esp. ch. x, 'The gates of the future'.
84   Bradley, F. H.: *Appearance and Reality: A Metaphysical Essay* (Oxford, 1893).
85   Brettschneider, Bertram D.: *The Philosophy of Samuel Alexander* (1964), esp. ch. I, 'Space, time, and space-time'. Cf. *75*.
86   Brillouin, L.: 'The arrow of time', in *TSP*, pp. 101–10.
87   Broad, D. C.: 'Time', in *Encyclopaedia of Religion and Ethics* (Edinburgh, 1921), XII, pp. 334–45.
88   Bünning, Erwin: *The Physiological Clock: Circadian Rhythms and Biological Chronometry*, 3rd rev. ed. (1973).
89   Burtt, Edwin A.: *The Metaphysical Foundations of Modern Physical Science*, 2nd rev. ed. (1932, repr. 1954), ch. III(*d*), V(*f*), and VII(4). Time in Galileo, Barrow, and Newton.
90   Čapek, Milič: 'Time', *DHI*, IV, pp. 389–98. With bibliography.
91   Čapek, Milič: 'Time in relativity theory: arguments for a philosophy of Becoming', in *VT*, pp. 434–54.
92   Caponigri, A. Robert: *Time and Idea: The Theory of History in Giambattista Vico* (1953).
93   Carr, H. Wildon: *Henri Bergson: The Philosophy of Change* (1911, repr. 1970). Also *The Philosophy of Change: A Study of . . . Bergson* (1914), and ' "Time" and "History" in contemporary philosophy, with special reference to Bergson and Croce', *PBA* (1917–18), pp. 331–49. Cf. *168*.
94   Cleugh, M. F.: *Time and its Importance in Modern Thought* (1937). A survey; with bibliography.
95   Cloudsley-Thompson, J. L.: 'Time sense of animals', in *VT*, pp. 296–311.
96   Costa de Beauregard, Olivier: 'Time in relativity theory: arguments for a philosophy of being', in *VT*, pp. 417–33.
97   der Leeuw, G. van: 'Primordial time and final time', in *MT*, pp. 324–50.
98   Dingle, Herbert: 'Time in relativity theory: measurement or coordinate?' in *VT*, pp. 455–72.
99   Dunne, J. W.: *The Serial Universe* (1934). Also his *An Experiment with Time* (1927).
100   Durell, Clement V.: *Readable Relativity: A Book for Non-Specialists* (1926).
101   Eddington, Sir Arthur: *The Nature of the Physical World* (1935; reprinted often), esp. chs. II–III. Also *Space, Time and Gravitation: An Outline of the General Relativity Theory* (Cambridge, 1920).
102   Einstein, Albert: 'On the electrodynamics of moving bodies' (1905), trans. in *The Principle of Relativity*, by H. A. Lorentz *et al.* (1923), pp. 37–65. The Special Theory of Relativity.
103   Einstein, Albert: 'The foundation of the General Theory of Relativity' (1916), as in previous entry, pp. 111–64.
104   Einstein, Albert: *Relativity: The Special and the General Theory—A Popular Exposition*, trans. Robert W. Lawson, 15th rev. ed. (1954).
105   Einstein, Albert: *The Meaning of Relativity*, 6th rev. ed. (1956).

106 Eisenstein, Elizabeth L.: 'Clio and Chronos: some aspects of history-book time', in *HT*, pp. 36–64.

107 Findlay, J. N.: 'Time: a treatment of some puzzles', in *Logic and Language*, 1st series, ed. Antony Flew (Oxford, 1968), ch. III.

108 Fischer, Roland: 'Biological time', in *VT*, pp. 357–82.

109 Fock, V.: *The Theory of Space, Time and Gravitation*, trans. N. Kemmer, 2nd rev. ed. (1964).

110 Fraasen, Bas. C. van: *An Introduction to the Philosophy of Time and Space* (1970).

111 Freeman, Eugene, and Wilfrid Sellars (eds.): *Basic Issues in the Philosophy of Time* (LaSalle, Ill., 1971). A collection of thirteen essays.

112 Gale, Richard M.: 'Some metaphysical statements about time', *JP*, LX (1963), pp. 225–37; also *The Language of Time* (1968).

113 Gale, Richard M. (ed.): *The Philosophy of Time: A Collection of Essays* (1967; 1968). With extensive bibliography.

114 Garnett, Christopher B., Jr.: *The Kantian Philosophy of Space* (1939).

115 Gelven, Michael: *A Commentary on Heidegger's 'Being and Time'* (1970). Cf. *127*.

116 Gentile, Giovanni: 'The transcending of time in history', in *Philosophy and History*, ed. Raymond Klibansky and H. J. Paton (Oxford, 1936), pp. 91–105.

117 Georgescu-Roegen, Nicholas: *The Entropy Law and the Economic Process* (Cambridge, Mass., 1971), esp. ch. v.

118 Gold, T.: 'The arrow of time', *American Journal of Physics*, XXX (1962), pp. 403–10.

119 Gold, T.: 'The arrow of time', in *Recent Developments in General Relativity* (1962), pp. 225–34.

120 Gold, T. (ed.): *The Nature of Time* (Ithaca, N.Y., 1967). A collection of thirteen essays.

121 Goldsborough, G. F.: *Recent Discussions on 'Time'* (1927). A lecture.

122 Gonseth, Ferdinand: *Time and Method: An Essay on the Methodology of Research*, trans. E. H. Guggenheimer (1972).

123 Grünbaum, Adolf: 'The status of temporal becoming', in *TSP*, pp. 67–87; *Philosophical Problems of Space and Time* (1963); also *Modern Science and Zeno's Paradoxes* (Middletown, Conn., 1967). Cf. *63*.

124 Hamner, Karl C.: 'Experimental evidence for the biological clock', in *VT*, pp. 281–95.

125 Hanna, Thomas (ed.): *The Bergsonian Heritage* (1962). With bibliography.

126 Heath, Louise R.: *The Concept of Time* (Chicago, Ill., 1936).

127 Heidegger, Martin: *Being and Time*, trans. John Macquarrie and Edward Robinson (1962).

128 Hintikka, Jaakko: *Time and Necessity: Studies in Aristotle's Theory of Modality* (Oxford, 1973).

129 Hoffmann, Banesch: 'Relativity', *DHI*, IV, pp. 74–92. With bibliography.

130 Husserl, Edmund: *The Phenomenology of Internal Time-Consciousness*, ed. Martin Heidegger, trans. James S. Churchill (Bloomington, Ind., 1964).

131 Ingarden, Roman: *Time and Modes of Being*, trans. Helen R. Michejda (Springfield, Ill., 1964).

132 Jammer, Max: *Concepts of Space: The History of Theories of Space in Physics* (Cambridge, Mass., 1964).

133 Jeans, Sir James: *The Mysterious Universe* (Cambridge, 1930); also 'Space and time', in his *Physics and Philosophy* (Cambridge, 1942), pp. 55–69.

134 Johnson, Martin: *Time, Knowledge, and the Nebulae: An Introduction to the Meanings of Time in Physics, Astronomy and Philosophy* . . . (1947).

135 Kalmus, Hans: 'Biological time scales', in *TSP*, pp. 147–51.

136 Knoll, Max: 'Transformations of science in our age', in *MT*, pp. 264–307.

137 Kracauer, Siegfried: 'Time and history', in *VT*, pp. 65–78.

138 Kümmel, Friedrich: 'Time as succession and the problem of duration', in *VT*, pp. 31–55.

139 Lanchester, F. W.: *Relativity: An Elementary Introduction of the Space-time Relations as Established by Minkowski* . . . (1935). Cf. *152*.

140 Lanczos, Cornelius: *Albert Einstein and the Cosmic World Order* (1965), esp. ch. III, 'The unification of space and time by Einstein and Minkowski'.

141 Lecomte du Noüy, Pierre: *Biological Time* (1937).

142 Lovejoy, Arthur O.: 'The problem of time in recent French philosophy', *PR*, XXI (1912), pp. 11–31, 322–43, 527–45.

143 Lucas, J. R.: *A Treatise on Time and Space* (1973).

144   Luce, Gay G.: *Body Time* (1972). With extensive bibliography.

145   Lynch, Kevin: *What Time is this Place?* (Cambridge, Mass., 1972).

146   Mach, Ernst: 'Newton's views of time, space, and motion', in *Readings in the Philosophy of Science*, ed. Herbert Feigl and May Brodbeck (1953), pp. 165–70.

147   MacIver, R. M.: *The Challenge of the Passing Years* (1962).

148   McTaggart, J. M. E.: 'The unreality of time' (1908), and 'The relation of time and eternity', in his *Philosophical Studies* (1934), chs. v–vi. Also 'Time', and 'Further considerations of time', in his *The Nature of Existence*, ed. C. D. Broad (Cambridge, 1927), vol. ii, chs. 33 and 51.

149   Marder, L.: *Time and the Space-Traveller* (1971). With bibliography.

150   Mehlberg, Henry: 'Philosophical aspects of physical time', in *TSP*, pp. 37–65.

151   Millikan, Robert A.: 'Time', in *TM*, pp. 3–19; repr. from his *Time, Matter and Values* (Chapel Hill, 1932).

152   Minkowski, H.: 'Space and time' (1908), as above (*102*), pp. 75–91.

153   Morgenstern, Irvin: *The Dimensional Structure of Time* (1960).

154   Munitz, Milton K.: *Space, Time and Creation: Philosophical Aspects of Scientific Cosmology* (Glencoe, Ill., 1957).

155   Nordenson, Harald: *Relativity, Time and Reality* (1969).

156   Nordman, Charles: *The Tyranny of Time: Einstein or Bergson?* (1925). Upholds the former.

157   Ornstein, Robert E.: *On the Experience of Time* (Harmondsworth, 1969). With bibliography.

158   Ouspensky, P. D.: *Tertium Organum*, trans. N. Bessaraboff and C. Bragdon, 3rd ed. (1968). Also *A New Model of the Universe*, trans. Anon., 2nd ed. (1934).

159   Pagel, Walter: 'J. B. van Helmont, *De tempore* and biological time', *Osiris*, viii (1948), pp. 346–417. With a translation of *De tempore* (1st ed., 1648).

160   Pears, D. F.: 'Time, truth, and inference', in *Essays in Conceptual Analysis*, ed. Antony Flew (1956), ch. xi. Cf. *148*.

161   Pepper, Stephen C., and others: *The Problem of Time*, in *UCPP*, xviii (1953). Eight lectures on philosophical aspects of time.

162   Pieper, Josef: *The End of Time: A Meditation on the Philosophy of History*, trans. Michael Bullock (1954).

163   Plessner, Helmuth: 'On the relation of time to death', in *MT*, pp. 233–63.

164   Prior, Arthur N.: *Time and Modality* (Oxford, 1957); *Past, Present and Future* (1967); and *Papers on Time and Tense* (1968).

165   Reichenbach, Hans: *The Direction of Time*, ed. Maria Reichenbach (Berkeley, Calif., 1956). Also *The Philosophy of Space and Time*, trans. Maria Reichenbach and John Freud (1957).

166   Reyna, Ruth: 'Metaphysics of time in Indian philosophy and its relevance to particle science', in *TSP*, pp. 227–39.

167   Russell, Bertrand: *The ABC of Relativity*, 3rd rev. ed., ed. Felix Pirani (1969).

168   Russell, Bertrand: *The Philosophy of Bergson* (Cambridge, 1914). A lecture, with a reply by H. W. Carr (see also above, *93*). For Russell's final adverse summing up of Bergson, see his *History of Western Philosophy* (1946), ch. xxviii.

169   Schlegel, Richard: *Time and the Physical World* (East Lansing, Mich., 1961). Also 'Time and thermodynamics', in *VT*, pp. 500–23, and 'Time and entropy', in *TSP*, pp. 27–35.

170   Schlick, Moritz: *Space and Time in Contemporary Physics*, trans. H. L. Brose, 3rd ed. (1920, repr. 1963).

171   Shapley, Harlow: 'On the lifetime of a galaxy', in *TM*, pp. 39–55.

172   Sklar, Lawrence: *Space, Time, and Spacetime* (Berkeley, Calif., 1974).

173   Smart, J. J. C.: 'The river of time', as above (*160*), ch. x.

174   Smart, J. J. C. (ed.): *Problems of Space and Time* (1964). A collection of essays; with extensive bibliography.

175   Stace, W. T.: *Time and Eternity: An Essay in the Philosophy of Religion* (Princeton, N.J., 1952).

176   Stallknecht, Newton P.: 'Time and possibility in Kant', in *The Heritage of Kant*, ed. G. T. Whitney and D. F. Bowers (Princeton, N.J., 1939), ch. iv.

177   Stambaugh, Joan: *Nietzsche's Thought of Eternal Return* (Baltimore, Md., 1972).

178   Starr, Chester G.: 'Historical and philosophical time', in *HT*, pp. 24–35.

179   Stebbing, L. Susan: 'Some ambiguities in discussions concerning time', as above (*116*), pp. 107–23.

180 Stromberg, Gustaf: 'Space, time and eternity', *Journal of the Franklin Institute*, CCLXXII (1961), pp. 134–44.
181 Swinburne, Richard: *Space and Time* (1968).
182 Synge, J. L.: 'The General Theory of Relativity', *Hermathena*, CVX (1973), pp. 57–71. An introductory lecture.
183 Szumilewicz, Irena: 'The direction of time and entropy', in *TSP*, pp. 181–92.
184 Taylor, Marc C.: 'Time's struggle with space: Kierkegaard's understanding of temporality', *HTR*, LXVI (1973), pp. 311–29.
185 Temmer, Mark J.: *Time in Rousseau and Kant* (Geneva, 1958).
186 Tillich, Paul: 'The struggle between time and space', in *The Theology of Culture*, ed. Robert C. Kimball (1964), pp. 30–9.
187 Walsh, W. H.: 'Kant on the perception of time', in *Kant Studies Today*, ed. Lewis W. Beck (LaSalle, Ill., 1969), pp. 160–80.
188 Watanabe, Satosi: 'Time and the probabilistic view of the world', in *VT*, pp. 527–63.
189 Whiteman, Michael: *Philosophy of Space and Time and the Inner Constitution of Nature* (1967).
190 Whitrow, G. J.: *The Natural Philosophy of Time* (1961). Cf. *16*.
191 Yaker, Henri, and H. Osmond and F. Cheek (eds.): *The Future of Time: Man's Temporal Environment* (1971). A collection of seventeen essays.
192 Zimmerman, E. J.: 'Time and quantum theory', in *VT*, pp. 479–99.

Where philosophy is concerned, see also the entries in Section II, above (pp. 252 ff.). On Bergson, Einstein, *et al.*, see also Section VIII*d* , below (pp. 264 ff.).

IV   PSYCHOLOGY AND SOCIOLOGY

In addition to the studies listed below, see the extensive bibliography in Doob (*198*), pp. 413–46, and the eleven essays on 'La Construction de temps humain' in the special issue of *Journal de Psychologie*, LIII (1956), pp. 257–472.

193 Aaronson, B. S.: 'Hypnotic alterations of space and time', *International Journal of Parapsychology*, x (1968), pp. 5–36.
194 Bonaparte, Marie: 'Time and the unconscious', *International Journal of Psycho-Analysis*, XXI (1940), pp. 427–68.
195 Boodin, John E.: *Time and Reality* ('Psychological Review' Monograph Supplements VI, iii, 1904).
196 Cohen, John: 'Time in psychology', in *TSP*, pp. 153–64; 'The experience of time', *Acta Psychologica*, x (1954), pp. 207–19; 'Psychological time', *Scientific American*, CCXI (1964), No. 5, pp. 116–24; and 'Subjective time', in *VT*, pp. 257–75.
197 De Grazia, Sebastian: *Of Time, Work and Leisure* (1962).
198 Doob, Leonard W.: *Patterning of Time* (New Haven, Conn., 1971). With extensive bibliography.
199 Fraisse, Paul: *The Psychology of Time*, trans. Jennifer Leith (1963).
200 Franz, M.-L. von: 'Time and synchronicity in analytic psychology', in *VT*, pp. 218–32.
201 Gurvitch, Georges: *The Spectrum of Social Time*, trans. Myrtle Korenbaum (Dordrecht, 1964).
202 James, William: 'The stream of thought', in his *The Principles of Psychology*, rev. ed. (1910), vol. I, ch. IV.
203 Jung, C. G.: 'On synchronicity', trans. R. F. C. Hull, in *MT*, pp. 201–11; also in *The Portable Jung*, ed. Joseph Campbell (1971), ch. XIV. Cf. *457*.
204 Kolaja, Jiri: *Social System and Time and Space: Introduction to the Theory of Recurrent Behavior* (Princeton, N.J., 1969).
205 Meerloo, Joost A. M.: *Along the Fourth Dimension: Man's Sense of Time and History* (1970); also 'The time sense in psychiatry', in *VT*, pp. 235–52.
206 Moore, Wilbert E.: *Man, Time, and Society* (1963).
207 Orme, J. E.: *Time, Experience and Behaviour* (1969).
208 Piaget, Jean: 'Time perception in children', in *VT*, pp. 202–16; also *The Child's Conception of Time*, trans. A. J. Pomerans (1969).
209 Progoff, Ira: *Jung, Synchronicity, and Human Destiny* (1973). Cf. *203*.

210   Sturt, Mary: *The Psychology of Time* (1925).
211   Wallis, Robert: *Time: Fourth Dimension of the Mind*, trans. B. B. and D. B. Montgomery (1968).
212   Woodrow, Herbert: 'Time perception', in *Handbook of Experimental Psychology*, ed. S. S. Stevens (1951), ch. XXXII. With further references.
213   Zipf, George K.: 'The repetition of words, time-perspective, and semantic balance', *Journal of General Psychology*, XXXII (1945), pp. 127–48.

V   ART AND ARCHITECTURE

214   Brett, Guy: *Kinetic Art: The Language of Movement* (1968).
215   Embler, Weller B.: 'Flight: a study of time and philosophy and the arts in the 20th century', *AS*, VIII (1971), pp. 306–23.
216   Giedion, Sigfried: *Space, Time and Architecture:The Growth of a New Tradition*, 5th rev. ed. (Cambridge, Mass., 1967). Cf. above, pp. 81 ff.
217   Giedion, Sigfried: *Architecture and the Phenomena of Transition: The Three Space Conceptions in Architecture* (Cambridge, Mass., 1971).
218   Golding, John: *Boccioni's 'Unique Forms of Continuity in Space'* (Newcastle upon Tyne, 1972).
219   Gombrich, E. H.: 'Moment and movement in art', *JWCI*, XXVII (1964), pp. 292–306.
220   Kirby, Michael: *The Art of Time: Essays on the Avant-Garde* (1969).
221   Lessing, G. E.: *Laocoön: An Essay on the Limits of Painting and Poetry* (1766), trans. Edward A. McCormick (1962). The classic expression of the spatial nature of the visual arts, and the temporal one of poetry (or music). The subjects of the former are objects 'which exist in space'; of the latter, objects 'which follow one another'.
222   Lowry, Bates: 'Time and motion', in his *The Visual Experience* (1967), ch. XIV.
223   Moholy-Nagy, László: *Vision in Motion*, rev. ed. (Chicago, Ill., 1947).
224   Neumann, Erich: 'Art and time', in *MT*, pp. 3–37.
225   Panofsky, Erwin: 'Father time', in his *Studies in Iconology: Humanistic Themes in the Art of the Renaissance* (1939), ch. III.
226   Popper, Frank: *Origins and Development of Kinetic Art*, trans. Stephen Bann (1968).
227   Praz, Mario: 'Spatial and temporal interpenetration', in his *Mnemosyne: The Parallel between Literature and the Visual Arts* (Princeton, N.J. 1970), ch. VII.
228   Saxl, Fritz: 'Veritas filia temporum', as above (*116*), pp. 197–222.
229   Souriau, Etienne: 'Time in the plastic arts', trans. Marjorie Kupersmith, *JAAC*, VII (1948–9), pp. 294–307.
230   Wittkower, Rudolf: 'Chance, time, and virtue', *Journal of the Warburg Institute*, I (1937–1938), pp. 313–21.
See also *145* and *463*.

VI   CINEMA

The recently increased attention to the cinema has augmented the number of studies concerned with time. The present list should therefore be regarded merely as suggestive of the discussions now under way.

231   Bluestone, George: 'Of time and space', in his *Novels into Film* (Baltimore, Md., 1957), ch. I(v). Also 'Time in film and fiction', *JAAC*, XIX (1961), pp. 311–15.
232   Eisenstein, Sergei: 'The filmic fourth dimension' (1922), in his *Film Form*, trans. Jay Leda (1949; 1951), pp. 64–71.
233   Gessner, Robert: 'Seven faces of time: an aesthetic for cinema', in *The Nature and Art of Motion*, ed. Gyorgy Kepes (1965), pp. 158–67. Also 'The faces of time', in his *The Moving Image* (1968), ch. X. See further *235*.
234   Hauser, Arnold: 'Space and time in the film', in his *The Social History of Art* (1951), II, pp. 940–8; repr. in *Film: A Montage of Theories*, ed. Richard D. MacCann (1966), pp. 187–98.
235   Jacobs, Lewis (ed.): 'Time and space', in *The Movies as Medium* (1970), ch. III. Includes the editor's 'The expression of time and space', Robert Gessner's 'The faces of time' (see also *233*), Maya Deren's 'Tempo and tension', J. H. Lawson's 'Time and space' (as in *237*), and Ivor Montagu's 'Rhythm'.

236 Kawin, Bruce F.: 'The continuous present', in his *Telling It Again and Again: Repetition in Literature and Film* (Ithaca, N.Y., 1972), ch. IV.

237 Lawson, John H.: 'Time and space', in his *Film: The Creative Process*, 2nd ed. (1967), ch. XXVI.

238 Luchting, Wolfgang A.: '*Hiroshima, mon amour*, time, and Proust', *JAAC*, XXI (1962–3), pp. 299–313.

239 Murray, Edmund: 'The stream-of-consciousness novel and film', in *The Cinematic Imagination: Writers and the Motion Pictures* (1972), chs. X–XII. On Joyce, Woolf and Faulkner.

240 Panofsky, Erwin: 'Style and medium in the motion pictures', *Critique*, I, iii (Jan.–Feb. 1947), pp. 5–28; revised from an essay first published in 1934. Also available in *Film*, ed. Daniel Talbot (1959), pp. 15–32; cf. *Film and the Liberal Arts*, ed. T. J. Ross (1970), pp. 375–94.

241 Robbe-Grillet, Alain: *Last Year at Marienbad*, trans. Richard Howard (1962). The text for the film by Alain Resnais, whose work is here said to be 'an attempt to construct a purely mental space and time'.

242 Slade, Mark: 'The screen as a clock without hands', in his *Language of Change* (Toronto, 1970), pp. 13–27.

243 Stephenson, Ralph, and J. R. Debrix: *The Cinema as Art*, rev. ed. (Harmondsworth, 1969). With four chapters on space, time, and space-time.

244 Ward, John: *Alain Resnais, or the Theme of Time* (1968).

See also *281* and *374*.

VII   MUSIC

In addition to the studies listed below, see Gisèle Brelet's *Le temps musical* (Paris, 1949) which, though the single most important study of musical time, is still not available in English. Relevant if highly technical is Karlheinz Stockhausen's discussion of 'How time passes' in *Die Reihe* (English trans., 1959), pp. 10–42.

245 Burrows, David: 'Music and the biology of time', *Perspectives of New Music*, fall–winter 1972, pp. 241–9.

246 Carpenter, Patricia: 'The musical object', *Current Musicology*, No. 5 (1967), pp. 56–87. With further references.

247 Dürr, Walther: 'Rhythm in music: a formal scaffolding of time', in *VT*, pp. 180–200.

248 Langer, Susanne: 'The image of time', in her *Feeling and Form* (1953), ch. VII.

249 Norton, Glyn P.: 'Retrospective time and the musical experience in Rousseau', *MLQ*, XXXIV (1937), pp. 131–45.

250 Schopenhauer, Arthur: 'On the metaphysics of music', in *The World as Will and Representation*, trans. E. F. J. Payne (1958), II, pp. 447–57.

251 Shattuck, Roger: 'Making time: a study of Stravinsky, Proust, and Sartre', *Kenyon Review*, XXV (1963), pp. 248–63. Cf. *515*.

252 Stambaugh, John: 'Music as a temporal form', *JP*, LXI (1964), pp. 265–80. Cf. Patricia Carpenter's comments, *JP*, LXII (1965), pp. 36–47.

253 Stravinsky, Igor: *Poetics of Music*, trans. Arthur Knodel and Ingolf Dahl (Cambridge, Mass, 1947; repr. 'Vintage Books', 1956), esp. pp. 31 ff. On 'ontological' and 'psychological' time. Cf. *251*.

254 Zuckerkandl, Victor: 'Time', in his *Sound and Symbol: Music and the External World*, trans. W. R. Trask (1956), chs. XI–XIII.

VIII   LITERATURE

(a) General studies

255 Allen, Dick: 'The poet looks at space—inner and outer', *AS*, VI (1969), pp. 185–93.

256 Bowling, Lawrence E.: 'What is the stream of consciousness technique?', *PMLA*, LXV (1950), pp. 333–45.

257 Burke, Kenneth: 'Kermode revisited', *Novel*, III (1969), pp. 77–82. A review of *266*.

258 Cohn, Dorrit: 'Narrated monologue: definition of a fictional style', *CL*, XIII (1966), pp. 97–112. Cf. *429*.

259 Daiches, David: *Time and the Poet* (Cardiff, 1965). A lecture.

260   De Man, Paul: 'The rhetoric of temporality', in *Interpretation: Theory and Practice*, ed. Charles S. Singleton (Baltimore, Md., 1969), ch. VI.

261   Dyck, Martin: 'Relativity in physics and in fiction', in *Studies in German Literature of the 19th and 20th Centuries*, ed. Siegfried Mews (Chapel Hill, 1970), pp. 174–85.

262   Fasel, Ida: 'Spatial form and spatial time', *Western Humanities Review*, XVI (1962), pp. 223–34. Cf. *427*.

263   Fetz, Howard W.: 'Of time and the novel', *Xavier University Studies*, VIII (1969), ii, pp. 1–17.

264   Hoffman, Frederick J.: *Freudianism and the Literary Mind*, 2nd ed. (Baton Rouge, 1957).

265   Hutchens, Eleanor N.: 'The novel as chronomorph', *Novel*, V (1972), pp. 215–22.

266   Kermode, Frank: *The Sense of an Ending: Studies in the Theory of Fiction* (1967).

267   Langbaum, Robert: *The Poetry of Experience: The Dramatic Monologue in Modern Literary Tradition* (1957).

268   Lessing, G. E.: *Laocoön* (as above, *221*).

269   Lukács, Georg: *The Theory of the Novel*, trans. Anna Bostock (1971), esp. pp. 120 ff., 150 f.

270   Maclay, Joanna H.: 'The aesthetics of time in narrative fiction', *Speech Teacher*, XVIII (1969), pp. 194–6.

271   Mendilow, A. A.: *Time and the Novel* (1952). Cf. above, pp. 69 ff.

272   Meyerhoff, Hans: *Time in Literature* (Berkeley, Calif., 1955). Cf. *481*.

273   Miller, J. Hillis: 'The literary criticism of Georges Poulet', *MLN*, LXXVIII (1963), pp. 471–88, and 'Geneva or Paris? The recent work of Georges Poulet', *UTQ*, XXXIX (1969–1970), pp. 212–28. A framework for Poulet (above, pp. 144 ff.).

274   Newbolt, Sir Henry: 'Poetry and time', *PBA* (1917–18), pp. 487–503.

275   Newton-De Molina, David: ' "Tempus edax rerum": a note on an aspect of Frank Kermode's theory of fictions', *Critical Quarterly*, XII (1970), pp. 352–81. Cf. *266*.

276   Patrides, C. A.: *The Grand Design of God: The Literary Form of the Christian View of History* (1972). Cf. *326*.

277   Priestley, J. B.: *Man and Time* (1964). See also his *Three Time-Plays* (1947)—i.e. *Dangerous Corner*, first produced in 1932, *Time and the Conways* and *I Have Been Here Before*, both in 1937—on 'split time, serial time, and circular time'.

278   Paul, David: 'Time and the novelist', *PRev*, XXI (1954), pp. 636–49.

279   Salinger, Herman: 'Time in the lyric', as above (*261*), pp. 157–73.

280   Spanos, William V.: 'Modern drama and the Aristotelian tradition: the formal imperatives of absurd time', *CLit*, XII (1971), pp. 345–72. See also his 'Modern literary criticism and the spatialization of time: An existential critique', *JAAC*, XXIX (1970), pp. 87–104. Cf. *522*.

281   Spiegel, Alan: 'Flaubert to Joyce: evolution of a cinematographic form', *Novel*, VI (1973), pp. 229–43.

282   Struve, Gleb: '*Monologue intérieur*: the origins of the formula and the first statement of its possibilities', *PMLA*, LXIX (1954), pp. 1101–11. See also above, p. 18, note 33.

283   Stutterheim, C. F. P.: 'Time in language and literature', in *VT*, pp. 163–79.

284   Sutton, Walter: 'The literary image and the reader: a consideration of the theory of spatial form', *JAAC*, XVI (1957–8), pp. 112–23. A rejoinder to Frank (*427*).

285   Timpe, Eugene F.: 'The spatial dimension: a stylistic typology', in *Patterns of Literary Style*, ed. Joseph Strelka (University Park, Pa., 1971), pp. 179–97.

286   Vonalt, Larry P.: 'Of time and literature', *SR*, LXXVII (1969), pp. 164–70. Reviews *266*, etc.

(b) On classical, medieval and Renaissance literature

287   Auerbach, Lawrence: 'The use of time in plays of four periods of the drama', *Dissertation Abstracts*, XXI (1960), pp. 698–9. Stage time in classical, Elizabethan, realistic and post-realistic periods.

288   Barnet, Sylvan: 'Prodigality and time in *The Merchant of Venice*', *PMLA*, LXXXVII (1972), pp. 26–30.

289   Blissett, William: 'This wide gap of time: *The Winter's Tale*', *ELR*, I (1971), pp. 52–70.

290   Brisman, Leslie: *Milton's Poetry of Choice and its Romantic Heirs* (Ithaca, N.Y., 1973).

291   Broude, Ronald: 'Time, truth, and right in [Kyd's] *The Spanish Tragedy*', *SP*, LXVIII (1971), pp. 130–45.

292   Buland, Mable: *The Presentation of Time in the Elizabethan Drama* (1912).

293 Burke, Peter: *The Renaissance Sense of the Past* (1970).

294 Bush, Douglas: 'English poetry: time and man', in his *Prefaces to Renaissance Literature* (Cambridge, Mass., 1965), ch. IV.

295 Charney, Maurice: ' "This mist, my friend, is mystical": place and time in Elizabethan Plays', in *The Rarer Action*, ed. Alan Cheuse and Richard Koffler (New Brunswick, N.J., 1970), pp. 24–35.

296 Cirillo, Albert R.: 'Noon–midnight and the temporal structure of *Paradise Lost*', *ELH*, XXIX (1962), pp. 372–95; ' "Hail holy light" and divine time in *Paradise Lost*', *JEGP*, LXVIII (1969), pp. 45–56; and 'Time, light, and the phoenix: the design of *Samson Agonistes*', in *Calm of Mind*, ed. Joseph A. Wittreich (Cleveland, 1971), pp. 209–33.

297 Colie, Rosalie L.: 'Time and eternity: paradox and structure in *Paradise Lost*', *JWCI*, XXIII (1960), pp. 127–38; repr. in her *Paradoxia Epidemica: The Renaissance Tradition of Paradox* (Princeton, N.J., 1966), ch. V.

298 Cope, Jackson I.: 'Time and space as Miltonic symbol', *ELH*, XXVI (1959), pp. 497–513; repr. in his *The Metaphoric Structure of 'Paradise Lost'* (Baltimore, Md., 1962), ch. III.

299 Davidson, Clifford: 'The triumph of time', *Dalhousie Review*, L (1970), pp. 170–81. On *Hamlet*.

300 Dorius, R. J.: 'A little more than a little', *SQ*, XI (1960), pp. 13–26; repr. as 'Prudence and excess in *Richard II* and the histories', in *Discussions of Shakespeare's Histories*, ed. R. J. Dorius (Boston, Mass., 1964), pp. 24–40.

301 Driver, Tom F.: *The Sense of History in Greek and Shakespearean Drama* (1960).

302 Driver, Tom F.: 'The Shakespearian clock: time and the vision of reality in *Romeo and Juliet* and *The Tempest*', *SQ*, XV (1964), pp. 363–70.

303 Durr, Robert A.: 'Spenser's calendar of Christian time', *ELH*, XXIV (1957), pp. 269–93.

304 Frye, Northrop: *Fools of Time: Studies in Shakespearean Tragedy* (Toronto, 1967).

305 Grant, Patrick: 'Time and temptation in *Paradise Regained*', *UTQ*, XLIII (1973), pp. 32–47.

306 Grivelet, Michel: 'Shakespeare's "war with time": the Sonnets and *Richard II*', *SS*, XXIII (1970), pp. 69–78.

307 Heninger, S. K., Jr.: 'Concepts of deity and of time', in his *Touches of Sweet Harmony: Pythagorean Cosmology and Renaissance Poetics* (San Marino, Calif., 1974), pp. 201–33.

308 Hunt, John Dixon: 'Grace, art and the neglect of time in *Love's Labour's Lost*', in *Shakespearian Comedy*, ed. Malcolm Bradbury and David Palmer (Stratford-upon-Avon Studies XIV, 1972), ch. IV. Cf. above, pp. 225 ff.

309 Jordan, Richard D.: *The Temple of Eternity: Thomas Traherne's Philosophy of Time* (Port Washington, N.Y., 1972).

310 Jorgensen, Paul A.: ' "Redeeming time" in *Henry IV*', *Tennessee Studies in Literature*, V (1960), pp. 101–9; repr. in his *Redeeming Shakespeare's Words* (Berkeley, Calif., 1962), ch. IV.

311 Kaula, David: 'The time sense of *Antony and Cleopatra*', *SQ*, XV (1964), pp. 211–23.

312 Knights, L. C.: 'Time's subjects: the Sonnets and *King Henry IV, Part 2*', in his *Some Shakespearean Themes* (1959), ch. III.

313 Kolve, V. A.: 'Medieval time and English place', in his *The Play called Corpus Christi* (Stanford, 1966), ch. V.

314 Lapp, John C.: 'The unities of time and place', in his *Aspects of Racinian Tragedy* (Toronto, 1955), ch. II. Cf. *323*.

315 Lewalski, Barbara K.: 'Time and history in *Paradise Regained*', in *The Prison and the Pinnacle*, ed. Balachandra Rajan (Toronto, 1973), pp. 49–81.

316 Lindenbaum, Peter: 'Time, sexual love, and the uses of pastoral in *The Winter's Tale*', *MLQ*, XXXIII (1972), pp. 3–22.

317 Lyons, Bridget G.: 'Milton's "Il Penseroso" and the idea of time', in her *Voices of Melancholy* (1971), pp. 149–61.

318 MacLean, Hugh: 'Time and horsemanship in Shakespeare's histories', *UTQ*, XXXV (1966), pp. 229–45.

319 McNeir, Waldo F.: '*The Tempest*: space-time and spectacle-theme', *Arlington Quarterly*, II (1970), iv, pp. 29–58.

320 Mahood, M. M.: 'Wordplay in *Macbeth*', in *Approaches to 'Macbeth'*, ed. Jay L. Halio (Belmont, Calif., 1966), pp. 54–63; repr. from her *Shakespeare's Wordplay* (1957), pp. 130–41.

321 Miner, Earl: 'Forms of perception: time and place', in his *The Metaphysical Mode from*

Donne to Cowley (Princeton, N.J., 1969), ch. II; also 'The ruins and remedies of time', in his *The Cavalier Mode from Jonson to Cowley* (Princeton, N.J., 1971), ch. III.

322   Montgomery, Robert L.: 'The dimensions of time in *Richard II*', *Shakespeare Studies*, IV (1968), pp. 73–85.

323   Mourgues, Odette de: 'The multivalency of time and space', in her *Racine; or, The Triumph of Relevance* (Cambridge, 1967), ch. II. Cf. *314*.

324   Mueller, Martin: 'Time and redemption in *Samson Agonistes* and *Iphigenie auf Tauris*', *UTQ*, XLI (1972), pp. 227–45.

325   Nelson, Lowry, Jr.: 'Time as a means of structure', in his *Baroque Lyric Poetry* (New Haven, Conn., 1961), part II.

326   Patrides, C. A.: 'The Renaissance view of time: a bibliographical note', *Notes and Queries*, n.s., X (1963), pp. 409–10. Also 'Renaissance estimates of the year of creation', *Huntington Library Quarterly*, XXVI (1963), 315–22. Cf. *276*.

327   Peterson, Douglas L.: *Time, Tide and Tempest: A Study of Shakespeare's Romances* (San Marino, Calif., 1973), esp. pp. 14–51.

328   Pocock, J. G. A.: 'Time, history and eschatology in the thought of Thomas Hobbes', in *The Diversity of History*, ed. J. H. Elliott and H. G. Koenigsberger (1970), ch. VI; repr. in his *Politics, Language and Time* (1972), ch. v.

329   Quinones, Ricardo J.: 'Views of time in Shakespeare', *Journal of the History of Ideas*, XXVI (1965), pp. 327–52. Also *The Renaissance Discovery of Time* (Cambridge, Mass., 1972). Cf. above, pp. 38 ff.

330   Reeves, Marjorie: *The Influence of Prophecy in the Later Middle Ages: A Study in Joachimism* (Oxford, 1969).

331   Robinson, James E.: 'Time and *The Tempest*', *JEGP*, LXIII (1964), pp. 255–67.

332   Rodgers, Catherine (Myers): *Time in the Narrative of 'The Faerie Queene'* (Salzburg, 1973).

333   Romilly, Jacqueline de: *Time in Greek Tragedy* (Ithaca, N.Y., 1968).

334   Rosenblatt, Jason P.: 'Structural unity and temporal concordance: the war in heaven in *Paradise Lost*', *PMLA*, LXXXVII (1972), pp. 31–41.

335   Russ, Jon R.: 'Time's attributes in Shakespeare's Sonnet 126', *English Studies*, LII (1971), pp. 318–23.

336   Seng, Peter J.: 'Songs, time, and the rejection of Falstaff', *SS*, XV (1962), pp. 31–40.

337   Smalley, Beryl: 'The Bible and eternity: John Wycliff's dilemma', *JWCI*, XXVII (1964), pp. 73–89.

338   Spencer, Benjamin T.: '*2 Henry IV* and the theme of time', *UTQ*, XIII (1944), pp. 394–9.

339   Stapleton, Laurence: 'Perspectives of time in *Paradise Lost*', *PQ*, XLV (1966), pp. 734–48. Also 'Milton's conception of time in *The Christian Doctrine*', *HTR*, LVII (1964), pp. 9–21.

340   Stein, Arnold: *George Herbert's Lyrics* (Baltimore, Md., 1968), pp. 39 ff., etc.

341   Stewart, Stanley: 'Time and [Herbert's] *The Temple*', *Studies in English Literature*, VI (1966), pp. 97–110. Also 'Time', in his *The Enclosed Garden: The Tradition and the Image in 17th-Century Poetry* (Madison, Wis., 1966), ch. IV.

342   Stoll, Elmer E.: 'Time and space in Milton', in his *From Shakespeare to Joyce* (1944), ch. XX.

343   Stroud, Ronald: 'The bastard to the time in *King John*', *Comparative Drama*, VI (1972), pp. 154–66.

344   Tanselle, G. Thomas: 'Time in *Romeo and Juliet*', *SQ*, XV (1964), pp. 349–61.

345   Tennenhouse, Leonard: '*Beowulf* and the sense of history', *Bucknell Review*, XIX (1971), iii, pp. 137–46.

346   Thornton, Harry and Agathe: *Time and Style: A Psycho-Linguistic Essay in Classical Literature* (Dunedin, 1962).

347   Toliver, Harold E.: 'Shakespeare and the abyss of time', *JEGP*, LXIV (1965), pp. 234–54. Also 'Falstaff, the prince, and the history play', *SQ*, XVI (1965), pp. 63–80.

348   Turner, Frederick: *Shakespeare and the Nature of Time: Moral and Philosophical Themes in Some Plays and Poems of Shakespeare* (Oxford, 1971).

349   Underwood, Horace H.: 'Time and space in the poetry of Vaughan', *SP*, LXIX (1972), pp. 231–41.

350   Walter, G. F.: 'John Donne's changing attitudes to time', *Studies in English Literature*, XIV (1974), pp. 79–89.

351   Wilson, Robert R.: 'The deformation of narrative time in *The Faerie Queene*', *UTQ*, XLI (1971), pp. 48–62. Also 'Images and "Allegoremes" of time in the poetry of Spenser', *ELR*, IV (1974), pp. 56–82.

352 Zwicky, Laurie: 'Kairos in *Paradise Regained*: the divine plan', *ELH*, XXXI (1964), pp. 271–7.

See also 7, 77, *225, 276, 499.*

(c) On eighteenth- and nineteenth-century literature

353 Alford, John A.: 'Wordsworth's use of the present perfect', *MLQ*, XXXIII (1972), pp. 119–29.

354 Baird, Theodore: 'The time-scheme of *Tristram Shandy* and a source', *PMLA*, LI (1936), pp. 803–20. See above, p. 17, note 18.

355 Baker, Jeffrey: 'Time and judgment in "The ruined cottage": a reading of Wordsworth's *Excursion*, book I', *Antigonish Review*, I (1970), iii, pp. 3–23.

356 Blackstone, Bernard: ' "The loops of time": spatio-temporal patterns in [Byron's] *Childe Harold'*, *Ariel*, II (1971), iv, pp. 5–17.

357 Buckley, Jerome H.: *The Triumph of Time: A Study of the Victorian Concepts of Time, History, Progress and Decadence* (Cambridge, Mass., 1966). Cf. above, pp. 57 ff.

358 Butler, E. M.: 'The element of time in Goethe's *Werther* and Kafka's *Prozess'*, *GLL*, XII (1958–9), pp. 248–58.

359 Cash, Arthur H.: 'The Lockean psychology of *Tristram Shandy'*, *ELH*, XXII (1955), pp. 125–35.

360 Cohen, Ralph: 'Thomson's poetry of space and time', in *Studies in Criticism and Aesthetics, 1660–1800*, ed. Howard Anderson and J. S. Shea (Minneapolis, 1967), pp. 176–92.

361 Curtis, F. B.: 'Blake and the "moment of time": an 18th century controversy in mathematics', *PQ*, LI (1972), pp. 460–70.

362 Curtis, James M.: 'Notes on spatial form in Tolstoy', *SR*, LXXVIII (1970), pp. 517–30. Also 'Spatial form as the intrinsic genre of Dostoevsky's novels', *MFS*, XVIII (1972), pp. 135–54.

363 Dobie, Ann B.: 'Early stream-of-consciousness writing: *Great Expectations'*, *NCF*, XXV (1971), pp. 405–16.

364 Doerksen, Victor G.: ' "Was auch der Zeiten Wandel sonst hinnehmen mag": the problem of time in Mörike's epistolary poetry', in *Deutung und Bedeutung*, ed. Brigitte Schludermann *et al.* (The Hague, 1973), pp. 134–51.

365 Gerhardi, Gerhard C.: 'Psychological time and revolutionary action in *Le Rouge et le noir'*, *PMLA*, LXXXVIII (1973), pp. 1115–25.

366 Gilbert, Elliot L.: ' "A wondrous contiguity": anachronism in Carlyle's prophecy and art', *PMLA*, LXXXVII (1972), pp. 432–42.

367 Grimes, Ronald L.: 'Time and space in Blake's major prophecies', in *Blake's Sublime Allegory*, ed. Stuart Curran and Joseph A. Wittreich (Madison, Wis., 1973), pp. 59–81.

368 Harper, J. W.: ' "Eternity our due": time in the poetry of Robert Browning', in *Victorian Poetry*, ed. Malcolm Bradbury and David Palmer (Stratford-upon-Avon Studies XV, 1972), ch. III.

369 Harvey, W. J.: 'George Eliot's treatment of time', in his *The Art of George Eliot* (1961), ch. V.

370 Hirsch, E. D., Jr.: 'Time', in his *Wordsworth and Schelling: A Typological Study of Romanticism* (New Haven, Conn., 1960), ch. V.

371 Holtz, William: 'Time's chariot and *Tristram Shandy'*, *MQR*, V (1966), pp. 197–203.

372 Lehman, B. H.: 'Of time, personality, and the author: a study of *Tristram Shandy'*, in *Studies in the Comic*, University of California Publications in English, VIII (1941), pp. 233–250; repr. in *Essays on the Eighteenth Century Novel*, ed. Robert D. Spector (Bloomington, Ind., 1965), ch. IX.

373 Lindenberger, Herbert: 'Time-consciousness', in his *On Wordsworth's 'Prelude'* (Princeton, N.J., 1963), chs. V–VI.

374 Lodge, David: 'Thomas Hardy and cinematographic form', *Novel*, VII (1974), pp. 246–54. Cf. *281.*

375 Lynen, John F.: *The Design of the Present: Essays on Time and Form in American Literature* (New Haven, Conn., 1969). On Edwards, Franklin, Irving, Cooper, Poe, Whitman; see also *472.*

376 Marken, Ronald: ' "Eternity in an hour"—Blake and time', *Discourse*, IX (1966), pp. 167–83.

377   Maskell, Duke: 'Locke and Sterne, or Can philosophy influence literature?', *EC*, XXIII (1973), pp. 22–40.

378   Mayoux, Jean-Jacques: 'Variations on the time-sense in *Tristram Shandy*', in *The Winged Skull*, ed. Arthur H. Cash and John M. Stedmont (Kent, Ohio, 1971), pp. 3–18.

379   Osborne, L. MacKenzie: 'The "chronological frontier" in Thomas Hardy's novels', *Studies in the Novel*, IV (1972), pp. 543–55.

380   Parker, Dorothy: 'The time scheme of *Pamela* and the character of B', *TSLL*, XI (1969), pp. 695–704.

381   Parkin, Rebecca P.: 'The role of time in Alexander Pope's *Epistle to a Lady*', *ELH*, XXXII (1965), pp. 490–50.

382   Passler, David L.: *Time, Form and Style in Boswell's 'Life of Johnson'* (New Haven, Conn., 1971).

383   Radandt, Friedhelm: 'Transitional time in Keller's *Züricher Novellen*', *PMLA*, XCIX (1974), pp. 77–84.

384   Raleigh, John H.: *Time, Place, and Idea: Essays on the Novel* (Carbondale, Ill., 1968), esp. ch. III, 'The English novel and the three kinds of time', and ch. VII, 'Dickens and the sense of time'.

385   Ross, Malcolm: 'Time, the timeless, and the timely: notes on a Victorian poem '[i.e. *In Memoriam*], *Proceedings and Transactions of the Royal Society of Canada*, 4th series, IX (1971), pp. 219–34.

386   Sallé, Jean Claude: 'Sterne's distinction of time and duration', *RES*, n.s., VI (1955), pp. 180–2. Cf. *354*.

387   Sherbo, Arthur: 'Time and place in Richardson's *Clarissa*', *Boston University Studies in English*, III (1957), pp. 139–46; and 'The time-scheme in [Fielding's] *Amelia*', *ibid.*, IV (1960), pp. 223–8.

388   Shuter, William: 'History as palingenesis in Pater and Hegel', *PMLA*, LXXXVI (1971), pp. 411–21.

389   Stallman, R. W.: 'Hardy's hour-glass novel' [i.e. *The Return of the Native*], as below (*525*), pp. 53–63.

390   Stevick, Philip: 'Fielding and the meaning of history', *PMLA*, LXXIX (1964), pp. 561–8.

391   Stone, Harry: 'Dickens and interior monologue', *PQ*, XXXVIII (1959), pp. 52–65.

392   Sutherland, J. A.: 'The handling of time in *Vanity Fair*', *Anglia*, LXXXIX (1971), pp. 349–356.

393   Weber, Jean-Paul: 'Edgar Poe or the theme of the Clock', in *Poe*, ed. Robert Regan (Englewood Cliffs, N.J., 1967), pp. 79–97.

See also *1, 15, 43, 276, 278, 290, 324, 401, 409, 474, 479, 486, 499*, etc. On Sterne see further *406, 482, 511*.

(*d*) On modern literature

In addition to the studies listed below, consult the bibliographies on individual authors in *MFS* — e.g. Joyce: IV (1958), pp. 71–99, and XV (1969), pp. 107–82; Kafka: VIII (1962), pp. 80–100; Conrad: X (1964), pp. 81–106; Thomas Wolfe: XI (1965), pp. 315–28; Faulkner: XIII (1967), pp. 115–61; Durrell: XIII (1967), pp. 417–21; *et al.* See also the extensive bibliographies in Frederick J. Hoffman and Olga W. Vickery, eds., *William Faulkner: Three Decades of Criticism* (East Lansing, 1960), pp. 393–428; Leslie A. Field, ed., *Thomas Wolfe: Three Decades of Criticism* (1969), pp. 273–93; Guiguet on Virginia Woolf (below, *440*), pp. 465–82; etc.

394   Alexander, Ian W.: 'Valéry and Yeats: the rehabilitation of time', *Scottish Periodical*, I, 1 (summer 1947), pp. 77–106.

395   Anders, Günther: 'Being without time: on Beckett's play *Waiting for Godot*', in *Samuel Beckett: A Collection of Critical Essays*, ed. Martin Esslin (Englewood Cliffs, N.J., 1965), pp. 140–51.

396   Barnes, Hazel E.: *Sartre* (1973), esp. pp. 20 ff., 66 ff.

397   Beach, Joseph Warren: *The Twentieth Century Novel: Studies in Technique* (1932).

398   Beckett, Samuel: *Proust* (1931).

399   Bergsten, Saffan: *Time and Eternity: A Study of the Structure and Symbolism of T. S. Eliot's 'Four Quartets'* ('Studia Litterarum Upsaliensia' I, Stockholm, 1960).

400   Beznos, Maurice J.: 'Aspects of time according to the theories of relativity in Marcel Proust's *A la recherche du temps perdu*', *Ohio University Review*, X (1968), pp. 74–102.

401   Bisson, L. A.: 'Proust, Bergson, and George Eliot', *Modern Language Review*, XL (1945), pp. 104–14.

402   Blodgett, Harriet: 'Joyce's time mind in *Ulysses:* a new emphasis', *JJQ*, V (1967), pp. 22–9.

403   Bork, Alfred M.: 'Durrell and relativity', *Centennial Review*, VII (1963), 191–203

404   Brée, Germaine: *Marcel Proust and Deliverance from Time*, trans. C. J. Richards and A. D. Truitt, 2nd ed. (New Brunswick, N.J., 1969).

405   Brivic, Sheldon R.: 'Time, sexuality and identity in Joyce's *Ulysses*', *JJQ*, VII (1969), pp. 30–51.

406   Brown, Robert C.: 'Laurence Sterne and Virginia Woolf: a study in literary continuity', *University of Kansas City Review*, XXVI (1959), pp. 153–9. Cf. *482*.

407   Champigny, Robert: 'Proust, Bergson and other philosophers', in *Proust*, ed. René Girard (Englewood Cliffs, N.J., 1962), pp. 122–31; trans. from *PMLA*, LXXIII (1958), pp. 129–35.

408   Church, Margaret: *Time and Reality: Studies in Contemporary Fiction* (Chapel Hill, 1963). On Joyce, Virginia Woolf, Aldous Huxley, Mann, Kafka, Thomas Wolfe, Faulkner, and Sartre. Cf. above, pp. 179 ff.

409   Clendenning, John: 'Time, doubt and vision: notes on Emerson and T. S. Eliot', *American Scholar*, XXXVI (1967), pp. 125–32.

410   Cohn, Dorrit: 'Kafka's eternal present: narrative tense in "Ein Landarzt" and other first-person stories', *PMLA*, LXXXIII (1968), pp. 144–50. Cf. *258*.

411   Cook, Albert: 'Proust: the invisible stilts of time', *MFS*, IV (1958), pp. 118–26.

412   Cowley, Malcolm: 'The man who abolished time: Thornton Wilder and the spirit of anti-history', *Saturday Review*, XXXIX (6 October 1956), pp. 13–14, 50–2.

413   Daiches, David: 'Time and sensibility', *MLQ*, XXV (1964), pp. 486–92. A review of *408*.

414   Dassin, Joan: 'The dialectics of recurrence: the relation of the individual to myth and legend in Thomas Mann's *Joseph and his Brothers*', *Centennial Review*, XV (1971), pp. 362–90.

415   Dorenkamp, Angela G.: 'Time and sacrament in *The Anathemata* [of David Jones]', *Renascence*, XXIII (1971), pp. 183–91.

416   Dunn, Albert A.: 'The articulation of time in *The Ambassadors*', *Criticism*, XIV (1972), pp. 137–50.

417   Durrell, Lawrence: 'Space time and poetry', in his *Key to Modern Poetry* (1952), ch. II.

418   Edel, Leon: 'The mind's eye view', in his *The Modern Psychological Novel* (1955; rev. ed., 1964), pp. 75–93. On Joyce's *Ulysses*.

419   Erickson, John D.: 'The Proust–Einstein relation: a study in relative point of view', in *Marcel Proust: A Critical Panorama*, ed. Larkin B. Price (Urbana, Ill., 1973), pp. 247–76.

420   Ewton, Ralph W., Jr.: 'The chronological structure of Thomas Mann's *Die Geschichten Jaakobs*', *Rice University Studies*, L (1964), iv, pp. 27–40.

421   Fagan, Edward R.: 'Disjointed time and the contemporary novel', *Journal of General Education*, XXIII (1971), pp. 151–60.

422   Farrelly, James: '*Gerontion*: time's eunuch', *University of Dayton Review*, VI (1969), ii, pp 27–34.

423   Fjelde, Rolf: 'Time, space, and Wyndham Lewis', *Western Review*, XV (1950–1), pp. 201–12. Cf. *467*.

424   Fleischauer, John F.: 'Simultaneity in Nabokov's prose style', *Style*, V (1971), pp. 57–69.

425   Follett, Wilson: 'Time and Thomas Mann', *The Atlantic*, CLXI (June 1938), pp. 792–4. A review of *Joseph in Egypt*.

426   Foster, Steven: 'Relativity and *The Waste Land*: a postulate', *TSLL*, VII (1965), pp. 77–97.

427   Frank, Joseph: 'Spatial form in modern literature', *SR*, LIII (1945), 221–40, 433–56, 643–53; repr. in his *The Widening Gyre* (New Brunswick, N.J., 1963), pp. 3–62. A shorter version is available in *Criticism*, ed. Mark Schorer *et al.* (1948), pp. 379–92. See above, p. 18, note 35.

428   Fraser, G. S.: *Lawrence Durrell: A Study*, rev. ed. (1973), ch. VI (iii).

429 Friedman, Melvin: *Stream of Consciousness: A Study in Literary Method* (New Haven, Conn., 1955). The best study. Cf. *450*.

430 Friedman, Melvin: 'The novels of Samuel Beckett: an amalgam of Joyce and Proust', *CL*, XII (1960), pp. 47–58.

431 Frohock, W. M.: 'Thomas Wolfe: time and the national neurosis', in his *The Novel of Violence in America*, 2nd ed. (Dallas, Tex., 1957), ch. III.

432 Garber, Frederick: 'Time and the city in Rilke's *Malte Laurids Brigge*', *CLit*, XI (1970), pp. 324–39.

433 Genette, Gérard: 'Time and narrative in *A la recherche du temps perdu*', trans. Paul De Man, in *Aspects of Narrative*, ed. J. Hillis Miller (1971), pp. 93–118.

434 German, Howard, and Sharon Kaehele: 'The dialectic of time in [Virginia Woolf's] *Orlando*', *CE*, XXIV (1962), pp. 35–41.

435 Goldthorpe, Rhiannon: 'The presentation of consciousness in Sartre's *La Nausée* and its theoretical basis: reflection and facticity', *French Studies*, XXII (1968), pp. 114–31.

436 Grenander, M. E., B. J. Rahn and F. Valvo: 'The time-scheme in *The Portrait of a Lady*', *AL*, XXXII (1960), pp. 127–35.

437 Gross, Beverly: 'Narrative time and the open-ended novel', *Criticism*, VIII (1966), pp. 362–76.

438 Gross, Harvey: *The Contrived Corridor: History and Fatality in Modern Literature* (Ann Arbor, Mich., 1971). On Henry Adams, Yeats, Pound, Malraux, Mann, and especially T. S. Eliot.

439 Grubbs, Henry A.: 'Sartre's recapturing of lost time', *MLN*, LXXIII (1958), pp. 515–22.

440 Guiguet, Jean: 'Time and space', in his *Virginia Woolf and her Works*, trans. Jean Stewart (1965), pp. 382–98.

441 Hamilton, Gary D.: 'The past in the present: a reading of [Faulkner's] *Go Down, Moses*', *Southern Humanities Review*, V (1971), pp. 171–81.

442 Harkness, Bruce: 'Conrad on Galsworthy: the time scheme of *Fraternity*', *MFS*, I (1955), ii, pp. 12–18.

443 Harper, Ralph: 'Remembering eternity: St Augustine and Proust', *Thought*, XXXIV (1959), pp. 569–606.

444 Hartley, Lodwick: 'Of time and Mrs Woolf', *SR*, XLVII (1939), pp. 235–41. On her novel *The Years*.

445 Harvey, Larwence E.: *Samuel Beckett: Poet and Critic* (Princeton, N.J., 1970), pp. 50 ff., 170 ff., 186 ff., 200 ff., 315 ff.

446 Henn, T. R.: '[Yeats's] *A Vision* and the interpretation of history', in his *The Lonely Tower*, 2nd ed. (1965), ch. XII.

447 Hirsch, David H.: 'T. S. Eliot and the vexation of time', *SRev*, n.s., III (1967), pp. 608–24.

448 Horowitz, Floyd R.: 'The Christian time sequence in Henry James's *The American*', *CLA Journal*, IX (1966), pp. 234–45.

449 Hudspeth, Robert N.: 'Conrad's use of time in *Chance*', *NCF*, XXI (1966–7), pp. 283–9.

450 Humphrey, Robert: *Stream of Consciousness in the Modern Novel* (Berkeley, Calif., 1954). Cf. *429*.

451 Hunter, Frederick J.: 'The value of time in modern drama', *JAAC*, XVI (1957–8), pp. 194–201.

452 Hutchinson, James D.: 'Time: the fourth dimension in Faulkner', *SDR*, VI (1968), ii, iii, pp. 91–103.

453 Ianni, L. A.: 'Lawrence Ferlinghetti's fourth person singular and the theory of relativity', *WSCL*, VIII (1967), pp. 392–406.

454 Isaacs, J. : 'The stream of consciousness', in his *An Assessment of Twentieth-Century Literature* (1951), ch. III. Introductory.

455 Jacobs, Robert G.: 'H. G. Wells, Joseph Conrad, and the relative universe', *Conradiana*, I (1968), pp. 51–5.

456 Jameson, Fredric: 'The rhythm of time', in his *Sartre: The Origins of a Style* (New Haven, Conn., 1961), ch. III; repr. in *Sartre*, ed. Edith Kern (Englewood Cliffs, N.J., 1962), pp. 104–20.

457 Jung, C. G.: '*Ulysses*: a monologue' (1932), in *The Spirit in Man, Art, and Literature*, trans. R. F. C. Hull (1966), pp. 109–32; from his *Collected Works*, vol. XV. Cf. *203*.

458 Karl, Frederick R.: 'Time in Conrad', in his *A Reader's Guide to Joseph Conrad* (1960), ch. III.

459 Karst, Roman: 'Franz Kafka: word-space-time', *Mosaic*, III, iv (1970), pp. 1–13.

460   Klawitter, Robert: 'Henri Bergson and Joyce's fictional world', *Comparative Literature Studies*, III (1966), pp. 429–37.

461   Klotz, Marvin: 'The triumph over time: narrative form in William Faulkner and William Styron', *Mississippi Quarterly*, XVII (1963–4), pp. 10–20.

462   Kohler, David: 'Time in the modern novel', *CE*, X (1948), pp. 15–24.

463   Korg, Jacob: 'Modern art techniques in *The Waste Land*', *JAAC*, XVIII (1959–60), pp. 456–63.

464   Kumar, Shiv K.: *Bergson and the Stream of Consciousness Novel* (Glasgow, 1962).

465   Kumar, Shiv K.: 'Space-time polarity in *Finnegans Wake*', *Modern Philology*, LIV (1957), pp. 230–3.

466   Leech, Clifford: 'The shaping of time: *Nostromo* and [Malcolm Lowry's] *Under the Volcano*', in *Imagined Worlds*, ed. Maynard Mack and Ian Gregor (1968), pp. 233–42.

467   Lewis, Wyndham: *Time and Western Man* (1927). The celebrated attack on 'time-mind'.

468   Lindsay, Jack: 'Time in modern literature', in *Festschrift zum achtzigsten Geburstag von George Lukács*, ed. Frank Benseler (Neuwied, 1965), pp. 491–501.

469   Litz, A. Walton: *The Art of James Joyce* (1964), esp. pp. 53–62 ('The "image" ').

470   Lowrey, Perrin: 'Concepts of time in *The Sound and the Fury*', in *English Institute Essays 1952*, ed. A. S. Downer (1954), pp. 57–82.

471   Lynch, William F.: 'Time', in his *Christ and Apollo: The Dimensions of the Literary Imagination* (1960), ch. II. Also 'Dissociation in time' (*ibid.*, pp. 169–77), repr. in *T. S. Eliot: 'Four Quartets'*, ed. Bernard Bergonzi (1969), pp. 247–53.

472   Lynen, John F.: 'Selfhood and the reality of time: T. S. Eliot', as above (*375*), ch. VI.

473   MacCallum, Reid: 'Time lost and regained', in his *Imitation and Design* (Toronto, 1953), pp. 132–61. On T. S. Eliot.

474   McElderry, B. R., Jr.: '[Thomas] Wolfe and Emerson on "Flow" ', *MFS*, II (1956), pp. 77–8.

475   Macksey, Richard: 'Architecture of time: dialectics and structure', in *Proust*, ed. René Girard (Englewood Cliffs, N.J., 1962), pp. 104–21.

476   Manierre, Virginia: 'How long is a minute?' *New Mexico Quarterly*, XXVI (1956), pp. 238–48. On Thomas Mann.

477   Martz, Louis L.: 'The wheel and the point: aspects of imagery and theme in Eliot's later poetry', *SR*, LV (1947), pp. 126–47.

478   Mein, Margaret: *Proust's Challenge to Time* (Manchester, 1962).

479   Mein, Margaret: *A Foretaste of Proust: A Study of Proust and his Precursors* (1974).

480   Melchiori, Giorgio: 'The moment as a time unit in fiction', in his *The Tightrope Walkers: Studies of Mannerism in Modern English Literature* (1956), pp. 175–87.

481   Meyerhoff, Hans: 'The time of life', *Forum* [Houston], III (1962), IX, pp. 13–17. See esp. 272.

482   Moglen, Helene: 'Laurence Sterne and the contemporary vision', as above (*378*), pp. 59–75.

483   Moloney, Michael F.: 'The enigma of time: Proust, Virginia Woolf, and Faulkner', *Thought*, XXXII (1957), pp. 69–85.

484   Monteiro, George: 'Bankruptcy in time: a reading of Faulkner's *Pylon*', *TCL*, IV (1958), pp. 9–20.

485   Morse, Donald E.: 'Meaning of time in Auden's *For the Time Being*', *Renascence*, XXII (1970), pp. 162–8.

486   Moyer, Patricia: 'Time and the artist in Kafka and Hawthorne', *MFS*, IV (1958), pp. 295–306.

487   Murray, Edward: *Arthur Miller, Dramatist* (1967), esp. pp. 24 ff. On the 'three time-sequences' in *Death of a Salesman*.

488   Nelson, Cary: *The Incarnate Word: Literature as Verbal Space* (Urbana, Ill., 1973).

489   Nettels, Elsa: '*The Ambassadors* and the sense of the past'. *MLQ*, XXXI (1970), pp. 220–35,

490   Neufeldt, Leonard N.: 'Time and man's possibilities in [Faulkner's] *Light in August*', *GR*, XXV (1971), pp. 27–40.

491   Noon, William T.: 'Modern literature and the sense of time', *Thought*, XXXIII (1958), pp. 571–603; repr. in *The Theory of the Novel*, ed. Philip Stevick (1967), pp. 280–313.

492   O'Brien, R. A.: 'Time, space, and language in Lawrence Durrell', *Waterloo Review*, VI (1961), pp. 16–24.

493   Olson, Paul R.: *Circle of Paradox: Time and Essence in the Poetry of Juan Ramón Jiménez* (Baltimore, Md., 1967).

494   Parry, Idris: 'Space and time in Rilke's Orpheus sonnets', *Modern Language Review*, LVIII (1963), pp. 524–31.

495   Patrides, C. A.: 'The renascence of the Renaissance: T. S. Eliot and the pattern of time', *MQR*, XII (1973), pp. 172–96. The 'blueprint' for the introduction and the essay above, pp. 1 ff., 159 ff.

496   Pearlman, Daniel D.: *The Barb of Time: On the Unity of Ezra Pound's Cantos* (1969).

497   Perlis, Alan D.: '[Faulkner's] *As I Lay Dying* as a study of time', *SDR*, X (1972), i, pp. 103–10.

498   Pouillon, Jean: 'Time and destiny in Faulkner', in *Faulkner*, ed. R. P. Warren (Englewood Cliffs, N.J., 1966), pp. 79–86; trans. J. Merriam from his *Temps et roman* (Paris, 1946), pp. 238–60.

499   Poulet, Georges: *Studies in Human Time*, trans. Elliott Coleman (Baltimore, Md., 1956). Cf. above, pp. 144 ff.

500   Pritchett, V. S.: 'Time frozen: *A Fable*', *PRev*, XXI (1954), pp. 557–61; repr. in *Faulkner*, ed. R. P. Warren (Englewood Cliffs, N.J., 1966), pp. 238–42.

501   Prusok, Rudi: 'Science in Mann's *Zauberberg*: the concept of space', *PMLA*, LXXXVIII (1973), pp. 52–61. See also Michael Ossar, 'Relativity theory in *Der Zauberberg*', *ibid.*, 1191–2.

502   Radke, Judith J.: 'The theater of Samuel Beckett: "Une durée à animer" ', *YFS*, XXIX (1962), pp. 57–64.

503   Read, Phyllis J.: 'The illusion of personality: cyclical time in Durrell's *Alexandria Quartet*', *MFS*, XIII (1967), pp. 389–99.

504   Rechtien, Brother John, S.M.: 'Time and eternity meet in the present', *TSLL*, VI (1964), pp. 5–21. On Beckett's *Godot*.

505   Richter, Harvena: *Virginia Woolf: The Inward Voyage* (Princeton, N.J., 1970), esp. ch. X, 'Three modes of time'.

506   Robson, Vincent: 'The psychosocial conflict and the distortion of time: a study of Diver's disintegration in [Fitzgerald's] *Tender is the Night*', *Language and Literature*, I (1972), ii, pp. 55–64.

507   Rodrigues, Eusebio L.: 'Time and technique in *The Sound and the Fury*', *Literary Criterion* [University of Mysore], VI (1965), iv, pp. 61–7.

508   Rose, Alan: 'The spatial form of [Henry James's] *The Golden Bowl*', *MFS*, XII (1966), pp. 103–16.

509   Rosenberg, John: *Dorothy Richardson: The Genius they Forgot* (1973). On the reputed pioneer of the 'stream of consciousness'.

510   Rothman, Nathan L.: 'Thomas Wolfe and James Joyce: a study in literary influence', in *A Southern Vanguard*, ed. Allen Tate (1947), pp. 52–77.

511   Rubin, Louis D., Jr.: 'Joyce and Sterne: a study in affinity', *Hopkins Review*, III (1950), ii, pp. 14–22.

512   Rubin, Louis D., Jr.: 'Thomas Wolfe in time and place', *Hopkins Review*, VI (1953), pp. 117–32; also his *Thomas Wolfe: The Weather of his Youth* (Baton Rouge, 1955), ch. II.

513   Sale, Roger: 'The narrative technique of *The Rainbow*', *MFS*, V (1959), pp. 29–38.

514   Scott, Nathan, Jr.: 'Mimesis and time in modern literature', in his *The Broken Center: Studies in the Theological Horizon of Modern Literature* (New Haven, Conn., 1966), ch. III.

515   Shattuck, Roger: *Proust's Binoculars: A Study of Memory, Time, and Recognition in 'A la recherche du temps perdu'* (1963; 1964). Cf. *251*.

516   Sheppard, R. W.: 'Rilke's *Duineser Elegien*: a critical appreciation in the light of Eliot's *Four Quartets*', *GLL*, XX (1966–7), pp. 205–18.

517   Slade, Joseph W.: 'The functions of eternal recurrence in Thomas Mann's *Joseph and his Brothers*', *Symposium*, XXV (1971), pp. 180–97.

518   Slochower, Harry: 'Marcel Proust: revolt against the tyranny of time', *SR*, LI (1943), pp. 370–81.

519   Smidt, Kristian: 'Bergson, and the problem of time', in his *Poetry and Belief in the Work of T. S. Eliot*, rev. ed. (1961), pp. 165–81.

520   Smith, Grover: *T. S. Eliot's Poetry and Plays* (Chicago, Ill., 1956), esp. ch. XII and XVIII.

521   Smyth, W. F.: 'Lawrence Durrell: modern love in chamber pots and space time', *Edge*, No. 2 (spring 1964), pp. 105–16.

522   Spanos, William V.: *The Christian Tradition in Modern British Verse Drama: The Poetics of Sacramental Time* (New Brunswick, N.J., 1967), esp. ch. VII, 'T. S. Eliot's plays of contemporary life: the redemption of time'. Cf. *280*.

523 Spears, Monroe K.: '*For the Time Being*', in *Auden*, ed. M. K. Spears (Englewood Cliffs, N. J., 1964), pp. 160–71.

524 Spencer, Sharon: *Space, Time and Structure in the Modern Novel* (1971).

525 Stallman, R. W.: 'Time and [Conrad's] *The Secret Agent*', and 'Gatsby and the hole in time', in his *The Houses that James Built* (East Lansing, Mich., 1961), pp. 111–50. The same volume contains the original version of the essay reprinted above, pp. 126 ff.

526 Stallman, R. W.: 'Time and the unnamed article in *The Ambassadors*', *MLN*, LXXII (1957), pp. 27–32. See also Leon Edel's reply, LXXIII (1958), pp. 177–9, and Stallman's rejoinder, LXXVI (1961), pp. 20–3.

527 Starkie, Enid: 'Bergson and literature', as above (*125*), pp. 74–99.

528 Stein, William Bysshe: '*Almayer's Folly*: the terrors of time', *Conradiana*, I (1968), No. 1, pp. 27–34. Also 'Conrad's east: time, history, action, and *Maya*', *TSLL*, VII (1965), pp. 265–83. Cf. above, pp. 114 ff.

529 Steinberg, Erwin R.: *The Stream of Consciousness and Beyond in 'Ulysses'* (Pittsburgh, Pa., 1973).

530 Stern, Madeleine B.: 'Counterclockwise: flux of time in literature', *SR*, XLIV (1936), pp. 338–65.

531 Stone, Edward: 'From James to Balderston: relativity and the 20s', *MFS*, I (1955), ii, pp. 2–11.

532 Struc, Roman S.: '*The Magic Mountain*: time and timelessness', *Research Studies* (Washington State University), XXXIX (1971), pp. 83–95.

533 Swiggart, Peter: 'Moral and temporal order in *The Sound and the Fury*', *SR*, LXI (1953), pp. 221–37; 'Time in Faulkner's novels', *MFS*, I (1955), ii, pp. 25–9; 'Rage against time', in his *The Art of Faulkner's Novels* (Austin, Tex., 1962), ch. VI.

534 Szanto, George H.: 'Structure as process: the temporal point of view', in his *Narrative Consciousness* (Austin, Tex., 1972), ch. VII. On Robbe-Grillet.

535 Tembeck, Robert: 'Dialectic and time in [Sartre's] *The Condemned of Altona*', *MD*, XII (1969), pp. 10–17.

536 Tomlin, E. W. F.: 'Reflections on *Time and Western Man*', in the special issue of *Agenda* reprinted as *Essays on Wyndham Lewis* (1971), pp. 97–108. Cf. 467.

537 Tompkins, Jane P.: 'The redemption of time in [Henry James's] *Notes of a Son and Brother*', *TSLL*, XIV (1973), pp. 681–90.

538 Torrance, Robert M.: 'Modes of being and time in the world of *Godot*', *MLQ*, XXVIII (1967), pp. 77–95.

539 Toyokuni, Takashi: 'A modern man obsessed by time: a note on "The man who loved islands"', *D. H. Lawrence Review*, VII (1974), pp. 78–82.

540 Underwood, Henry, J. Jr.: 'Sartre on *The Sound and the Fury*: some errors', *MFS*, XII (1966), 477–9. Stresses one 'error': Sartre 'overstates' his thesis. Cf. above, pp. 203 ff.

541 Uscatescu, George: 'Time and destiny in the novels of Mircea Eliade', as above (70), pp. 397–406. Cf. 37–8.

542 Vande Kieft, Ruth M.: 'Faulkner's defeat of time in *Absalom, Absalom!*', *SRev*, VI (1970), pp. 1100–9.

543 Verheul, Kees: *The Theme of Time in the Poetry of Anna Axmatova* (The Hague, 1971).

544 Vickery, Olga W.: 'Faulkner and the contours of time', *GR*, XII (1958), pp. 192–201.

545 Wagner, Geoffrey: 'Wyndham Lewis and James Joyce: a study in controversy', *SAQ*, LVI (1957), pp. 57–66.

546 Ward, Anne: 'Speculations on Eliot's time-world: an analysis of *The Family Reunion* in relation to Hulme and Bergson', *AL*, XXI (1949), pp. 18–34.

547 Warren, Robert Penn: 'Faulkner: the south, the Negro, and time', in *Faulkner*, ed. R. P. Warren (Englewood Cliffs, N.J., 1966), pp. 251–71.

548 Webb, Eugene: *The Plays of Samuel Beckett* (Seattle, 1972), pp. 34–9. Time in *Godot*.

549 Weigand, Hermann J.: *Thomas Mann's Novel 'Der Zauberberg'* (1933), ch. II, 'Organization'.

550 Weinstein, Norman: *Gertrude Stein and the Literature of the Modern Consciousness* (1970), pp. 37–8. On the reputed 'tension' in Stein between actual and narrative time.

551 Weitz, Morris: 'T. S. Eliot: time as a mode of salvation', *SR*, LX (1952), pp. 48–64; repr. in *T. S. Eliot: 'Four Quartets'*, ed. Bernard Bergonzi (1969), pp. 138–52.

552 Weitz, Morris: '*A la recherche du temps perdu*: philosophy as artistic vision', in his *Philosophy in Literature* (Detroit, 1963), ch. IV.

553 Whitaker, Thomas R.: *Swan and Shadow: Yeats' Dialogue with History* (Chapel Hill, 1964).

554 Whitehead, Lee M.: 'The moment out of time: Golding's *Pincher Martin*', *CLit*, XII (1971), pp. 18–41.

555 Widmer, Kingsley: 'Timeless prose', *TCL*, IV (1958), pp. 3–8. On e. e. cummings's *The Enormous Room*.

556 Williams, George: '*Four Quartets* and history', in his *A Reader's Guide to T. S. Eliot* (1953), ch. VIII.

557 Wilson, Jame S.: 'Time and Virginia Woolf', *Virginia Quarterly Review*, XVIII (1942), pp. 267–76.

558 Woolf, Virginia: 'Modern fiction' (1919), in *The Common Reader* (1st Series), 2nd ed. (1925), esp. pp. 189–90.

559 Zink, Karl E.: 'Flux and the frozen moment: the imagery of stasis in Faulkner's prose', *PMLA*, LXXI (1956), pp. 285–301.

560 Ziolkowski, Theodore: 'The discordant clocks', in his *Dimensions of the Modern Novel: German Texts and European Contexts* (Princeton, N.J., 1969), ch. VI.

561 Ziolkowski, Theodore: 'Franz Kafka: *The Trial*', as in the previous entry, ch. II(1).

562 Ziolkowski, Theodore: 'Hermann Broch and relativity in fiction', *WSCL*, VIII (1967), pp. 365–76.

See also 78, 84, *125*, *227*, *231*, *238–9*, *251*, *262*, *271–2*, *276–7*, *281*, *358*.